Optical Code Division Multiple Access

A Practical Perspective

This book is a comprehensive guide to optical fiber communications, from the basic principles to the latest developments in OCDMA for next-generation Fiber-to-the-Home (FTTH) systems. Part I starts with the fundamentals of light propagation in optical fibers, multiple access protocols, and their enabling techniques. Part II is dedicated to the practical characteristics of current and next-generation passive optical network (PON) technology. It covers the key building blocks of OCDMA devices such as optical encoders and decoders, signal impairment due to noise, and data confidentiality, a unique property of OCDMA. This is followed by a discussion of hybrid system architectures with TDM and WDM and practical aspects such as system cost, energy efficiency and long-reach PONs. Featuring the latest research, with cutting-edge coverage of system design, optical implementations, and experimental demonstrations in testbeds, this text is ideal for students, researchers and practitioners in the industry seeking to obtain an up-to-date understanding of optical communication networks, particularly optical access networks.

Ken-ichi Kitayama has been a Professor in the Department of Electrical, Electronic, and Information Engineering, at the Graduate School of Engineering, Osaka University, Japan since 1999. He received the B.E., M.E., and Dr.Eng. degrees in communication engineering from Osaka University in 1974, 1976, and 1981, respectively. In 1976 he joined the NTT Laboratories. In 1982–1983, he spent a year as a visiting Research Fellow under the supervision of Professor Shyh Wang and Professor T. Kenneth Gustafson at the University of California, Berkeley. From 1995 to 1999, he was with the Communications Research Laboratory, CRL, the Ministry of Posts and Telecommunications (currently known as the National Institute of Information and Communications Technology, NICT), Japan.

Optical Code Division Multiple Access

A Practical Perspective

KEN-ICHI KITAYAMA

Osaka University, Japan

CAMBRIDGE
UNIVERSITY PRESS

CAMBRIDGE
UNIVERSITY PRESS

University Printing House, Cambridge CB2 8BS, United Kingdom

Published in the United States of America by Cambridge University Press, New York

Cambridge University Press is part of the University of Cambridge.

It furthers the University's mission by disseminating knowledge in the pursuit of
education, learning and research at the highest international levels of excellence.

www.cambridge.org
Information on this title: www.cambridge.org/9781107026162

© Ken-ichi Kitayama 2014

This publication is in copyright. Subject to statutory exception
and to the provisions of relevant collective licensing agreements,
no reproduction of any part may take place without the written
permission of Cambridge University Press.

First published 2014

Printing in the United Kingdom by TJ International Ltd. Padstow Cornwall

A catalog record for this publication is available from the British Library

Library of Congress Cataloguing in Publication data
Kitayama, Ken-Ichi.
Optical code division multiple access: a practical perspective / Ken-ichi Kitayama.
   pages   cm
Includes bibliographical references and index.
ISBN 978-1-107-02616-2 (hardback)
1. Optical fiber communication.   2. Code division multiple access.   I. Title.
TK5103.592.F52K57   2014
621.382′75 – dc23   2013034909

ISBN 978-1-107-02616-2 Hardback

Cambridge University Press has no responsibility for the persistence or accuracy of
URLs for external or third-party internet websites referred to in this publication,
and does not guarantee that any content on such websites is, or will remain,
accurate or appropriate.

To my wife, Michiyo, my daughters, Hiromi and Midori

Contents

Preface

Between 2000 and the end of 2011, almost two thirds of subscribers to the plain old telephone service (POTS) in Japan switched to the broadband service, and half of the subscribers to the broadband service are now connected to the Fiber-to-the-Home (FTTH) system. The rate of increase in new FTTH subscriptions is already showing signs of leveling off. This is the case not only in Japan but also in the rest of the world where a similar trend will be observed in the near future as the FTTH system becomes widespread.

This rapid growth of the broadband service, along with the emerging need for backhauling the huge data traffic of mobile phones and inter-data center traffic, will eventually trigger the problem of "capacity crunch" in long-haul optical fiber transmission systems. The capacity crunch happens when the data traffic to be transferred from one end of the network to the other end overflows the total capacity of optical fiber cables deployed on the planet. Only incessant innovation of the transmission system and optical network technologies can solve this problem. Technologies of optical fiber transmission systems currently under intense development include digital coherent transmission and space division multiplexing (SDM). There are two approaches to SDM: one is via a multicore fiber which has a number of cores embedded in the cladding of the fiber, and the other is mode division multiplexing (MDM) via a multimode fiber which supports more than one waveguide mode. The digital coherent transmission has the capability of equalizing the impairments of multi-level phase-shift-keying modulated optical signals, incurred by chromatic and polarization dispersions and non-linear effects, with the aid of powerful digital signal processing (DSP). This is a result of rapid progress in silicon complementary metal oxide semiconductor (CMOS) LSI technology, underpinned by Moore's law. In this way, the limitations in the transmission distance and the bit rate have been overcome, although the system is not perfect.

Such advanced technologies developed for backbone networks will be adopted in metro/access networks after a few or ten years as their costs go down. For example, the first commercial 10 Gb/s system was deployed in a long-haul transmission line between Tokyo and Osaka in Japan in 2003, and now the commercial 10 Gb/s passive optical network (PON) is almost ready to be deployed. Thus, current optical access networks will soon evolve to the next-generation (NG) PON with higher bit rate and longer reach. A forum of the world telecommunications industries has studied a roadmap of the NG-PON: one system is NG-PON1 aiming at evolutionary growth, which supports "brown field" deployment, and the other is NG-PON2 aiming at revolutionary growth, supporting a "green field" deployment. Note that in the brown field deployment newly introduced technology has to coexist with ongoing PONs, while the green field deployment can be disruptive, with little requirement for coexistence with other PONs. At the time of writing, the objective of NG-PON2 has been altered from revolutionary to evolutionary. However, this will not be the end of the NG-PON scenario because there remain plenty of green fields on the globe where a revolutionary technology might be deployed from the beginning. Therefore, further advanced NG-PON in a true sense is likely to appear on the scene as post NG-PON2.

Optical code division multiple access (OCDMA) has been emerging as a promising technology of choice for the NG-PON. OCDMA has unique capabilities such as fully asynchronous transmission, low latency access, and soft capacity on demand. Another advantage is inherent data confidentiality. Messages are encoded at the transmitter and can be recovered only by the authorized subscriber, who knows the optical code.

This book should be valuable to both students at universities and mid-career professionals in the telecom industry. I reference very recent research progress as much as possible, which will be useful to researchers in this field. This book can be added to collections as a popular item in university libraries and R&D centers of industry. Parts I and II in this book serve different purposes. Part I is devoted to the fundamental technology underlying optical fiber communications. Part II offers an extensive coverage of a wide variety of technologies, relevant to OCDMA-PON systems. Those who are new to this field would be better to start with Part I and then proceed to Part II. Part II is for those who have a solid background in optical communication and networks but need to understand the practical perspectives and future-proof technology of FTTH systems. To aid in teaching and learning the material, selected problems are provided at the ends of the chapters in Part I.

The book consists of ten chapters with the following organization. Chapter 1 offers an introduction to optical fiber communications and networks. In Part I, Chapter 2 describes the basics of PONs and multiple access techniques. Chapter 3 describes how light propagates in an optical fiber and what the guided modes are. In Chapter 4 the model and building blocks of optical transmission systems as well as the methodology of performance evaluation are described. Chapter 5 deals with enabling techniques of OCDMA. In Part II, Chapter 6 defines OCDMA, including its roots, system classification, and noise unique to OCDMA. In Chapter 7 various techniques of optical encoding and decoding are described. Chapter 8 covers data confidentiality which is inherently provided with OCDMA. Chapter 9 offers an extensive coverage of experimental demonstrations done by the author's group in the OCDMA testbeds, including hybrid systems of TDM-OCDMA, WDM-OCDMA, and WDM-TDM-OCDMA PONs, followed by space-OCDMA. Chapter 10 includes a comparison of the cost and the power consumption of various PON systems, applications of optical code labeling in optical networks, and practical aspects of PON such as testing and equipment, and safety issues with respect to exposure of the human body.

Acknowledgments

My first and foremost gratitude is due to my colleagues at the Communications Research Laboratory, CRL, the Ministry of Posts and Telecommunications (currently, renamed the National Institute of Technology, NICT), Naoya Wada, Xu Wang presently of Heriot-Watt University, Moriya Nakamura presently of Meiji University, Hedeyuki Sotobayashi presently of Aoyama Gakuin University, Nobuaki Kataoka formerly with NICT, Toshio Kuri, Yoshinari Awaji, and Satoshi Shimizu for long-term collaboration since 1995. I would like to express my sincere thanks to Gabriella Cincotti, Rome Tre University, also for long-term collaboration. It was my good fortune to work under my boss at CRL, Tadashi Shiomi, presently at Yokohama National University. Without his generous support, research activity in OCDMA would never have taken off so smoothly.

I also thank Akira Himeno, Masayuki Okuno, and Takashi Saida of NTT Electronics Corp., Takashi Mizuochi, Jun-ichi Nakagawa, Naoki Suzuki, and Satoshi Yoshima of Mitsubishi Electronic Corp., Yoshihiro Terada, Akira Sakamoto, Koji Omichi, Ryo Maruyama, Shoichiro Matsuo, Nobuo Kuwaki, and Ryozo Yamauchi of Fujikura, Akihiko Nishiki of OKI Electric Industry Corp., Naoto Yoshimoto, Shunji Kimura, Shin Kaneko, Noriki Miki, and Hideaki Kimura of NTT Access Network Service Systems Laboratories, Yosihiro Tomiyama of Aisthesis Corp., and Yasuyuki Kato, Tadatoshi Tanifuji, and Masahiro Ikeda formerly of NTT Labs for their collaboration. I would like to thank my colleagues at Osaka University, Akihiro Maruta, Yuki Yoshida, and Takahiro Kodama, for their inspiring discussions and continuous support.

I am grateful to the following individuals who have provided me with technical materials: Yoshio Itaya, Honchul Ji, Kunio Kokura, Yoichi Maeda, Hideo Miyahara, Kunio Mori, Toshikazu Sakano, Hiromichi Shinohara, Shin-ichi Todoroki, Yoshito Shuto, Moshe Tur, and Shuichi Yanagi. I have also greatly benefitted from the help of my students at Osaka University. I thank Masaya Nakazawa, Ryosuke Matsumoto, and Daisule Hisano for executing computer plots of figures. The figures were drawn by Ai Yamamoto with inimitable care.

And last but not least, my sincere thanks are due to my mentor, the Honorary Professor Nobuaki Kumagai, who guided me in such an exciting research field of optical communications at Osaka University when I was an undergraduate student.

Finally, my special thanks are extended to Phil Mayler, Publishing Director, Mia Balashova, Publishing Assistant, and Rachel Cox, Production Editor of Cambridge University Press for their tremendous effort in organizing the logistics of the book, including the editing and promotion that make this publication happen.

Ken-ichi Kitayama
Osaka, Japan

1   Introduction

1.1   Broadband service trend

Internet traffic has been expanding exponentially with the rapid growth in cloud computing data, smart phone voice and videoconferencing, and streaming contents of IP TV, video sharing applications such as YouTube, and 3D TV, as shown in Fig. 1.1 [1]. In 2012 global IP traffic was expected to exceed 40 exa-bytes per month, of which consumer IP traffic is the largest portion. The growth in Internet traffic is driving large carriers to prepare to provide enough bandwidth to meet market demand. According to the forecast [2], the compound annual growth rate (CAGR) of Internet traffic is 32% in six years in the period 2010–2015. On the other hand, the profits of telecommunication carriers are staying constant or even decreasing as the profitable old telephone service is being switched to the voice over IP (VoIP) service. Therefore, there is strong motivation for the carriers and service providers to reduce their capital expenditure (CAPEX) by fully exploiting the capacity of their existing installed optical fiber cables without the costly deployment of new cables.

The number of broadband subscribers is increasing rapidly all over the world, exceeding 600 million in 2010. The broadband services are provided either by a wired infrastructure such as Fiber-to-the-Home (FTTH), digital subscriber line (DSL), and cable modem (CATV) or fixed wireless access (FWA). In particular, the increase in FTTx subscribers has been remarkable in recent years. The line bit rate ranges from a few tens of kb/s to Gb/s, and even higher, and the broadband service includes so-called "triple play" of data communication, voice over IP (VoIP), and video or TV broadcasting. As the broadband service penetrates deep into residential areas in developing countries in Asia, Eastern Europe, and South America, the traffic volume will continue growing exponentially, and this has a tremendous impact on the backbone network.

From Fig. 1.2 one can see that the number of broadband subscribers in Japan has increased rapidly to 30 million at the end of 2010, with almost two thirds of fixed telephone subscribers having switched to the broadband service [3]. The broadband subscribers are broken up into 14.4 million FTTH, 11.6 million DSL, and 4.1 million CATV. It is interesting to observe that the DSL penetrated faster than FTTH in the earlier stage, but later the number of DSL subscribers decreased while the number of FTTH subscribers was picking up slowly, with the crossover occurring at the end of 2008. The monthly charges of the broadband service in Japan are plotted as a function of the bit

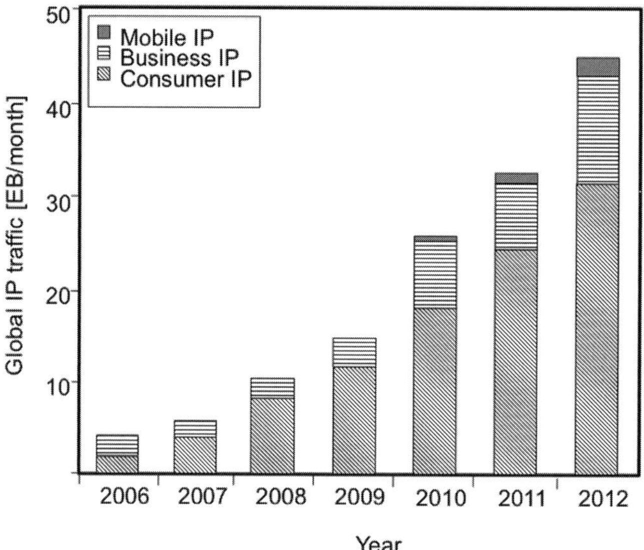

Figure 1.1 Global Internet traffic growth in the world [2]. © Reprinted by permission of Cisco.

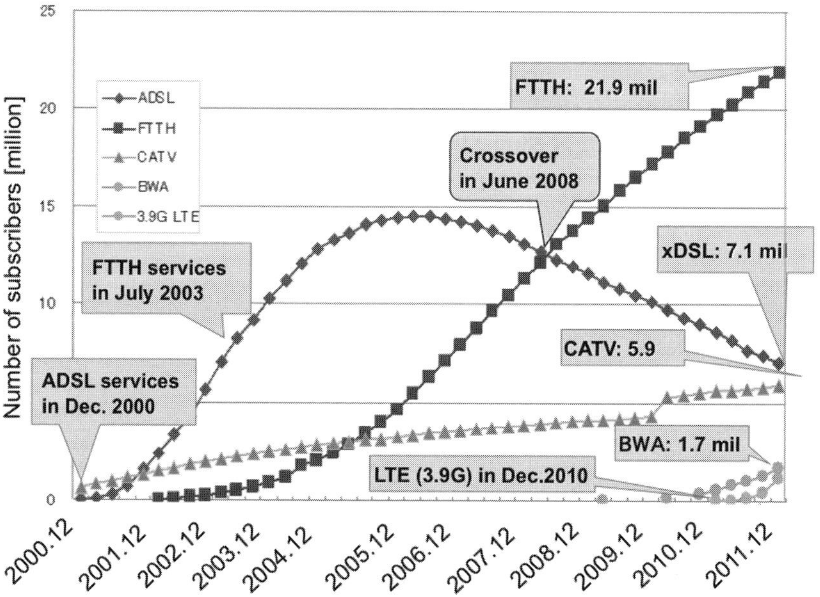

Figure 1.2 Number of subscribers of the broadband service in Japan [3]. © Reprinted by permission of the Ministry of Internal Affairs and Communications.

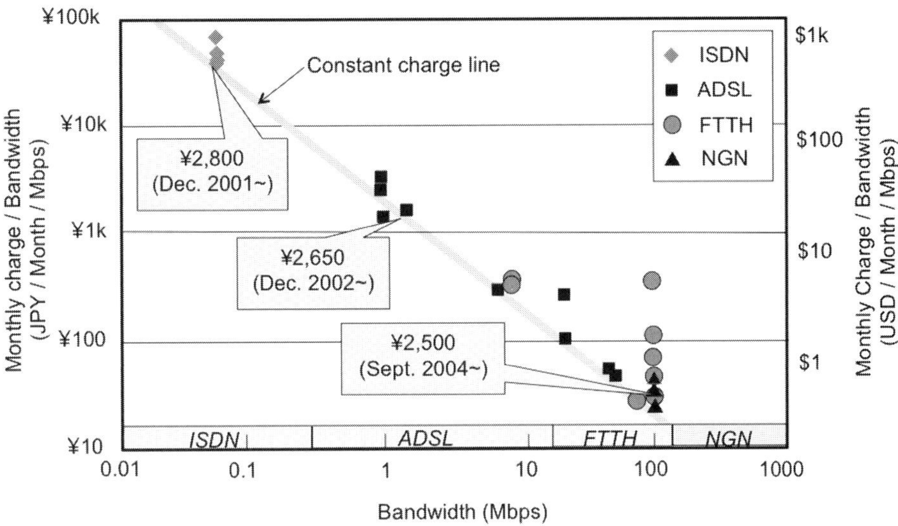

Figure 1.3 Monthly charge versus bandwidth of the broadband service in Japan [4], by courtesy of T. Sakano.

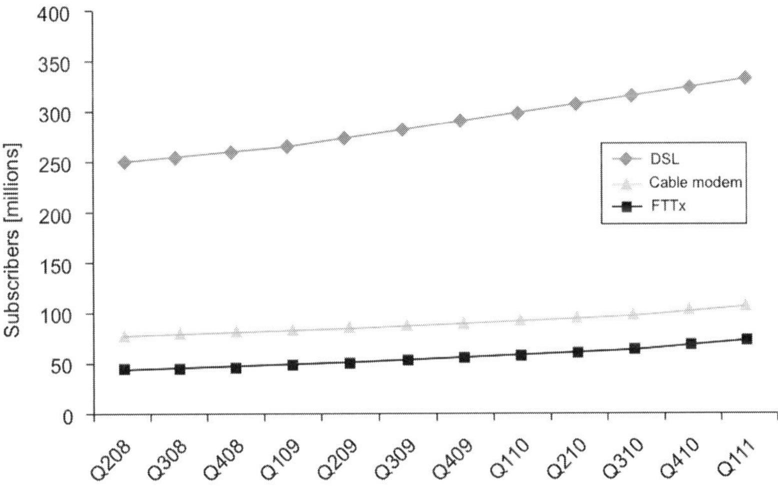

Figure 1.4 Broadband subscribers in the world for 2006–2009 [5]. © POINT topic by permission.

rate in Fig. 1.3 [4]. The charge has been decreasing rapidly to below 1 USD/Mbps as the bit rate has increased higher than 100 Mb/s.

By contrast, this trend is not the case in developing countries. As a consequence, the overall number of DSL subscribers in the world is still dominant over the number of FTTH subscribers, as seen in Fig. 1.4 [5]. In the top ten countries in 2010, China has the largest number of subscribers, 127 billion, followed by 857 million in the USA and Japan, 261 million in Germany, and 209 million in France, UK, Russia, South Korea, Brazil, and Italy.

1.2   Historical perspective of optical fiber communications

Without the two inventions of the laser and the optical fiber, the optical fiber communications technology would not exist today. These two inventions were awarded with Nobel Prizes in Physics.

> The Nobel Prize in Physics 1964 was divided, one half awarded to Charles Hard Townes, the other half jointly to Nicolay Gennadiyevich Basov and Aleksandr Mikhailovich Prokhorov. The citation reads *"for fundamental work in the field of quantum electronics, which has led to the construction of oscillators and amplifiers based on the maser-laser principle."*
>
> The Nobel Prize in Physics 2009 was divided, one half awarded to Charles Kuen Kao. The citation reads *"for groundbreaking achievements concerning the transmission of light in fibers for optical communication."*

Looking back at the history of glass, shown in Fig. 1.5 [6], the glass made by the ancient Egyptians in 1 BC was sufficiently transparent in the region of visible wavelengths. In 1966 Dr. Kao, the 2009 Nobel Prize laureate for Physics, first proposed the concept of cladded glass fibers as a *"new form of communication medium"* and in his seminal paper he predicted that the fiber loss could be reduced to as low as 20 dB/km [7]. Today's float glass used for window glass has typically a loss of a few dB/mm. The past forty years have witnessed a significant reduction in loss of silica-based optical fibers and the transition of the transmission window into the near infrared, as shown in Fig. 1.6 [8]. Since a record loss of 17 dB/km for silica fiber reported in 1970, it has taken only fifteen years for the attenuation to be reduced to 0.2 dB/km [9].

1.3   Optical transmission systems

There are two basic multiplexing and demultiplexing techniques, time division multiplexing (TDM) and wavelength division multiplexing (WDM). Multiplexing allows a number of channels to be established in an optical fiber, and it is motivated by the economical reason that the installation of new optical fiber cables is costly, and there is a need to exploit fully the potential transmission capacity of existing fibers. Figure 1.7 shows schematically current mainstream techniques, TDM and WDM, along with optical division multiplexing (OCDM). T_B denotes the one-bit time duration. In TDM a time slot is divided into the channel count, N slots, and one data-bit pulse is squeezed into the time duration T_B/N. Since WDM uses wavelengths as the channels, the time duration of the slot is maintained at T_B. In OCDM a channel is assigned with a unique temporal waveform. OCDM is a multiplexing and demultiplexing technique, and it will be discriminated from optical code division multiple access (OCDMA). More details of OCDMA will be described in Section 1.5.

TDM and WDM are compared in Fig. 1.8. In TDM, shown in Fig. 1.8(a), three time slots are multiplexed in the time T_B using a switch at the transmitter, and after

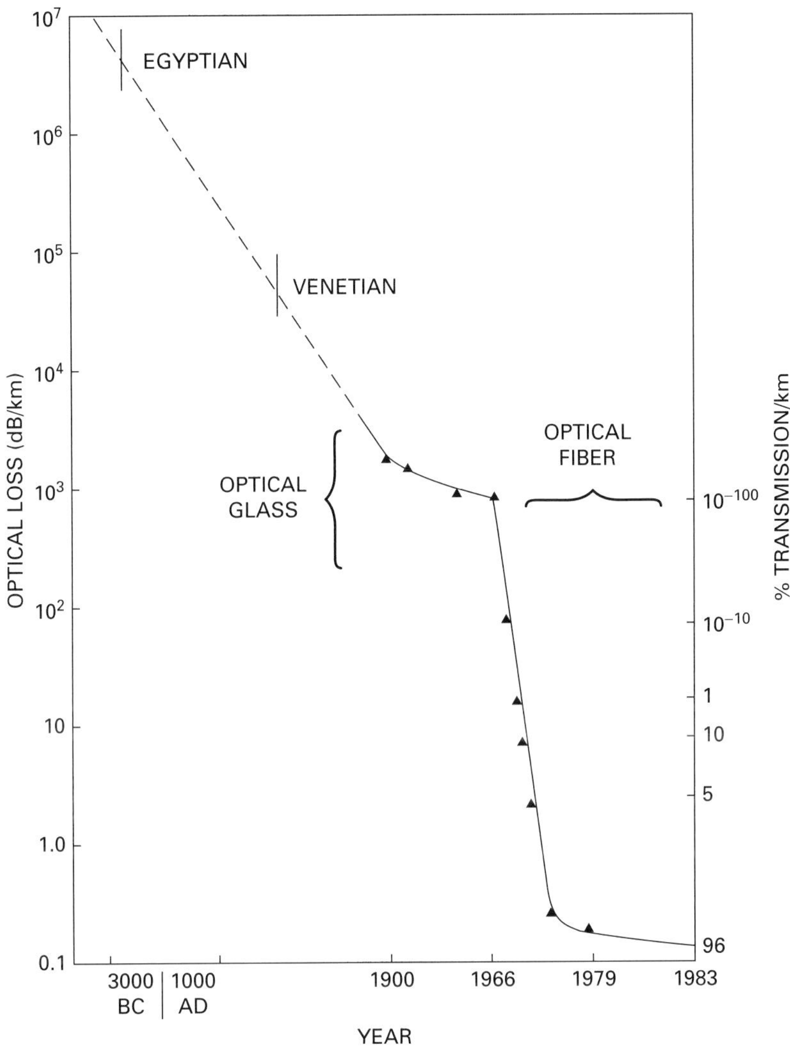

Figure 1.5 Historical loss reduction of glass versus year [6]. © Reprinted by permission of IEEE.

electrical-to-optical (EO) conversion the multiplexed optical signal is transmitted in an optical fiber. At the receiver the TDM signal is demultiplexed into a separate channel using a switch after optical-to-electrical (OE) conversion. Obviously, strict time synchronization between the transmitter and receiver is required, and the electrical switch has to operate at a speed of $T_B/3$, three times faster than the one-bit time duration. As the electrical switch is used for multiplexing and demultiplexing, this type of TDM is sometimes referred to as ETDM, in order to distinguish it from optical TDM (OTDM). Multiplexing and demultiplexing of OTDM are performed using a high-speed optical device. In WDM, shown in Fig. 1.8(b), data from three channels after EO conversion into different wavelengths are multiplexed using a wavelength multiplexer at the transmitter

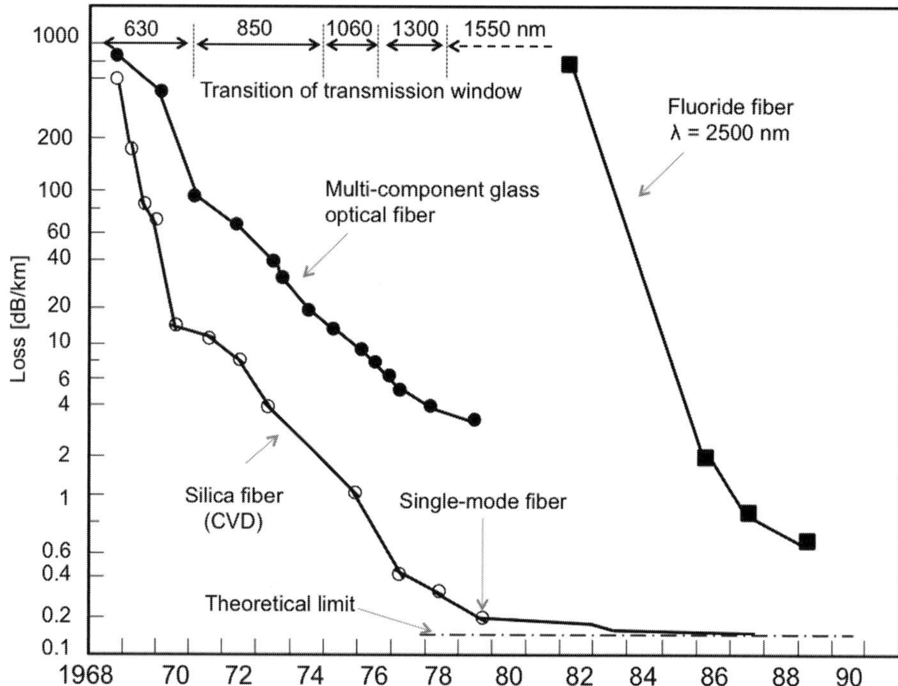

Figure 1.6 Silica optical fiber loss versus year [8]. © Reprinted by permission of Ohmsha.

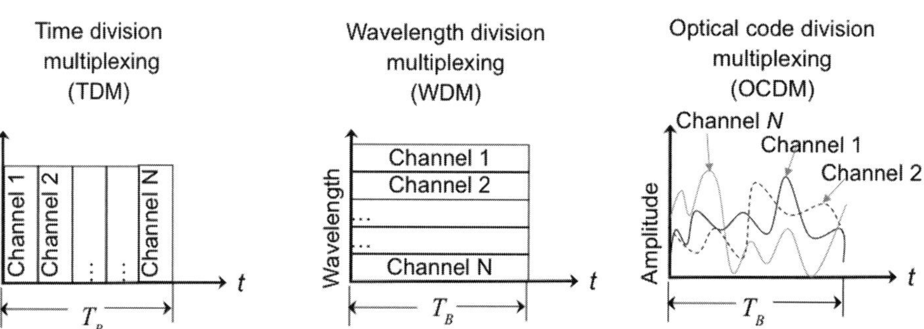

Figure 1.7 Multiplexing techniques: TDM and WDM, and OCDM.

and transmitted in an optical fiber. At the receiver the WDM signal is demultiplexed using a wavelength demultiplexer into a separate channel and converted into electrical signals. Multiplexing and demultiplexing of WDM are performed optically using passive optical devices, and therefore the data rate is not restricted by the speed of the multiplexer and demultiplexer, unlike TDM.

To cope with demands for the increase in bandwidth, optical fiber communications technology has been continuously evolving and has gone through three generations over three decades. The evolution of the transmission performance of optical fiber communications, the product of total transmission capacity per fiber and the distance,

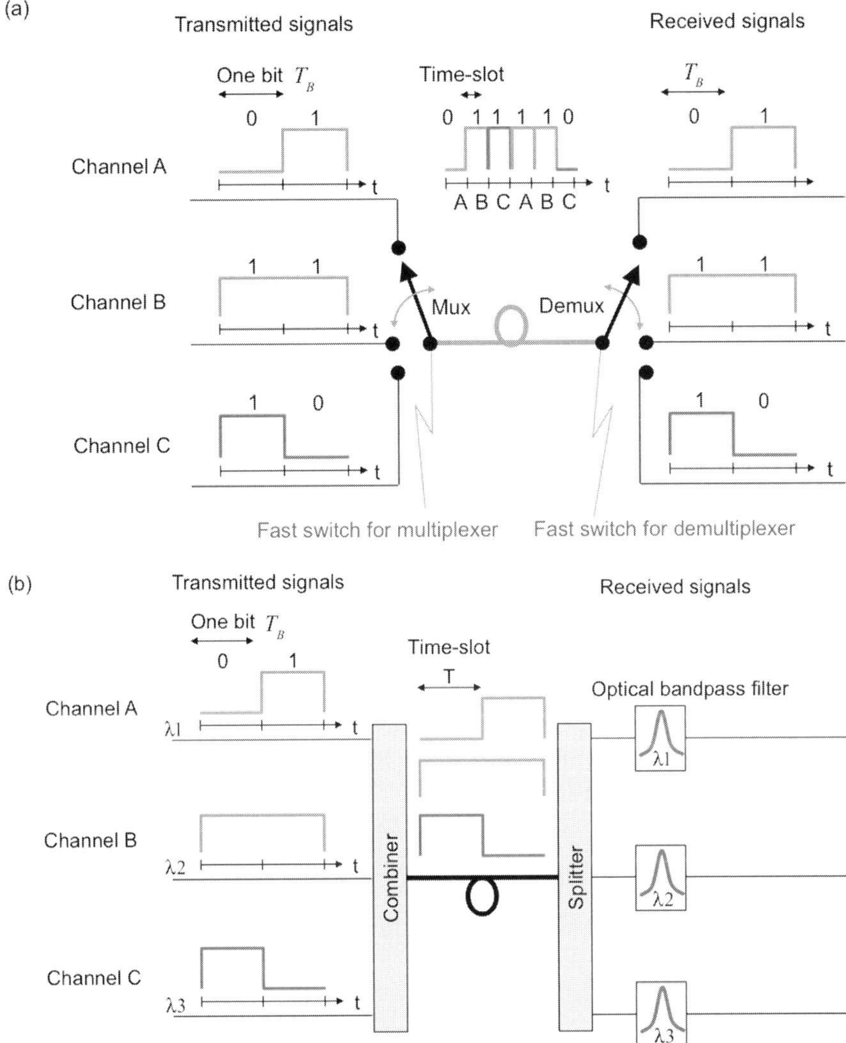

Figure 1.8 (a) Multiplexing and demultiplexing of TDM and (b) multiplexing and demultiplexing of WDM.

is shown in Fig. 1.9 [10]. Note that the solid and white circles, respectively, denote commercial systems in operation and experimental demonstrations in laboratories. It is interesting to see that commercial system deployment follows five or six years behind the laboratory experiments. A commercial optical fiber transmission system using a graded-index multimode fiber, F-100M and F-32M, at a wavelength of around 1300 nm, was the first system to be launched in 1981 in Japan. It was followed by deployment of the single-mode fiber in F-400M at a wavelength of 1300 nm in 1983, and the operation wavelength was shifted to 1550 nm in 1984, resulting in extension of the repeater span from 20 km to 80 km. During three decades the bit rate per wavelength has increased by a factor

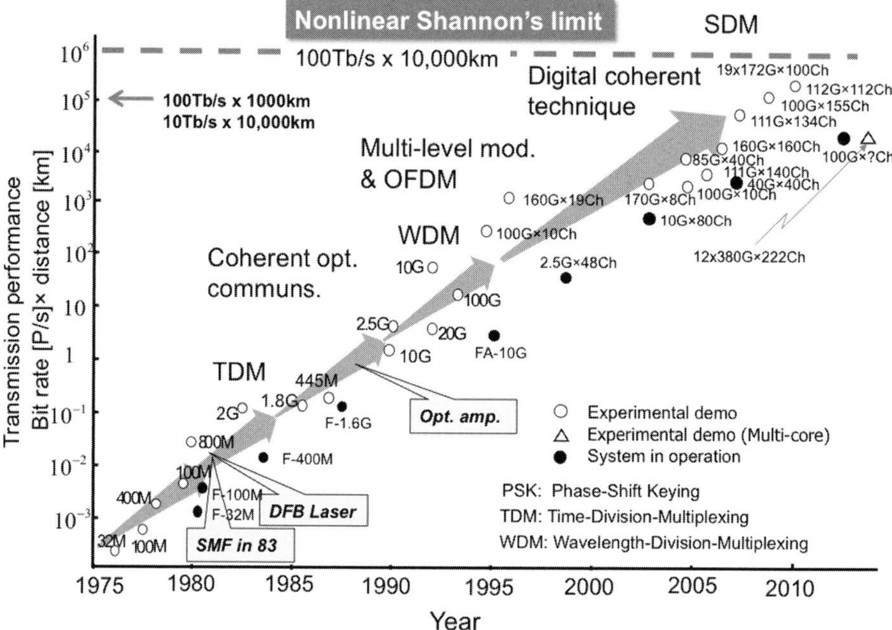

Figure 1.9 Evolution of optical fiber transmission technology: product of transmission capacity and distance versus year. Updated after [10].

of 10^3 from 100 Mb/s to 100 Gb/s, and the product of the bit rate and the transmission distance has increased by a factor of 10^6 as seen in Fig. 1.9. Some of the major laboratory experiments have exceeded 10^5 Pb/s × km, which typically range between 100 Tb/s over 1000 km and 10 Tb/s over 10,000 km. On the other hand, the deployment of commercial systems in public telecommunication networks has evolved steadily up to 40 Gb/s with the total capacity per fiber of a few Tb/s. The commercial deployment of a digital coherent 100 Gb/s long-haul system has been underway since early 2013.

For possible solutions to the emerging issue of capacity crunch, other multiplexing techniques such as space division multiplexing (SDM) have been revisited. As shown in Fig. 1.10 [11], there are two approaches. One approach is via a multicore fiber with a number of cores embedded in cladding of the fiber [12]. A record total transmission capacity up to 1.0 Pb/s using 11-core fiber, 380 Gb/s, 222-WDM has been reported, but the transmission distance is limited to 53 km as shown by the triangle in Fig. 1.9 [13]. However, unprecedented challenges remain in fiber splicing, optical amplification, crosstalk reduction between cores, and light coupling. The other approach is mode division multiplexing (MDM) via a multimode fiber which supports more than one waveguide mode. The modes can be exploited as independent channels, as shown in Fig. 1.11. In practice, for long distance optical communication, mode coupling between the independent channels is inevitable. According to information theory, however, the coupling between the channels does not necessarily cause a loss of capacity as long as this coupling can be described as a unitary transformation. For simplicity, compare two-mode transmission in a fiber with single-mode transmission at the same total input power. The

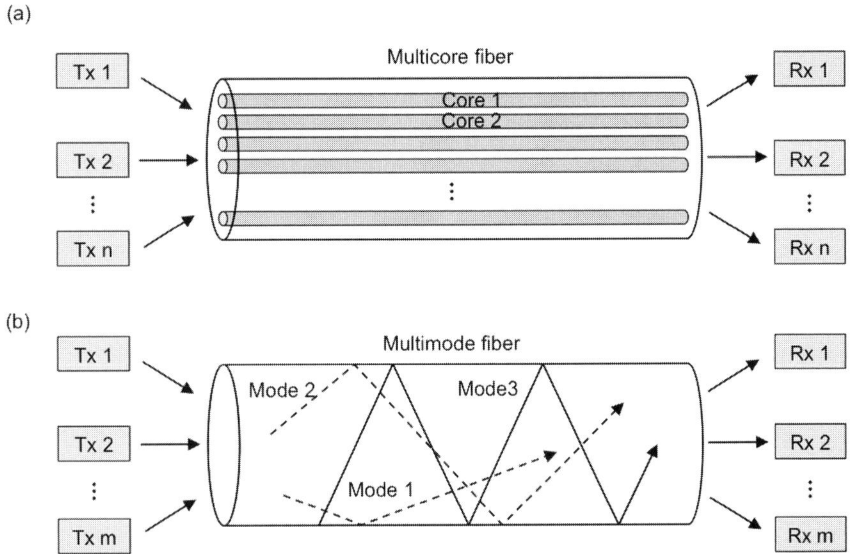

Figure 1.10 Space division multiplexing (SDM): (a) multicore fiber and (b) multimode fiber [11]. © Reprinted by permission of NTT.

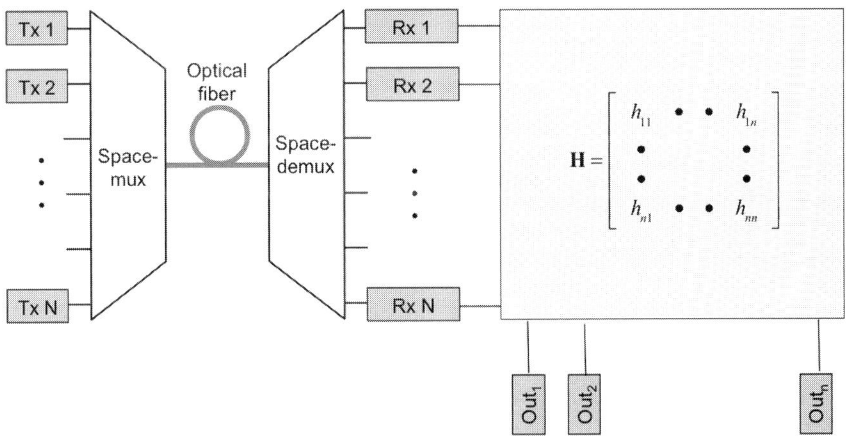

Figure 1.11 Space division multiplexing (SDM) transmission system using MIMO technique [12]. © Reprinted by permission of IEEE.

spectral efficiency of the two-mode transmission is $2\log_2(1 + \frac{1}{2}S/N)$ and to some extent gains over the efficiency $\log_2(1 + S/N)$ of the single-mode transmission. Here, S/N denotes the signal-to-noise ratio. This concept of parallel transmission is well established in wireless communication, where multiple-input multiple-output (MIMO) techniques have been widely used to increase the reach and capacity of wireless links [14]. Note that there is a difference between parallel transmission and multiple transmission. In two-channel parallel transmission the information data are split in half, and the two signals are transmitted separately on two channels. In multiple transmission the same information data are duplicated and transmitted separately on two channels, and therefore no gain

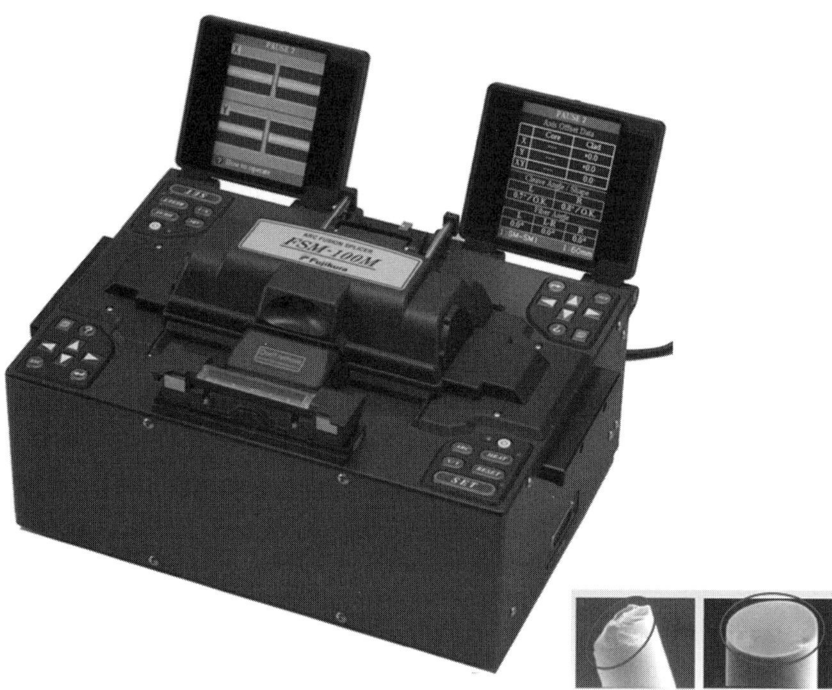

Figure 1.12 Fiber fusion splice machine. © Reprinted by courtesy of Fujikura Ltd.

in spectral efficiency is obtained. Recently, MIMO has been introduced in optical fiber communications. An experiment has demonstrated 20 GSymbol/s quadrature-phase-shift-keying (QPSK) signals transmission over a 6×6 MIMO channel over a 1200 km differential-group delay compensated few-mode or two-mode fiber [15]. More details of few-mode (two-mode) fibers are described in Section 3.4.

Let us review the progress in optical fiber and device technologies since the mid-1970s. The first generation was characterized by electrical time division multiplexing (ETDM) in the early 1980s. A single-frequency laser such as a distributed feedback (DFB) laser diode and distributed Bragg reflector (DBR) laser diode along with a single-mode fiber (SMF) were key enablers to realize multi-gigabit rate transmissions. Believe it or not, it was doubtful at that time whether a single-mode fiber having only about 5 μm radius core could be connected with low loss. Having achieved an optical fiber loss below 0.2 dB/km, a low-loss splice was regarded as a stringent requirement. This problem was solved using a well-engineered automatic fusion splice machine and a fiber cutter. The fusion splice machine melts the end faces of the fibers at a temperature of about 1500 °C by arc discharge between electrodes, and splices by pushing forward two fibers. It required skilful fine manual position alignment in three directions between the fibers. However, an automatic fusion splice machine was soon developed in which the positioning mechanism uses a video monitor. This machine, shown in Fig. 1.12, works automatically and allows the worker in the field plant to cut and position fibers without any particular skill. In this way a fusion splice can be completed in 15 seconds. The typical value of splice loss of the single-mode fiber is 0.03 dB.

Figure 1.13 WDM grid.

In the second generation in the late 1990s, an optical amplifier had a tremendous impact on the cost reduction per channel as well as the extension of the transmission distance. In particular, an erbium-doped fiber amplifier (EDFA) played a key role in extending the transmission distance. The EDFA has a large gain of 30 dB and also covers the most transparent window of fused silica fiber, the conventional (C)- and long (L)-bands of the spectral region of 11 THz, ranging from 1530 nm to 1625 nm as shown in Fig. 1.13. The first fiber amplifier was demonstrated in early 1964 using Nd^{3+}-doped barium crown glass [16]. Another major advance was achieved with the operation of end-pumped glass-clad Nd^{3+}-doped silica fiber lasers [17]. A decade later, low threshold, high-slope-efficiency Nd^{3+}-doped fiber lasers were demonstrated [18]. One problem was the pumping wavelength. First, the 515 nm line of an argon laser was used as the pumping wavelength, and later developments included the second line at 980 nm and the third line at 1480 nm. Soon laser diodes became available for pumping both wavelengths. In parallel, WDM was introduced to increase further the transmission capacity per fiber by using a number of colored channels. The EDFA together with WDM technology has had a crucial economic impact by reducing drastically the cost of a long-haul transmission system, because a number of WDM channels can be amplified simultaneously with a single EDFA. For example, an EDFA can amplify 80 WDM channels in the C-band with a frequency interval of 50 GHz. More details of the EDFA will be described in Section 5.5.

In the third generation, the challenge has been to increase further the transmission capacity. Remember that there is a strict requirement that the signal bandwidth be squeezed into the 50 GHz frequency interval to maintain existing WDM systems with the least modification. This will keep the capital expenditure (CAPEX) as low as possible for the telecom carriers. For better spectral efficiency, wireless communication technologies came into play in optical fiber communications, such as multi-level phase- and/or amplitude-shift keying modulation formats as well as MOMO. A commercial dense WDM (DWDM) 40 Gb/s RZ-DQPSK system was deployed in 2007 in the trunk line between Tokyo and Osaka in Japan. As shown in Fig. 1.14 [19], by comparing 0.25 b/s/Hz of single-polarization OOK at 10 Gb/s with the frequency interval of 50 GHz, the spectral efficiency increases drastically up to 6 b/s/Hz on adopting a multi-level modulation format. In addition, it has become common practice to utilize polarization division multiplexing (PDM) or dual polarization (DP) to double the transmission capacity. The 100 Gb/s transmission system which will be deployed in 2013 in Japan uses the PDM QPSK modulation format. The quest for a higher bit rate, such as 400 Gb/s per wavelength, is already underway, in which multi-level higher than QPSK, for example,

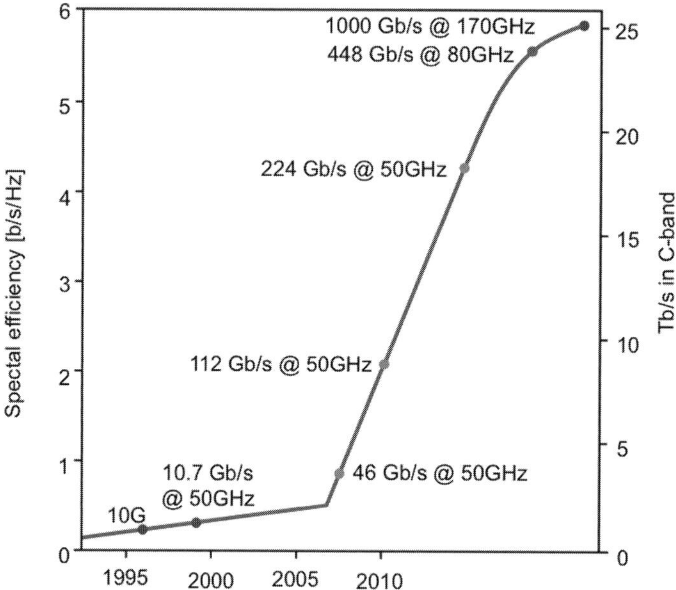

Figure 1.14 Spectral efficiency of commercial systems versus year [19]. © Reprinted by permission of IEEE.

16-QAM and 64-QAM, will be adopted. The price to be paid for higher multi-level modulation format is a higher signal-to-noise ratio, resulting in a shorter transmission distance. To quadruplicate the bit rate, keeping the symbol rate the same as for the 100 Gb/s system, a multicarrier modulation technique will be adopted. For 16-QAM modulation, two optical subcarriers with Nyquist interval around 30 Gbaud gives a bit rate of 400 Gb/s, and the spectral efficiency will be close to 6 b/s/Hz in the DWDM system. It is noteworthy that the symbol rate is limited mainly by the sampling speed of the analog-to-digital converter used for digital signal processing at the receiver. The key enabler for detecting the phase information of an optical carrier is a digital coherent transmission technique. This is because coherent receivers are capable of measuring phase and polarization as well as amplitude. Coherent optical communication attracted much attention in the mid-1980s [20, 21] because of its better receiver sensitivity compared with direct detection, leading to extension of the transmission reach in a long-haul transmission system. With the emergence of optical fiber amplifiers, however, interest in coherent optical communication decreased because the EDFA pre-amplifier was a much easier way to increase the receiver sensitivity than coherent detection.

The digital coherent transmission differs from its predecessor in that it is intended to equalize dynamically rather than statically the deterioration of multi-level modulated optical signals caused by chromatic and polarization dispersion with the aid of powerful digital signal processing (DSP). This is a result of the rapid progress in silicon complementary metal oxide semiconductor (CMOS) LSI technology, underpinned by Moore's law. In such a way the limitation to the transmission distance has been overcome, although it is not perfect. With the power of DSP, another problem of the carrier

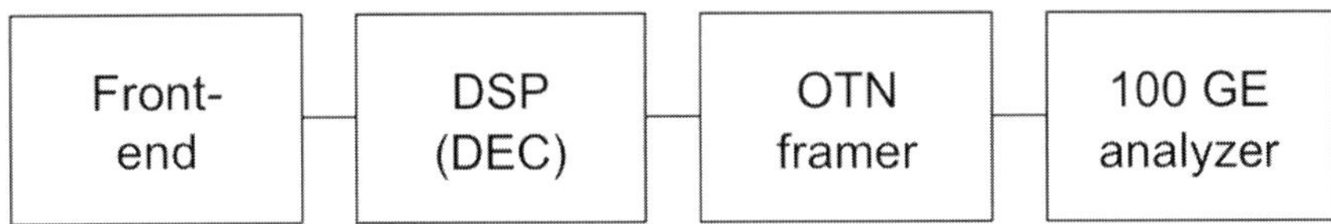

Figure 1.15 100 Gb/s digital coherent signal processor LSI: configuration and DSP module [22]. © Reprinted by permission of the Optical Society of America.

frequency offset between the light sources at the transmitter and the local oscillator has been solved. This tolerates a rigid requirement for the phase stability of the light sources of heterodyne and homodyne detection, which was one of the stumbling blocks in the previous coherent detection. At first, DSP at the symbol rate resorted to offline processing. Very recently, however, a milestone of 127 Gb/s PDM-QPSK Ethernet over OTN has demonstrated a chromatic dispersion compensation of 40,000 ps/nm by using real-time DSP at the receiver, amounting to more than 2000 km transmission in a standard single-mode fiber (SMF) at 1550 nm [22]. The receiver configuration along with a photograph of the DSP package is shown in Fig. 1.15. The DSP consists of a training sequence addition/removal framer, analog-to-digital converter, equalizers, framer synchronizer, and forward error correction (FEC) en/decoder. The 40 nm CMOS process has been adopted, and the DSP LSI has 100 million gates.

On increasing the transmission capacity per fiber, the total input power goes up. This raises two issues: one is non-linear effects of optical fibers, and the other is physical damage to optical fibers. C. E. Shannon forecast [23] that the capacity of a linear channel with additive noise would grow indefinitely with increasing signal power. However, this is not the case with non-linear channels of optical fiber transmission systems. The most predominant non-linear effect arises from the intensity dependent refractive index (Kerr effect) and results in a number of phenomena such as self-phase modulation, cross-phase modulation and inter-channel and intra-channel four-wave mixing. For DWDM transmission systems, intra-channel crosstalk between the subcarrier and inter-channel crosstalk between channels remain predominant, even if the linear impairments are compensated for by the DSP. The intra-channel non-linear effects have to be compensated for both by pre-distortion based on perturbation theory (PPD) at the transmitter and post-compensation using digital backpropagation (DBP) [24]. The polarization crosstalk due to XPM can be mitigated by a non-linear polarization crosstalk canceller (NPCC) [25].

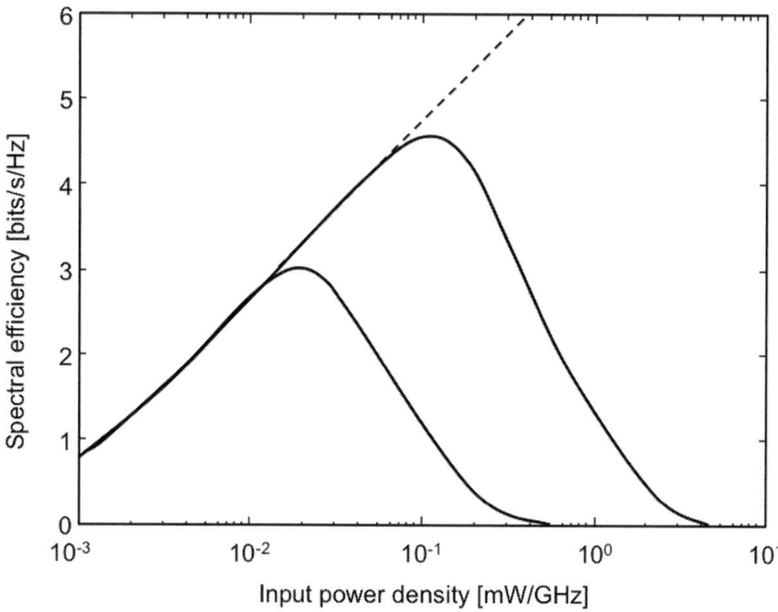

Figure 1.16 Spectral efficiency versus signal power spectral density [26]. © Reprinted by permission of *Nature*.

The theoretical analysis in Fig. 1.16 shows that the capacity of a non-linear channel does not grow indefinitely with increasing signal power, but has a maximal value [26]. This is a fundamental feature which distinguishes non-linear communication channels from linear communication channels. According to the theory [27], if the non-linear coefficient γ could be reduced by a factor of 1000 from the value of a conventional single-mode fiber, which is nearly achievable in an ideal hollow-core fiber [28], the spectral efficiency would increase by 80%. On the other hand, the reduction in fiber loss, even by a factor of one third conjectured in hollow-core fibers, has a very limited effect on the spectral efficiency.

In Fig. 1.17 the total input power in a fiber is plotted as a function of the product of the transmission capacity and the distance for various transmission experiments [11]. It is predicted that to realize a 1 Pb/s transmission system, a few watts of total input signal power will be required, and this exceeds the threshold power of physical damage to conventional optical fibers. One of the predominant types of damage is the so-called "fiber fuse." The laser light propagating in the fiber is strongly absorbed by the heated part when a fiber is heated locally to a temperature of about 1000 °C, and this heating induces a breakdown of the transparency of the fiber. The threshold of total input power for fiber fuse is 1.2~1.4 W. More details of fiber fuse will be described in Section 10.4.2.

1.4   Optical networks

A network consists of nodes and point-to-point links in between the nodes. A switch which connects a link with another link is located at a node. The connection between the

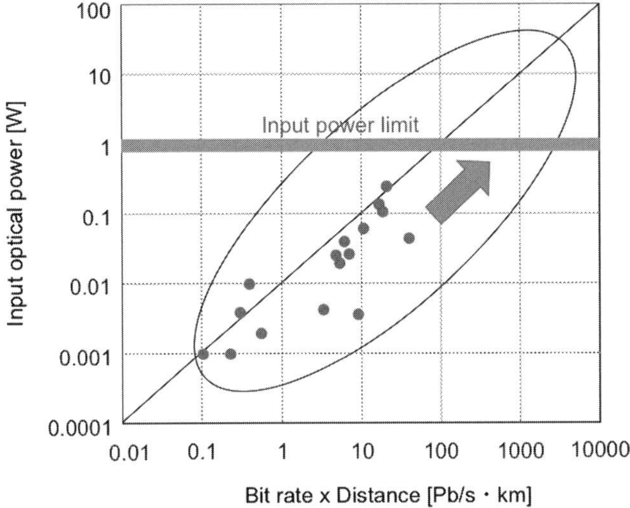

Figure 1.17 Total input power versus product of the transmission capacity and the distance [11]. © Reprinted by permission of NTT.

links may be either one-to-one or one-to-many, that is, so-called multipoint connection. A nationwide telecom network is broken up into wide area networks (WANs) or long-haul networks, metropolitan area networks (MANs), and local area networks as shown in Fig. 1.18. A WAN or long-haul network interconnects cities, extending over thousands of kilometers across a continent, while a MAN is an interoffice network of city size, typically a few tens of kilometers, in which central offices (COs) of the carrier interconnect. The access network extends from a CO to LANs in business districts, campuses and residential zones. In the access network, optical fibers or metal cables are laid from each telephone office to customers' homes in an area of typically ten or a few tens of kilometers. WANs and MANs are owned by telecom carriers, and LANs are privately owned networks. From a topological viewpoint, a WAN is logically a mesh network whether or not the nodes are physically connected with a pair of optical fiber cables. There will be direct connections between the COs, and there may be a logical connection as denoted by a dotted line in the figure. In reality the long-haul network of NTT consists of ring networks, interconnected through a gateway node.

The telecommunication network has been evolving from an opaque network to a transparent network by incorporating technological innovations. An opaque network is at an intermediate stage, in which metal cables in the point-to-point links are replaced with optical fiber cables before the optical switches replace electrical switches at the nodes. Therefore, the optical signal has to undergo OE and EO conversions each time it passes through the switch at the node. Currently, the public network of NTT in Japan, shown in Fig. 1.19 [10], is translucent, which means that part of the network has become transparent, and the rest of it remains opaque. In the metro ring network in large cities, a small-scale optical switch, a reconfigurable add/drop multiplexer (ROADM), has been deployed at nodes.

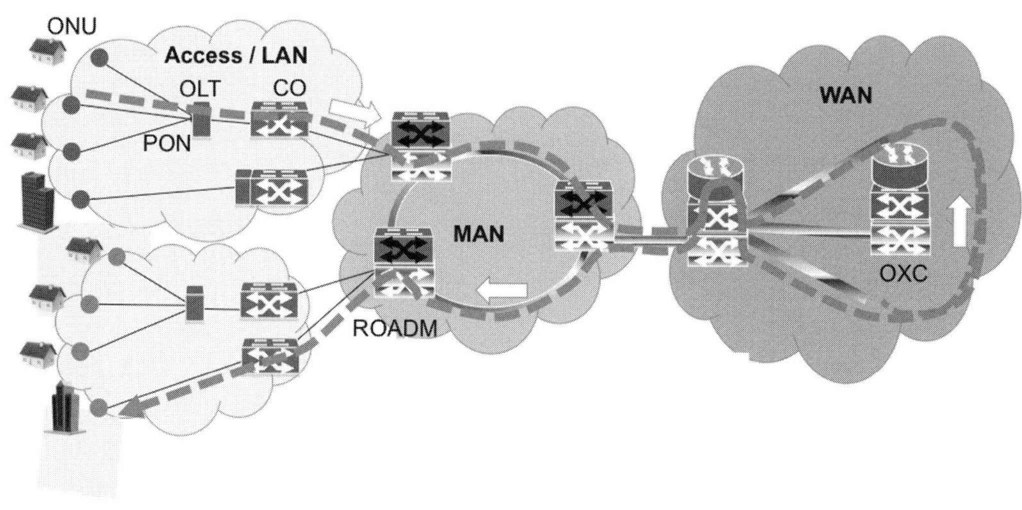

Figure 1.18 Optical networks: WAN, MAN, and Access/LAN.

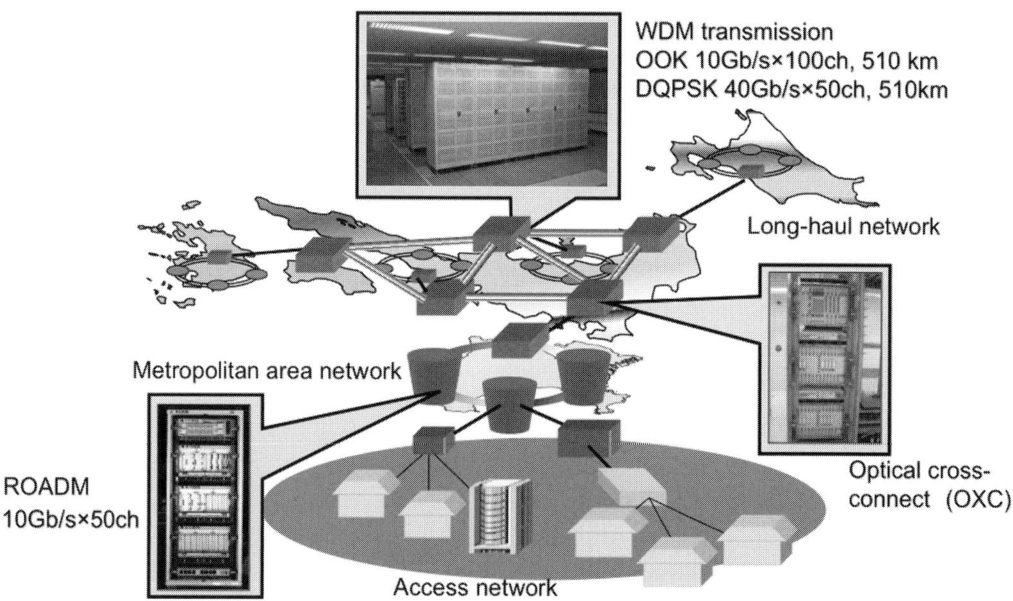

Figure 1.19 State-of-the-art optical network in Japan. Updated after [10].

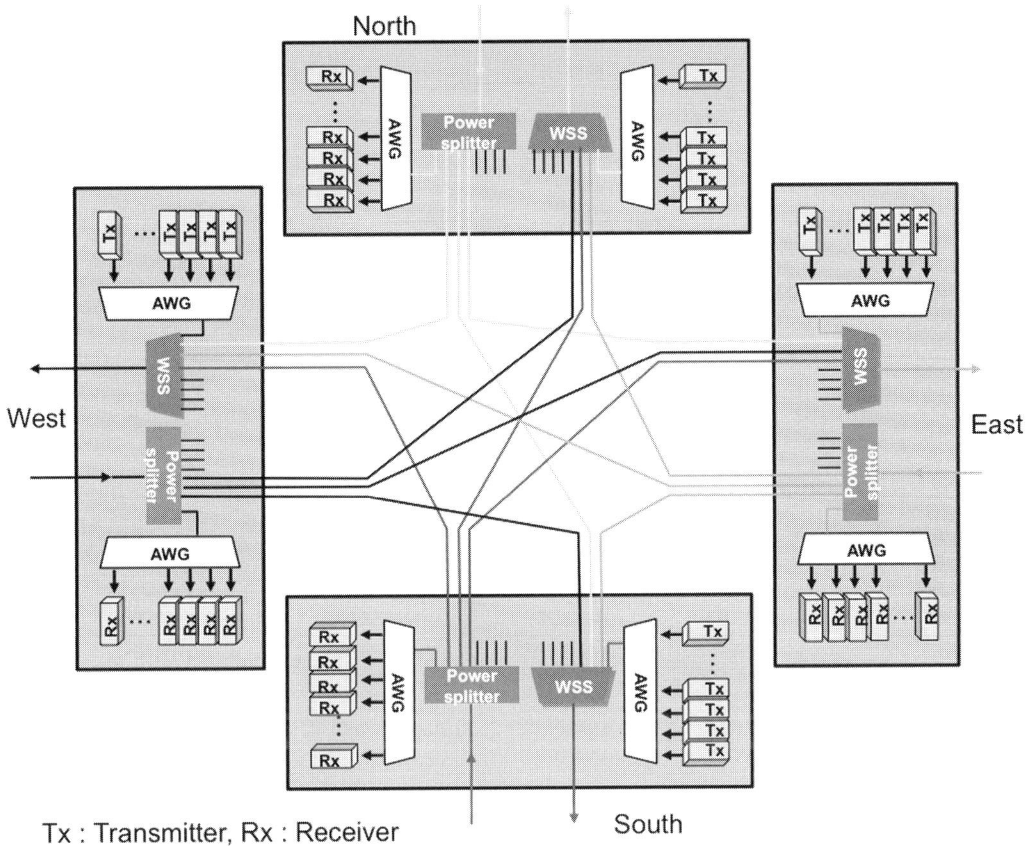

Figure 1.20 Architecture of colorless, directionless, and contentionless ROADM [27]. © Reprinted by permission of IEEE.

A typical architecture of a four-degree wavelength-selective switch (WSS) based ROADM is shown in Fig. 1.20 [29]. At the input port of the ROADM, the optical power splitter creates multiple copies of input WDM channels, and one copy is sent to each output port. At the output port, all copies of the WDM channels from each of the other input ports are injected into different WSS ports. The WSS possesses a single common port and multi-wavelength ports. The output of the WSS common port is amplified and launched onto the outbound transmission fiber. A copy of all WDM channels is directed to an arrayed waveguide grating (AWG) filter which demultiplexes individual channels for local optical link termination (drop). Similarly, another AWG filter multiplexes the input from a series of wavelength-specific add ports onto a single fiber connected to one of the ports of the WSS. By introducing optical switches into the network, a dynamic configuration of the network is enabled, and thus connections can be dynamically set up and torn down on demand. For example, traffic engineering can be performed in such a

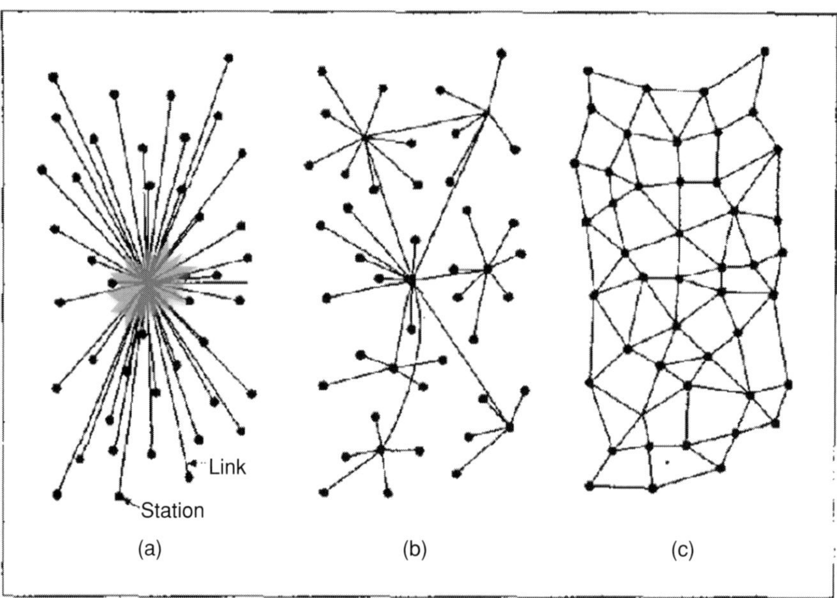

Figure 1.21 Three topologies of networks: (a) centralized, (b) decentralized, and (c) distributed [28]. © IEEE by permission.

way that more bandwidth or wavelength can be assigned where the traffic increases, and less wavelength can be assigned as the traffic volume becomes smaller. This will be the case during the daytime and during the night in the business districts and residential areas. However, the ROADM has limitations in its functionality because a fixed wavelength is assigned to specific ports, a fixed direction is assigned to the wavelength multiplexer and demultiplexer, and add/drop is partitioned to avoid wavelength contention. Flexibility in ROADM switching capability will be enhanced by adding the functionalities colorless, directionless, and contentionless (CDC) to the WSS. A downside of the CDC ROADM will be its high cost because it requires a large number of transmitters and receivers. By scaling up the ROADM an optical cross-connect (OXC) switch having a large port count can be configured. The OXC switch can switch a number of a optical signals from the input ports to output ports without OE–EO conversion. This switch is expected to be deployed at the nodes in long-haul networks in the near future. The evolution of opaque to transparent will create optical networks in a true sense. The benefits gained from an optical network are summed up as follows:

- minimum latency,
- larger bandwidth,
- reduction of power consumption.

Packet switching is a basis of the Internet. The concept of packet switching was invented by Paul Baran in the mid-1960s. The motivation for this concept was to enhance the survivability of military-owned networks against attack during the cold war era. Figure 1.21 shows original drawings of three different networks, centralized,

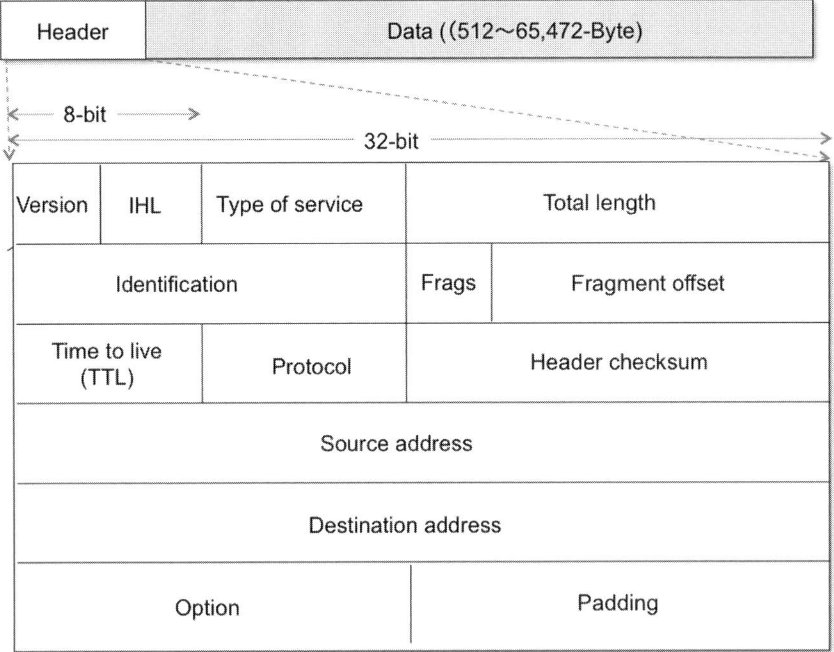

Figure 1.22 Structure of IPv4 header.

decentralized, and distributed networks [30]. The centralized network has all its nodes connected to a central switching node to allow simple switching, giving it a single point of high vulnerability, while the decentralized network comprises small centralized clusters. The distributed network is a network without any hierarchical structure. There is no single point which is vulnerable to paralysis, unlike the centralized topology. The network has to respond quickly to changes caused by damage and has the capability to deflect traffic autonomously from the point of damage and route to other surviving nodes. To enable this, the idea is to send digital data by segmenting the message into packets and allowing the packets to traverse through nodes hop by hop. For example, if half the network is instantly destroyed, the remainder of the network reorganizes itself and routes traffic effectively. The routing protocol travels along with packets by going to and from addresses so that according to the address the node can forward the packet to the next neighboring node. The concept of packet switching in a distributed network helped built the ARPANET, the origin of Internet Protocol (IP). The structure of a packet of the current IP version 4 (IPv4) is shown in Fig. 1.22. The header consists of 32-bit long IP addresses of the source and destination. An 8-bit time to live (TTL) field helps prevent packets from circulating in the Internet. Each router decrements the TTL count by one as the packet traverses the router, and the packet is no longer forwarded and is discarded when the time to live field hits zero. It is noteworthy that the next version of IP, IPv6, has 128-bit long addresses, which was developed to deal with the anticipated exhaustion of addresses of IPv4. It is difficult to imagine the total number of addresses,

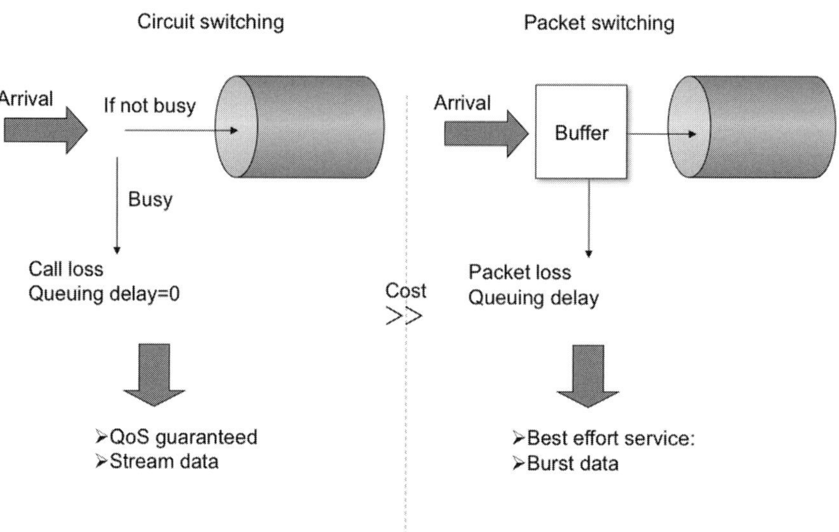

Figure 1.23 Circuit switching versus packet switching [39]. © Reprinted by courtesy of H. Miyahara.

2^{128}, approximately 340 undecillion ($=10^{36}$). A good way to visualize this value is to imagine the size of address space that one would fill with a corresponding number of 1 mm^3 grains of sand. This would be equivalent to 300 million planets Earth.

It is useful to look at the fundamental differences between packet switching and circuit switching as summarized in Fig. 1.23 [31]. Circuit switching, which the plain old telephone service is based upon, guarantees quality of service (QoS). This means that unless the circuit is not busy, the circuit is established, and the message can reach the receiver without loss of data bits. However, if the circuit is busy, call loss occurs, requiring a re-dial. In packet switching, by contrast, the message is stored in a buffer at each node and forwarded if the bandwidth is available for the next hop, and hence queuing delay occurs. If the buffer is full upon the arrival of the packet, the packet will be discarded, and packet loss occurs. As a consequence, the level of service is not guaranteed QoS but it is a best effort service. The cost is crucial. Packet switching can offer a less expensive service than circuit switching because of its better bandwidth utilization.

An example of the end-to-end connection is illustrated by the dashed line in Fig. 1.18. The data signal bearing a customer's message generated at the optical network unit (ONU) in the source access network goes into the MAN, traverses over the WAN to the other MAN, and reaches the destination access network. Suppose that the data signal is carried in Ethernet frames and transmitted over GE-PON. The Ethernet frames are aggregated in a CO. At the CO the traffic from ONUs is further aggregated and encapsulated into optical transport units (OTUs) at 10 Gb/s or even higher, say 100 Gb/s, and sent out over the WAN to the destination MAN, followed by delivery to the destination access network. The OTU is a frame defined in optical transport network (OTN) architecture. The OTN provides cost-effective transparent transport over WDM

Table 1.1 OTN/SONET/Ethernet transmission rates

Data rate (Gb/s)	OTN	SONET OC	Ethernet
2.488		48	
2.67	1		
9.689		192	
10.7	2		10GbE
43.0	3		40GbE
112.0	4		100GbE

OTN optical-channel transport unit; SONET synchronous optical network; OC optical carrier level.

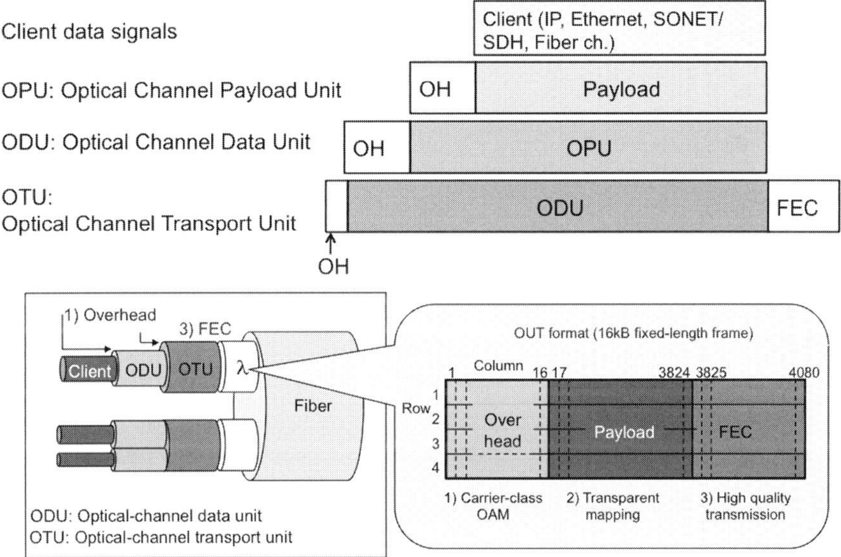

Figure 1.24 Frame structures of optical channel transport (OUT).

networks for a variety of client signals such as the Ethernet and synchronous optical network (SONET), which is standardized by ITU-T. The Ethernet frame can only be transmitted over 40 km or less, and the OTN enables long-haul transmission by adding necessary functions such as FEC and encapsulating an Ethernet frame in the OTN as shown in Fig. 1.24. As summarized in Table 1.1, four data rates are available which are all compatible with the data rates of existing Ethernet frames.

An emerging transmission technique with an elastic bandwidth or bit rate will have a significant impact on WDM networks. As shown in Fig. 1.25, an optical path is no longer of a fixed bandwidth but can be flexible on demand. For example, for a longer reach the bandwidth of the optical path has to be decreased, and for a shorter reach a higher bandwidth optical path can be established. This flexibility in optical path provisioning will enhance efficiencies in spectrum utilization, protection, and power consumption. The conventional paradigm of WDM will shift to a new "flexible grid network." Key optical components include gridless in addition to CDC, the so-called

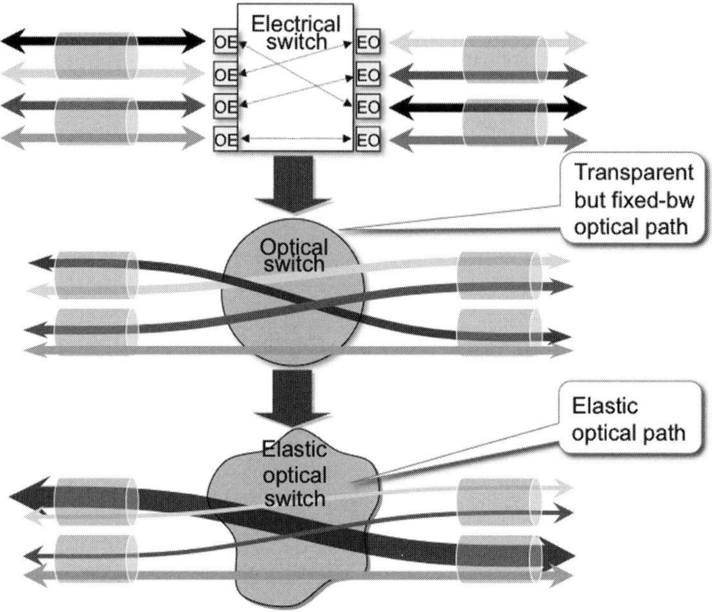

Figure 1.25 Evolution of the optical path from static bandwidth to elastic bandwidth.

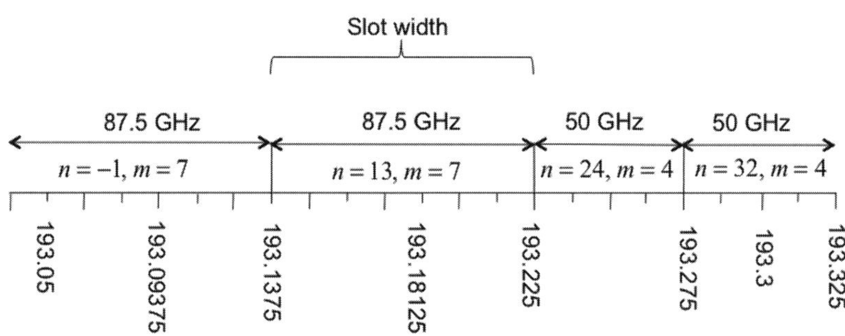

Figure 1.26 Flexible grid [30]. © Reprinted by permission of ITU-T.

CDCG ROADM, using bandwidth-tuneable WSS and bit rate-adaptive transmitter and receiver. The standardization of the flexible grid in ITU-T G694.1 was completed in 2012. As shown in Fig. 1.26, it recommends a slot width with slot width granularity of 12.5 GHz along with a nominal central frequency granularity of 6.25 GHz [32]. This flexible grid will allow a mixed bit rate transmission system to allocate frequency slots with different widths, leading to better exploitation of the spectrum.

Finally, to cope with the problem of global warming, one has to be conscious of the energy consumption of information and communication technology (ICT). The

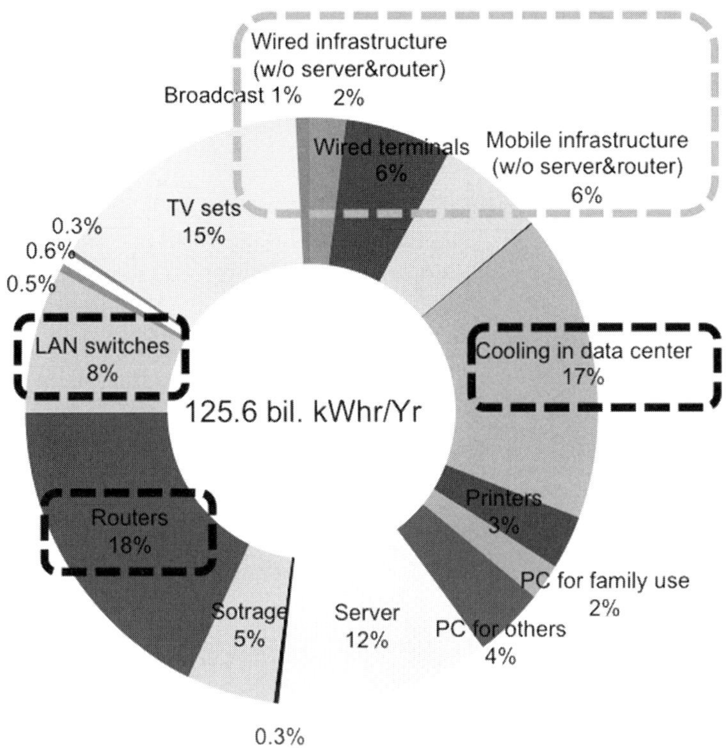

Figure 1.27 Power consumption of ICT in 2020 [31]. © Reprinted by permission of the Ministry of Internal Affairs and Communications in Japan.

breakdown of power consumption of ICT forecast in 2020 in Japan is shown in Fig. 1.27 [33]. Note that this forecast is based upon the assumption that no countermeasure of power saving is taken. The energy consumption of ICT is only 6% of the total energy consumption in 2012 in Japan, but it is forecast to reach 10% of the total 125.6 billion kWh in 2020 with the CAGR, the compound annual growth rate of 6%. In the breakdown, the LAN switches and routers along with their cooling and ventilation consume 43%, while the wired and wireless communications consume less power, 14%. This is a warning that one should pay serious attention to reducing the power consumption of ICT, particularly of L2/L3 switches, widely deployed in data centers, and Internet exchanges to residential areas. Therefore, it may be the case in the future that cutting-edge highly functional transmission and network equipment and components will not be acceptable to the market if they are not energy efficient.

1.5 Access networks

"Last mile" is a generic term for any final leg of broadband connectivity from a telecommunications provider to the subscriber in an access network. It is sometimes referred

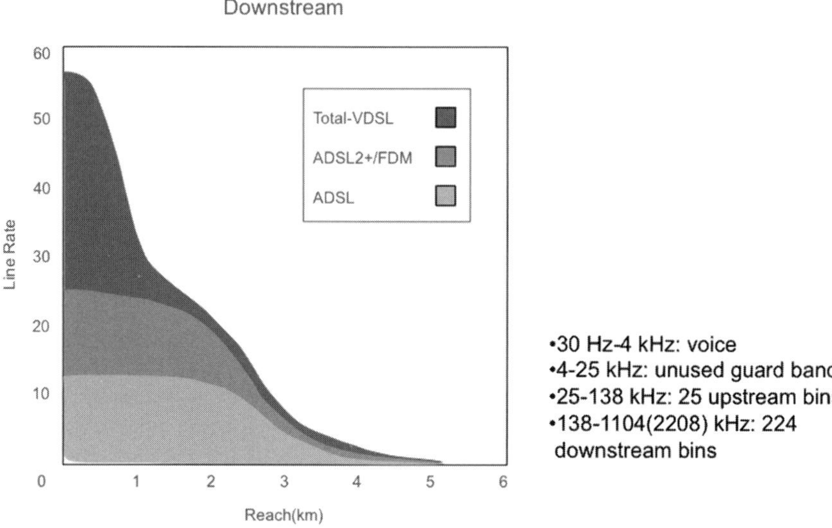

Figure 1.28 Bandwidth versus distance of digital subscriber line (DSL).

to as "first mile" because it is the first mile from the subscriber to the service provider. Access networks fall into three categories: wireless, copper, and fiber. Cost is the first priority for service providers. The cost of deploying the access system has to be minimized, while maximizing revenue from the service offerings. In order to reduce the cost, the digital subscriber line (DSL) uses installed twisted pairs of copper cable for telephony. Voice for telephony occupies up to 4 kHz of the bandwidth of copper cables, and DSL uses the higher frequency band. As seen in Fig. 1.28, the line rates of VDSL and ADSL decrease rapidly due to attenuation as the distance increases. DSL is capable of 30 Mb/s for a distance of 1 km, but it can only provide less than 10 Mb/s for a reach of 3 km. DSL uses a point-to-point architecture which is distinct from the wireless network which uses point-to-multipoint (P2MP) architecture, and the whole bandwidth is dedicated to each subscriber. Although DSL has a slight edge over optical fiber in terms of cost, the raw bandwidth capability of optical fiber overwhelms that of DSL. This is why optical fiber has penetrated widely, replacing deployed DSL.

Wireless has the lowest deployment cost because it has the lowest outside plant costs. WiFi and WiMAX[1] are widely deployed among the wireless family. WiFi has a range of only 100 m and a bit rate of 10–50 Mb/s. WiMAX is designed for fixed and mobile access networks, and its data rate is 70 Mb/s only if the distance is shorter than 5 km. They use point-to-multipoint (P2MP) architecture, and hence the bandwidth has to be shared with multiple subscribers. Therefore, wireless access lacks sufficient bandwidth to support high bandwidth video applications such as IPTV broadcasting.

The last but not least option to be considered for access networks is optical fibers. Fiber-to-the-Home (FTTH) adopts optical devices and components developed for

---

[1] WiFi and WiMAX are standards of IEEE802.11 and IEEE802.16, respectively.

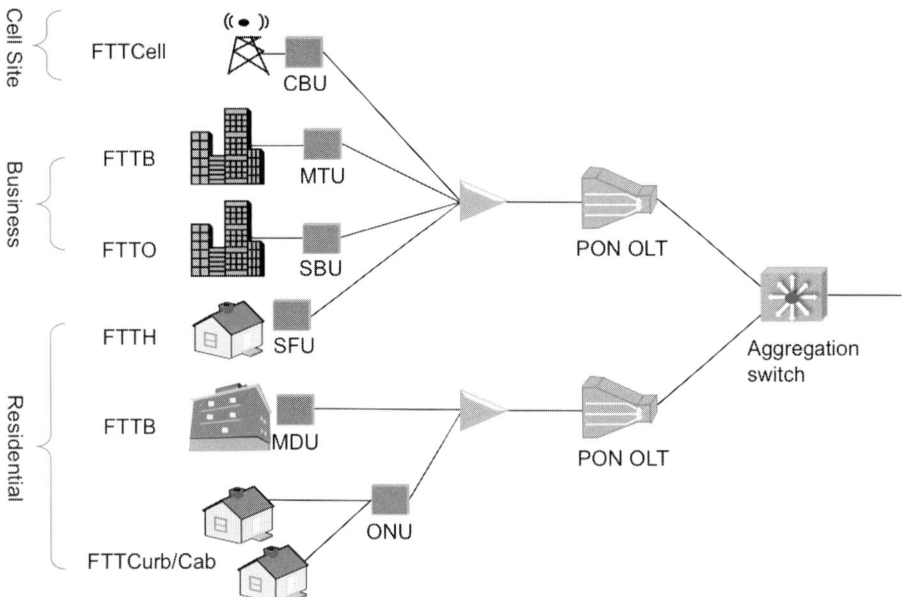

Figure 1.29 Architectures of various FTTx.

long-haul optical fiber transmission systems and high-volume products of LAN technology such as the Ethernet. FTTx is the access network which connects between the CO and subscribers with optical fibers. FTTx (Fiber-to-the-x) is the deployment of fiber cables to a specific location "x" with regard to the customer premises. In Fig. 1.29, x represents a node, curb or cabinet, building or basement, cell, or home. The difference between various "x" is in the degree of fiber penetration. In Fiber-to-the-Node (FTTN) the fiber is terminated in a street cabinet up to several kilometers away from the subscriber premises, with the final connection being copper. In Fiber-to-the-Curb/Cabinet (FTTC) the fiber is terminated in a street cabinet which is closer to the subscriber premises, typically within 300 m. In Fiber-to-the-Building/Fiber-to-the-Basement (FTTB) and Fiber-to-the-Office (FTTO), the fiber reaches the building, for example the basement, with a final copper connection to the individual space. Obviously, the deeper the penetration of fiber to the subscriber premises becomes, the higher the bit rate that can be offered. FTTB can offer the highest bit rate since the remaining segments use standard Ethernet or coaxial cable. The other FTTxs generally use VDSL for the final leg.

In addition, the next-generation FTTx architecture will include fiber-to-the-cell (FTT-Cell) backhaul applications via a cell-site backhaul unit (CBU) to the cellular base stations. Network convergence offers the prospect of optimizing the total cost of ownership (TCO) for network operators by eliminating heterogeneous and manifold network technology solutions in the access and aggregation domains. In this context, fixed access backhauling and mobile backhauling must be considered. Long term evolution (LTE) and LTE-advanced (LTE-A) are the relevant mobile network technologies which must be provided with backhauling solutions:

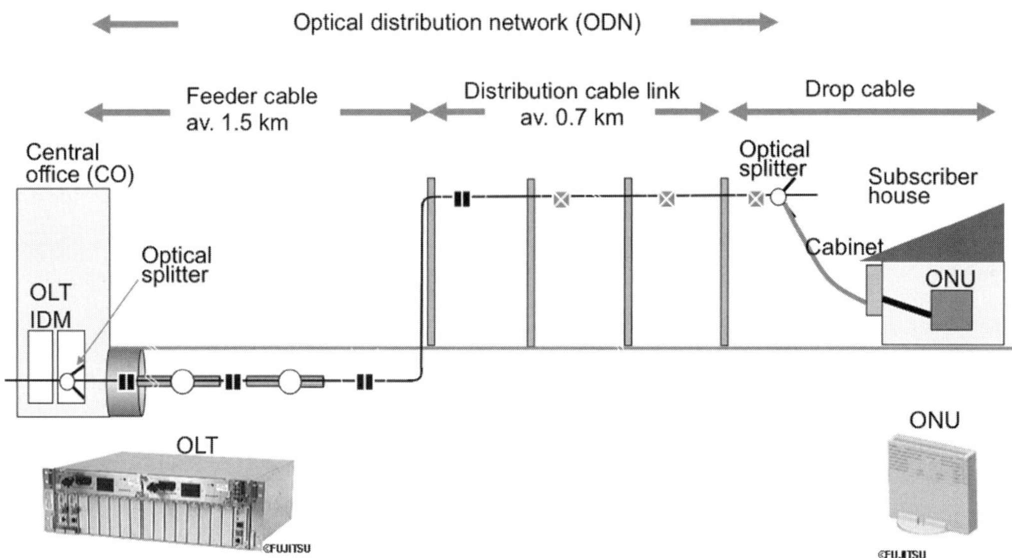

IDM: Integrated distribution module

Figure 1.30 Outside plant of FTTH, courtesy of H. Shinohara, NTT and S. Kinoshita, Fujitsu.

CBU: Cell-site backhaul unit
SFU: Single family unit
SBU: Single business unit
MTU: Multi tenant unit
MDU: Multi dwelling unit.

The passive optical network (PON) is the most important class of fiber access system. The typical outside plant shown in Fig. 1.30 consists of equipment and components located between the OLT (optical line terminal) in the CO, also referred to as the head-end, and the optical network units (ONUs) in the customer premises. Photographs of overviews of OLTs and ONUs are also shown in the inset. At the CO the public-switched telephone network (PSTN) and Internet services are interfaced with the optical distribution network (ODN) via the OLT. The gear and components include both optical and non-optical components of the network. The optical components make up the optical distribution network (ODN) and include splices, connectors, splitters, optical fiber cables, patch cords and possibly drop terminals with drop cables. The non-optical components include pedestals, cabinets, patch panels, splice closures and miscellaneous hardware. The optical fiber cables consist of the feeder cable from the OLT to the first splitter, distribution cable linking the splitter to the drop terminal near the subscribers, and the dedicated drop cables (<30 m) connecting to the individual ONUs. Details of the optical components are described in Chapter 5. Installation of outside plant equipment in an FTTx can be carried out in many ways, and each installation may be different, depending on factors such as the distance from the CO, residential density

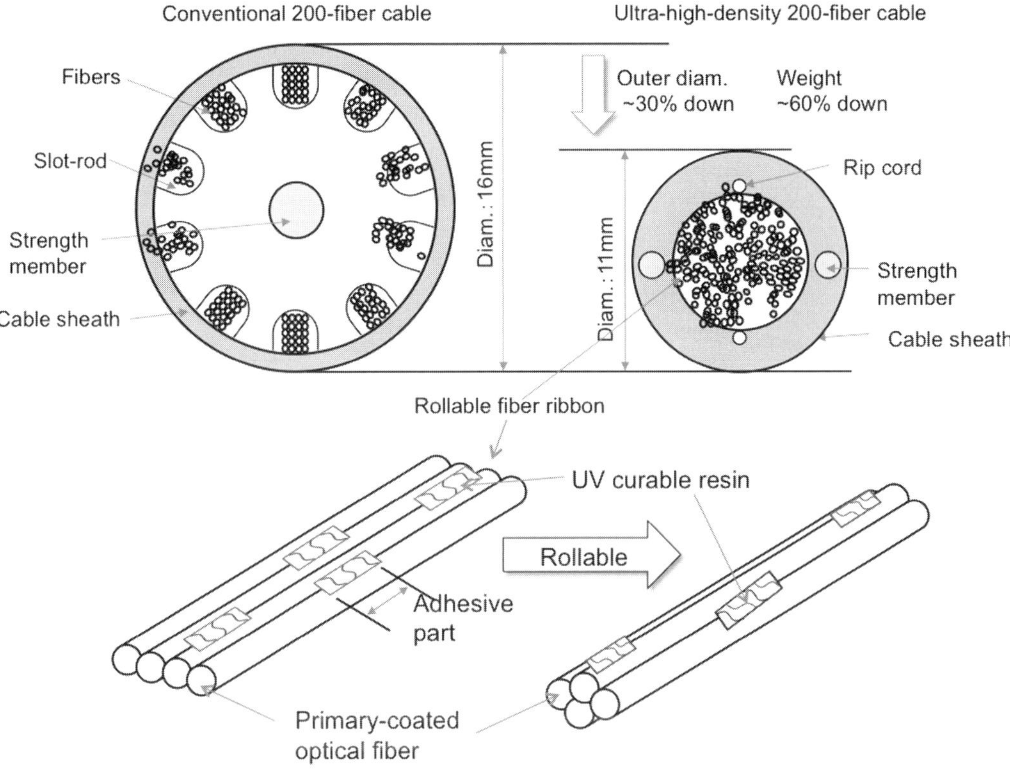

Figure 1.31 Structure of optical fiber cable for the feeder cable [32]. © Reprinted by permission of Elsevier.

and distribution urbanism etc. Optical fiber cables can be installed using either aerial or underground installation techniques. The placement of splitters and other passive components and cabinets used depends on geographical factors and on the PON topology.

In Fig. 1.31 cross-sectional views of conventional (on the left) and ultra-high-density optical fiber cables (on the right) for the feeder cable section and the distribution cable section are shown. Both cables have 200 optical fibers [34]. Compared with the conventional cable, the newly developed cable, which will soon be deployed in commercial PON systems, is significantly thinner and lighter. The inset of Fig. 1.31 shows 4~20 0.25 mm diameter primary-coated optical fibers aligned with an adhesive UV curable resin. This structure enables the fiber ribbon to be rolled up easily and accommodated very tightly in a cable. The ribbon can be spliced by mass fusion splicing machines or connected mechanically using a connector. Mechanical splices are the least expensive but have higher insertion loss and back-reflections than fusion splices. The fusion splice has very low loss (<0.1 dB) and almost no back-reflection but typically requires expensive and extensive fusion-splicing equipment and a well-trained technician; see the fusion splice machine shown in Fig. 1.12. The number of splices depends on the cable section lengths used, a typical section length being 2~6 km.

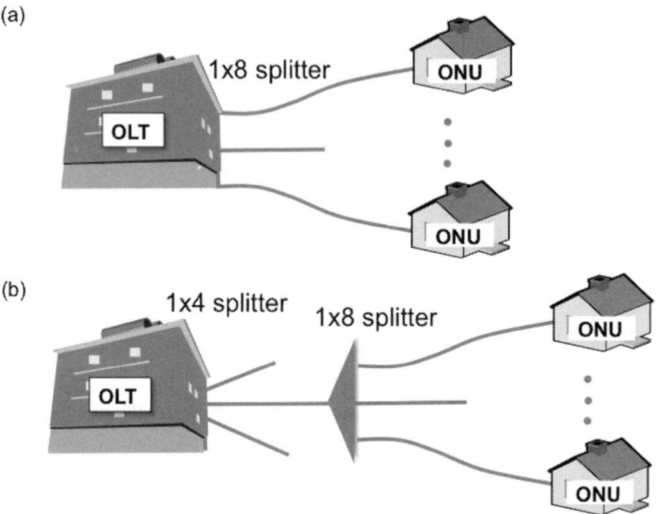

Figure 1.32 Passive optical network: (a) single star topology, and (b) double star topology.

There are two types of connections in the FTTH architecture: point-to-point and P2MP. Point-to-point architecture uses a dedicated fiber from the CO to each subscriber, referred to as a point-to-point network, while in the P2MP architecture a single fiber from the CO is shared with a number of subscribers. The PON has two main advantages.

- There is no requirement to install and maintain active components. Hence, operators can significantly reduce the power requirements for the outside plant. Furthermore, PON provides a more reliable access network solution due to the increased reliability of the optical fiber and passive splitters.
- Owing to P2MP architecture, the cost of deploying an optical access infrastructure is lower compared, for example, with point-to-point topologies because the cost of fiber installation between the CO and subscribers is shared among many subscribers. This contrasts with point-to-point access solutions.

The architecture of a PON is classified into two categories: single star and double star, shown in Fig. 1.32(a) and (b), respectively. The single star architecture is of point-to-point topology, while the double star architecture is in the form of P2MP. In the double star architecture there is a splitter between an OLT and a number of subscribers. If the splitter is a passive optical device, it is classified in the passive category. A passive optical splitter, a so-called star coupler, works in such a way that the power of the downstream signal is simply split and delivered to the subscribers, and the upstream signals from subscribers are combined and transmitted to the OLT. See Chapter 5 for more details of the star coupler.

The physical topology of an active network resembles a PON, in which the optical splitter is replaced with powered electronics such as an Ethernet switch, electronic-to-optical (EO) and optical-to-electronic (OE) converters, and an amplifier (Fig. 1.33). The logical topology is not P2MP but point-to-point because each subscriber is provided with

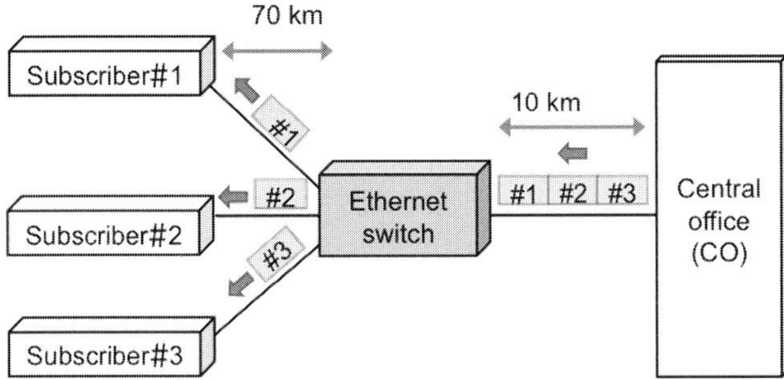

Figure 1.33 Active double star topology.

a dedicated link of full bandwidth. Owing to the electronics, the distance limitation is 80 km, regardless of the number of subscribers. However, it requires a large investment in the outside plant. Switches have to be installed in secured cabinets between the CO and the subscribers.

Scaling of PON systems in terms of the number of users, geographic coverage, and the transmission capacity has a significant impact on holding back both CAPEX and OPEX. Long-reach PON would be a cost-effective solution by consolidating several COs into a single CO. The transmission distance between the OLT and ONUs of 10G-EPON systems needs to be extended from the traditional 20 km range (60 km in the case of ITU-T G.987.4 extended reach XG-PON). An extension to over 100 km is desirable because it would allow several telephone central offices to be consolidated and would widen the coverage of PON. By extending geographic coverage, long-reach PON can combine optical access and metro networks into an integrated system. Without the long reach capability, the PON system cannot be deployed widely in the near future. The split ratio is another issue for the system economics. To attain the maximum split ratio based on the given power budget, the current 10G-EPON and XG-PON are standardized at a maximum 32- and 64-way split. Increasing the split ratio beyond 1:64 (e.g., 1:128 to 1:256) is attractive, especially for the service-independent introduction scenario, so that it improves the overall economics of 10G-EPON and XG-PON passively. This is summarized in Table 2.1.

1.6　Local area networks

The Ethernet is a predominantly LAN technology. It was originally developed by R. Metcalfe[2] at the Palo Alto Research Center, XEROX in 1973 for computer networking, and was introduced commercially by 3Com in 1980. The working group IEEE803.3 standardized the first 10BASE5 Ethernet at 10 Mb/s in 1982. This used coaxial cable as a shared medium. Since the use of a single cable also means that the bandwidth is shared,

---

[2]　A well-known assertion by Metcalf states that "The usefulness, or utility, of a network equals the square of the number of subscribers."

the network traffic can be very slow when many terminals are active simultaneously. Hence, the coaxial cables were soon replaced by twisted pairs of copper cables and optical fiber links to meet market demand for higher data rates. The data rate has been increased by a factor of ten from the original 10 Mb/s to 100 Mb/s and 1 Gb/s in 1998, followed by 10 Gb/s in 2003. Note that the Ethernet at x-Gb/s is traditionally referred to as x-GbE. In 2010, 100GbE along with 40GbE were standardized [35]. This rapid upgrading of the data rate is motivated by the ever increasing bit rate of the backbone network. It is interesting that the line rate follows the trend of CPU speed, doubling every twenty-four months, but this is only natural because the Ethernet provides the computer interface, the network interface card (NIC), which connects computer terminals to the LAN.

100GbE will enable the realization of end-to-end connections between LANs at 100 Gb/s by directly bridging over the WAN without any bit rate conversion. As shown in Fig. 1.24, a client signal of the Ethernet frame is encapsulated in the OTN frame and transported over the WAN. Systems communicating over the Ethernet segment a stream of data into individual packets called frames. A unique identifier, a 48 bit media access control (MAC) address, is assigned to each NIC so that any information sent by one terminal is received by all, even if that information is intended for just one destination. Each frame contains source and destination addresses and error-checking data so that damaged data can be detected and re-transmitted. Carrier sense multiple access with collision detection (CSMA/CD) protocol is used for the MAC protocol, which enables a number of computer terminals in the LAN to share the medium without collision. Since all communications happen on the same wire, collision occurs in a way that a message overlaps in time with the other messages using the shared medium. Each terminal eventually learns of the collision by CSMA/CD protocol and tries again later.

Problems

1.1 What makes the broadband service distinct from the plain old telephone service?

1.2 What are the most important inventions in optical fiber communications?

1.3 By comparing TDM and WDM, discuss their merits and demerits from the viewpoints of the line rate, capability of compound line rate, and cost of repeaters.

1.4 What are the benefits of replacing an electrical switch with an optical switch?

1.5 By comparing FTTH with DSL, discuss the merits and demerits of both systems.

Part I

2 Optical multiple access systems

2.1 Passive optical network

2.1.1 Technology roadmap

Various PON systems are currently in commercial operation. As the number of subscribers is growing rapidly throughout the world, as described in Section 1.1, and as there is a perennial demand for higher bit rates, the PON system is continuously evolving and is expected to experience further evolution in the future. Network operators are motivated to leverage the next-generation (NG) networks to deliver high bandwidth to mobile cell sites for 3G/4G backhaul as well. Two types of gigabit-class PON systems have been standardized: the gigabit Ethernet passive optical network (1G-EPON) by the Institute of Electrical and Electronics Engineers (IEEE) and the gigabit-capable passive optical network (G-PON) by the International Telecommunication Union, telecommunication standardization sector (ITU-T). The gigabit-class PON is currently under deployment worldwide. Hereafter, 1G-EPON is abbreviated as EPON. Prior to the gigabit-class PON, the broadband passive optical network (BPON) was rolled out. Having had major investment by operators in deploying G-PON and EPON, NG-PON must be able to protect the investment of the legacy PONs by ensuring seamless subscriber migration from existing PON architectures such as gigabit-class PON to NG-PON. G-PON and EPON will be detailed in Section 2.1.2. NG-PON is required to fully support various services for residential subscribers, business customers, and mobile backhaul through its high quality of service and high bit rate capability.

A technology roadmap toward NG-PON is shown in Fig. 2.1 [1]. There are three categories: NG-PON1, NG-PON2, and possibly NG-PON3. NG-PON1 aims at evolutionary growth, and it can support "brown field" deployment, where co-existence with G-PON/EPON on the same ODN is a must. ODN refers to the fiber plant and the splitter between the OLT and ONUs. The co-existence with legacy systems enables seamless upgrade of individual subscribers to NG-PON on a live ODN without disrupting the services of other subscribers. XG-PON in NG-PON1 represents a PON system with 10 Gb/s line rate, at least in the downlink direction (here X is the roman symbol for 10). Uplink line rate candidates include 2.5 and 10 Gb/s. On the other hand, NG-PON2 was originally supposed to support "green field" deployment, aiming at revolutionary growth, and it may be disruptive with no requirement in terms of co-existence with G-PON/EPON on

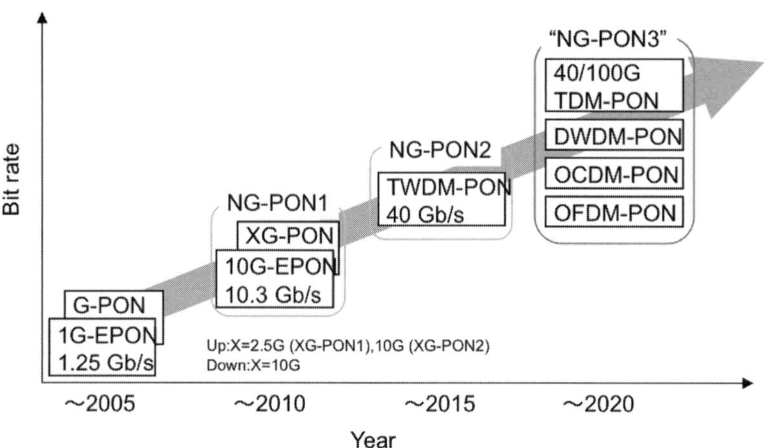

Figure 2.1 Technology roadmap toward the NG-PON [1]. © Reprinted by permission of IEEE and modified.

the same ODN. However, the objective of NG-PON2 has shifted from a revolutionary to an evolutionary technology, and co-existense capabilities with G-PON or XG-PON have been emphasized [2]. It mainly focuses on 40 Gb/s TDM-WDM (TWDM)-PON. As a consequence, a revolutionary approach in a true sense is likely to be revived as the post NG-PON2, NG-PON3. NG-PON3 will be open to a wide variety of candidate technologies, including 100G/400G TDM, dense WDM (DWDM), ultra-dense WDM (UDWDM). OCDMA, and OFDMA. It is too early to pick one of these as the best candidate because extensive research and development on components is needed.

OCDMA provides another dimension for multiple access, other than TDMA using time slots and WDMA using wavelength. The unique and the most attractive property of OCDMA is its asynchronous nature. OCDMA can support a fully asynchronous transmission, which allows all the users to access on demand without contention. In contrast, TDMA requires a tight control on synchronization, and WDMA needs a tight control on wavelength. As the optical encoding and decoding are performed in the optical domain, the low latency of data transmission is guaranteed, and neither the transmitter nor the receiver requires any complex electrical equipment. OCDMA can provide soft capacity on demand, in that users can be added to the network on demand by assigning new optical codes. Another advantage of OCDMA is the inherent nature of data confidentiality, because messages are encoded at the transmitter and can be recovered only by the authorized user, who knows the optical code.

2.1.2   Current PON

PON architecture has evolved from BPON to gigabit-capable G-PON and EPON. G-PON has been specified by the ITU-T G.984 series, including the requirements and basic architecture (G.984.1), the physical-medium-dependent (PMD) layer (G.984.2), the transmission convergence (TC) layer (G.984.3), and the management requirements

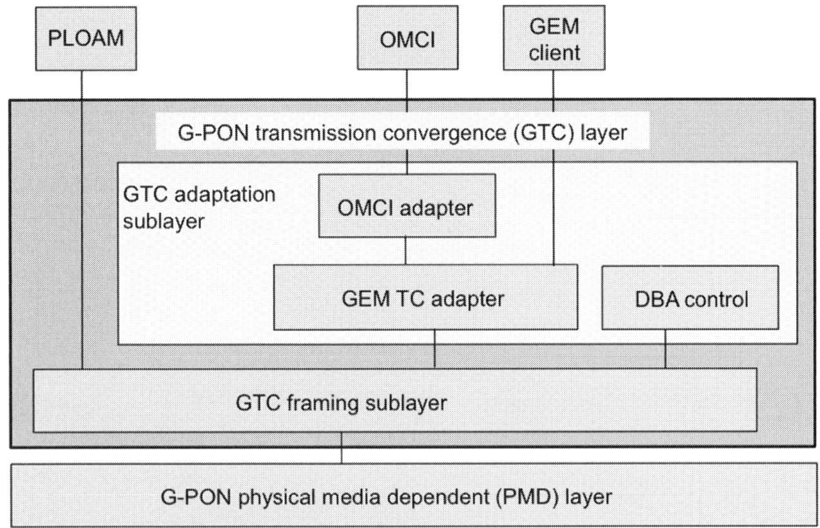

GTC:G-PON transmission convergence, GEM: G-PON encapsulation method, DBA: Dynamic bandwidth allocation, TC: transmission convergence, OMCI: ONT management and control interface, PLOAM: Physical layer operations, administration, and management

Figure 2.2 Protocol stack of G-PON transmission control (GTC) [8]. © Reprinted by permission of ITU-T G.984.3.

(G.984.4). On the other hand, IEEE 802.3, Ethernet Working Group, has specified two lower layers of the open systems interconnection (OSI) reference model such as the physical and data link layers. Here, the main characteristics of G-PON and EPON are reviewed briefly.

2.1.2.1   G-PON

The G-PON transmission control (GTC) layer adapts user data onto the physical media dependent (PMD) layer and provides basic management. The GTC is divided into two sublayers as depicted in Fig. 2.2 [8]. The PMD layer defines the bit rate of uplink and downlink signals and the optical parameters for various rate combinations. In the lower GTC framing sublayer, the frame structure is defined. Each GTC burst payload consists of a variable number of ATM and/or G-PON encapsulation method (GEM) frames. GTC uses fixed 125 μs framing for the downlink signal. It is possible to carry both Ethernet frames and TDM traffic in their native formats as well as asynchronous transfer mode (ATM) cells of BPON as shown in Fig. 2.3(a). The same frame structure is maintained for both uplink and downlink with asymmetric bit rate, only the amount of payload differs. GTC also provides transmission features, including encryption, forward error correction (FEC), and media access control (MAC) function. The physical layer operation administration and maintenance (PLOAM) carries a message-based protocol for PMD and GTC layer management.

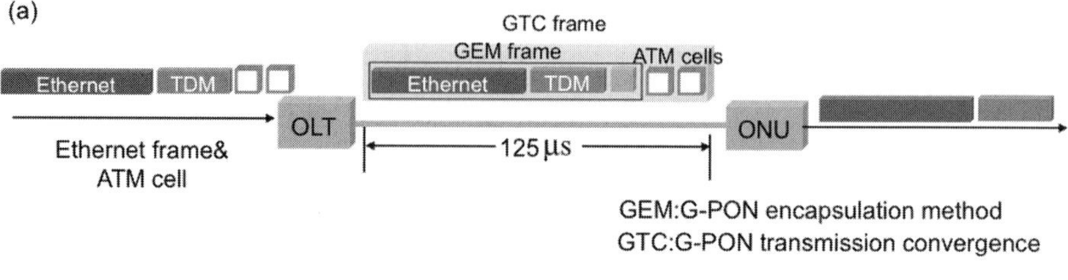

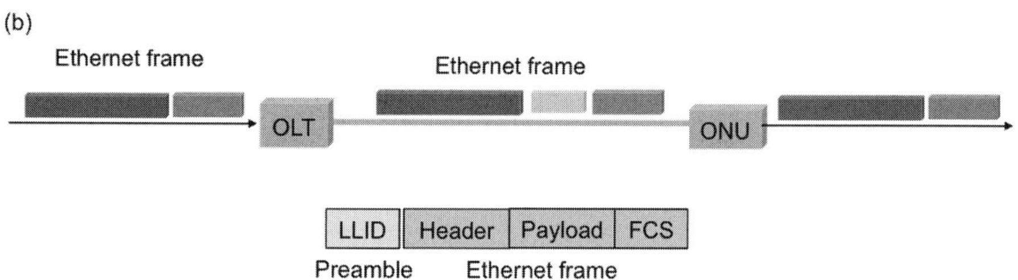

Figure 2.3 Frame formats of (a) G-PON and (b) EPON [13]. © Reprinted by permission of IEEE and modified.

The upper sublayer is the TC adaptation sublayer based upon GEM. GEM defines a protocol-dependent encapsulation for variable-size packets. It also defines the bandwidth allocation method. Bandwidth allocation can be the same in every frame according to a static allocation, or different in every frame in response to dynamic traffic fluctuations. The former option is the fixed bandwidth allocation (FBA), while the latter option is the basis of dynamic bandwidth allocation (DBA). DBA will also be described in more detail in Section 2.1.3.3. The management interface between the OLT and the ONU is specified so that the access carrier will be able to manage the ONUs using the ONU management and control interface (OMCI). OMCI comprises an ONU management information base (MIB) and the OLT management control channel protocol (OMCC).

G-PON has seven transmission-line rate combinations: 1.244 Gb/s in the downlink combined with 155 Mb/s; 622 Mb/s or 1.244 Gb/s in the uplink or 2.488 Gb/s in the downlink combined with 155 Mb/s; and 622 Mb/s, 1.244 Gb/s, or 2.488 Gb/s in the uplink. The split ratio ranges from 1:16, 1:32 up to 1:64. The transmission distance between the CO and ONU is up to 20 km, and the maximum differential logical reach is also 20 km. The operating wavelengths for the downlink and uplink directions on a single fiber system are in the ranges 1480–1500 nm and 1260–1360 nm, respectively, as shown in Fig. 2.4 [12]. The wavelength band 1550–1560 nm is reserved for the overlay of analog RF video distribution systems, and the testing of PON equipment using an optical time domain reflectometer (OTDR) will be conducted at the wavelength of 1625–1650 nm. The OTDR will be described in Section 10.3.2. The main specifications are summarized in Table 2.1. Both BPON (G.983.1) and G-PON (G.984.1) standards specify a two-fiber solution with dedicated upstream and downstream transmission fibers with a wavelength

Table 2.1 PON standards: key specifications

		BPON [3] (ITU-T G983 series)	G-PON [4] (ITU-T G984 series)	EPON [5] (IEEE802.3ah)	10G-EPON [6, 7] (ITU-T G987 series and IEEE802.3av)
Line rate	Downlink	2.49 Gb/s	2.49 Gb/s	1.25 Gb/s	10.3 Gb/s
	Uplink	1.24 Gb/s	1.24 Gb/s	1.25 Gb/s	10.3 Gb/s or 1.25 Gb/s
Signal coding (coding overhead)			NRZ scrambling ($<$10%)	8B/10B (20%)	64B/66B (3%)
Forward error correction (FEC)			Optional RS(255,239)	Mandatory RS(255, 223)	Mandatory RS(255, 223)
Wavelength	Downlink	1480–1580 nm (DFB)	1480–1500 nm (DFB)	1480–1500 nm (DFB)	1575–1580 nm (EML*)
	Uplink	1260–1360 nm (DFB)	1260–1360 nm (DFB) or 1290–1330 nm (reduced version)	1260–1360 nm (DFB)	1260–1280 nm (DFB)
Maximum distance		20 km	20 km	20 km	20 km
Split ratio		1:32	Max 1:64	1:16; 64 feasible	1:16; 64 feasible
Laser on-off		154 ns	25.6 ns	512 ns	512 ns
Burst preamble			35.2 ns	Automatic gain control Clock and data recovery	Automatic gain control Clock and data recovery
Frame size		ATM cell 53 bytes	Frame fragment 1518 bytes (maximum)	Ethernet frames 64–1518 bytes	Ethernet frames 64–1518 bytes
OH for bandwidth allocation			12 bytes	GATE/REPORT 64 bytes	GATE/REPORT 64 bytes

*EML electroabsorption modulator.

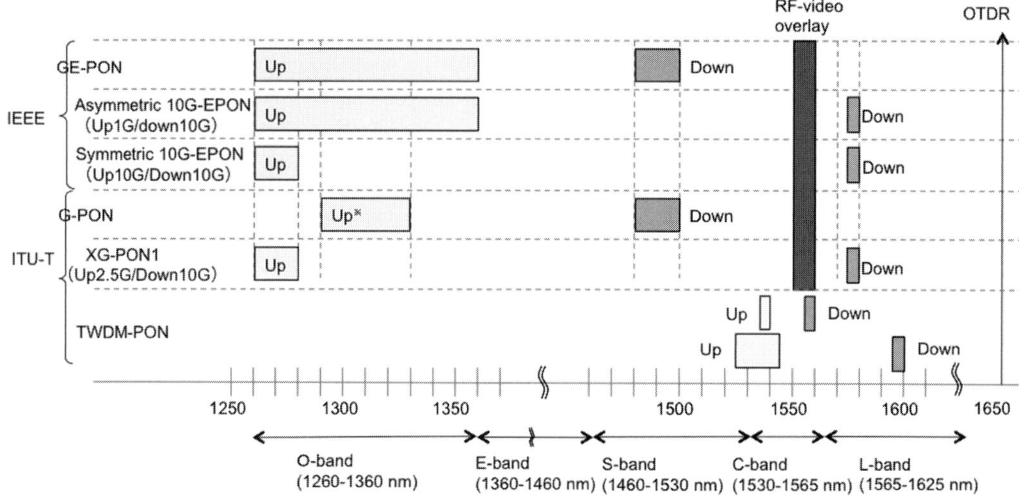

Figure 2.4 Wavelength allocations of BPON, G-PON, EPON, and 10G-EPON. Modified after [12].

of 1300 nm used in both directions in the two-fiber solution. However, few or probably no system has been deployed with the two-fiber solution.

2.1.2.2   EPON

Here, specifications of EPON are outlined. A communication system is broken up into a seven-layer Open Systems Interconnection (OSI) reference model, standardized by the International Organization of Standardization. The relation between OSI reference model and the IEEE802.3 layering diagram is shown in Fig. 2.5 [10]. The bottom four layers of the OSI model deal with the network and the top three deal with the application. The tools offered by the lower four layers relevant to networking are: provisioning by the physical layer of a link made from copper, optical fiber, or wireless technology; a data link protocol by the data link layer that asks for re-transmission if a message is received in error; and routing and addressing by the network layer. IEEE 802.3 focuses on two layers of the OSI model, the physical layer (PHY) and data link layer. In access networks there exists only one connection between the OLT and ONU with a single hop, therefore, there is no routing decision to be made, and the network layer is absent. However, since PON is a multiple access network like the local area network (LAN), the physical medium is shared with an OLT and a number of ONUs, and hence the MAC sublayer in the Ethernet is replaced with multipoint media access control (MPMC), which runs the multipoint control protocol (MPCP) to coordinate the access to the shared PON medium among ONUs. The MPCP handshake between the OLT and ONU is performed in the following way [11].

- Using the discovery GATE message, the OLT sends a request to all unregistered ONUs to transmit.

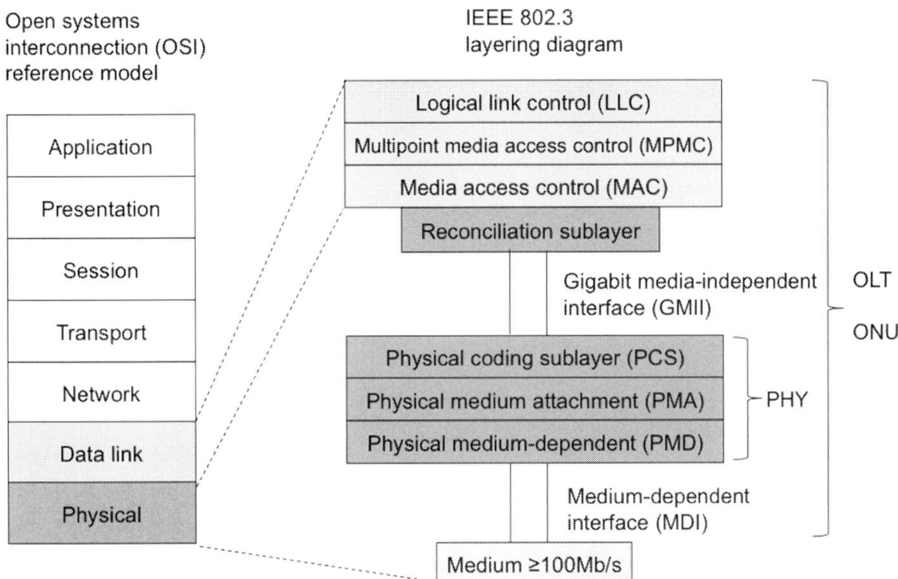

Figure 2.5 Open Systems Interconnection (OSI) reference model and 802.3 layering diagram [10].

- An unregistered ONU answers by using a REGISTER REQ registration request message.
- When received and approved, the OLT registers the ONU using the REGISTER message.
- The handshake ends with the ONU acknowledgment REGISTER_ACK.

The PMD layer specifies the physical characteristics of the optical transmitter and receiver. EPON technology provides bidirectional 1 Gb/s links using a wavelength of 1490 nm for the downlink and 1310 nm for the uplink, with 1550 nm reserved for future extensions or additional services, such as analog video broadcast as seen from Fig. 2.4. The typical distance between CO and subscriber is 20 km. The split ratio ranges from 1:16 to as high as 1:64. EPON is a burst mode protocol, and the traffic uses the same Ethernet packet format with standard interpacket gap (IPG) as any enterprise switch does. EPON does not use encapsulating framing on either the uplink or downlink. An uplink burst is simply a sequence of Ethernet packets with regular IPG between them. The packet preamble contains additional fields to the conventional Ethernet's header, shown in Fig. 10.9. In the EPON frame format shown in Fig. 2.3(b), the frame check sequence (FCS) detects errors using a cyclic redundancy check (CRC). The logical link ID (LLID) field defines the destination ONU in the downlink. An ONU filters the received frames based on the LLID in the frame's preamble and its own unique LLID value assigned by the OLT. A special value is reserved for broadcast messages sent to all ONUs. In uplink transmissions the LLID field marks the source ONU. Up to 32,768 logical ONUs can be supported via the reconciliation sublayer by using a 15-bit LLID. The physical coding sublayer (PCS) deals with line-coding. It adopts the 8B/10B line coding to produce a dc balanced output, which adds an overhead of 20%. The use

Table 2.2 Key parameters of downlink transmission

	PR20	PR30
Transmitter type	EML* + amplifier	EML
Maximum launch power [dB m]	+9	+5
Minimum launch power [dB m]	+5	+2
Extinction ratio [dB]	9	9
Penalties due to transmitter and dispersion	1.5	1.5
Receiver type	PIN-PD	APD
Receiver sensitivity [dB m]	−20.5	−28.5
Power budget [dB]	25.5	30.5
Channel insertion loss [dB]	10~24	15~29
Distance [km]	20	20

*EML electroabsorption modulator.

of FEC is optional, and it uses RS(255,239) block codes in the PCS layer. Although not defined in the IEEE 802.3ah specification, all EPON implementations incorporate encryption. Encryption is AES based with the exception of a special algorithm. The operating wavelengths for the downlink and uplink directions are the same as those of G-PON in Fig. 2.4. The key specifications are also summarized in Table 2.2.

2.1.2.3   10G-EPON and XG-PON

In NG-PON1 shown in Fig. 2.1, the target bit rate is 10 Gb/s-class. The IEEE 802.3av 10G-EPON standardization was completed in September 2009. NG-PON1 ensures that the vast investment in gigabit PON is meaningful, and subscribers who do not require a bit rate higher than 1 Gb/s will be able to continue receiving the gigabit service. There will be two migration scenarios, from EPON to 10G-EPON and from G-PON to 10G-EPON. This standard is a successor to the current EPON standard, IEEE 802.3ah and G-PON, ITU-T G984 series. The main differences between the 10G-EPON and EPON specifications are summarized in Table 2.1. The line codings for the 10G-EPON and EPON are 64B/66B and 8B/10B, respectively, and the 64B/66B line coding for 10G-EPON reduces the bit-to-band overhead from 20% in the EPON to 3%. 8B/10B line code maps 8-bit symbols to 10-bit symbols to maintain dc balance. It is noted that the forward error correction (FEC) is also mandatory for 10G-EPON. 10G-EPON supports both symmetric 10 Gb/s downlink and uplink, and asymmetric 10 Gb/s downlink and 1.25 Gb/s uplink data rates, while EPON provides only a 1.25 Gb/s symmetric data rate.

In order to realize smooth migration from existing G-PON and EPON to 10G-EPON, XG-PON has been standardized by ITU-T. The wavelength allocation of XG-PON is the same as for symmetric 10G-EPON as shown in Fig. 2.4. In a migration scenario from EPON to 10G-EPON, there are three combinations of different bit rates, and they have to be supported simultaneously using different wavelength assignments and/or TDMA as shown in Fig. 2.6 [12]. In the downlink, 1 Gb/s and 10 Gb/s signals are transmitted

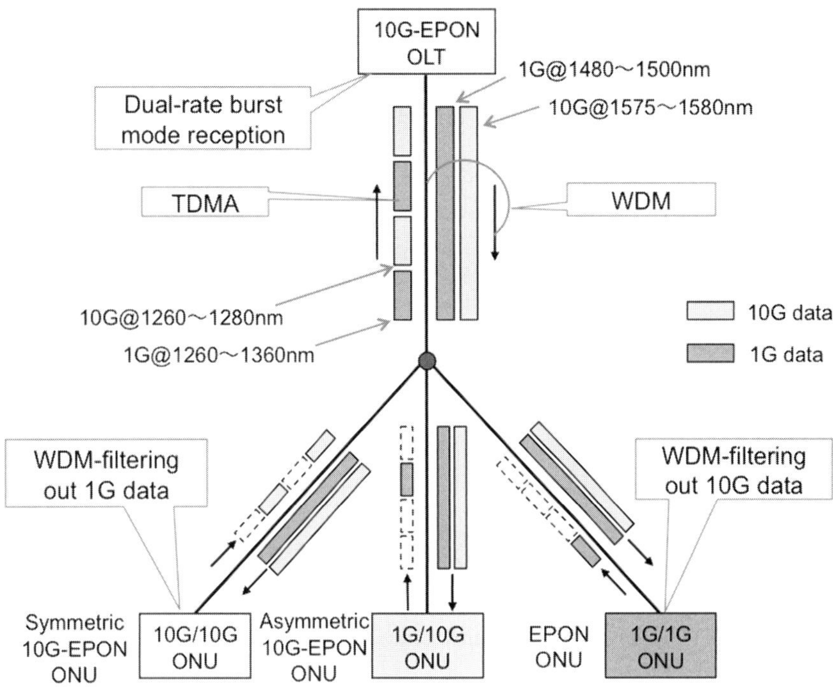

Figure 2.6 Technologies required for XG-PON, the co-existence of 10G-EPON with EPON [12]. © Reprinted by permission of NTT.

on different wavelength bands, 1480~1500 nm and 1575~1580 nm, respectively. In the uplink, TDMA is adopted since the same wavelength band, 1260~1360 nm, is used for 1 Gb/s and 10 Gb/s signals.

Another migration scenario is from G-PON to 10G-EPON, shown in Fig. 2.7. The line rate of the upstream signal is 2.5 Gb/s for XG-PON1. In this scenario, the OLT and several ONUs of G-PON are supposed to be deployed first, followed by the deployment of OLT and ONUs of XG-PON1. The WDM filter (WDM1) is installed to combine and separate G-PON and XG-PON1 signals into and out of the common ODN [1]. It is noted that the WDM1 has to be installed at the same time as XG-PON1 is added, in order to avoid service outage of the existing G-PON. Appendix I of ITU-T G.984.5 provides an example of WDM1 characteristics. The backward compatibility requirement imposes several technical constraints on the specifications: dual-rate burst-mode operation at the OLT receiver and dual-rate dynamic bandwidth allocation (DBA). The burst-mode receiver and DBA will be detailed in Section 5.6 and Section 2.1.3.3.

2.1.3 Practical aspects

2.1.3.1 Power budget

The power budget ensures the maximum split ratio and the maximum distance between the OLT and ONU at a given line rate and given launch power of the transmitter and

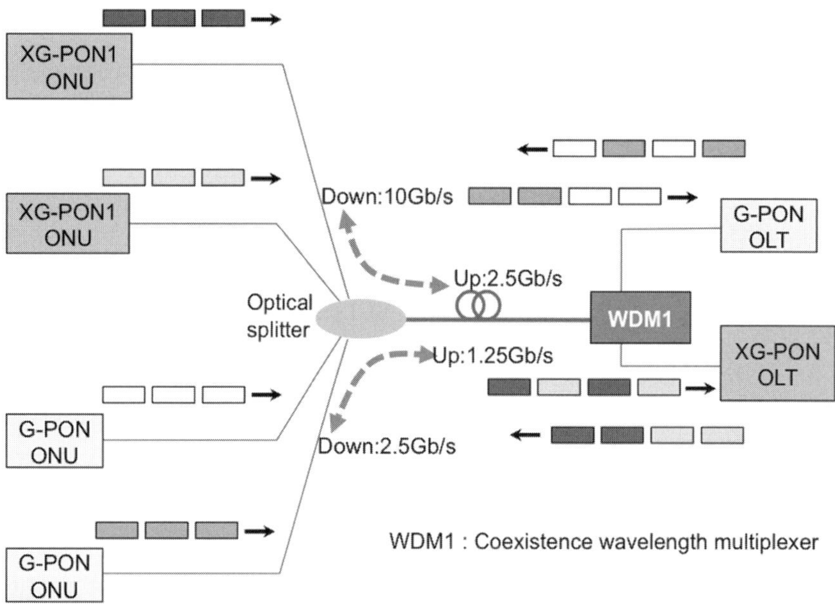

Figure 2.7 Migration scenario from existing G-PON to XG-PON1. Modified after [1].

receiver sensitivity. The power budget indicates the difference between output power and receiver sensitivity, and it includes the channel insertion loss of the ODN and the induced penalties. There are several classes of power budget in each type of PON. According to IEEE 802.3av, for example, three classes of power budget for 10G-EPON are specified to support split ratios of 1:16 and 1:32, and distances of at least 10 km and at least 20 km. Among these, we focus on two classes. One is the medium power budget class PR20 which supports the channel insertion loss of 24 dB of the symmetric bit rate with either the split ratio of 1:16 and distance of 20 km or the split ratio of 1:32 and distance of 10 km, and the other is the high-power class PR30, which supports the channel insertion loss of 29 dB with split ratio of 1:16 and distance of 20 km. Tables 2.2 and 2.3 present key characteristics of the optical transceivers for PR10, PR20, and PR30 in the downlink and uplink directions, respectively [13]. The parameters listed in Tables 2.2 and 2.3 are based on the condition that a bit error rate of 10^{-3} is required before FEC decoding. The power budget includes the channel insertion loss of the ODN and the induced penalties denoted by the transmitter and dispersion penalty in the tables.

The optical fiber which is widely deployed in the ODN is fully compliant with ITU-T G.652, known as standard single-mode fiber, S-SMF. The zero dispersion wavelength is located at 1310 nm, and the typical chromatic dispersion at 1550 nm is 17 ps/nm/km. The nominal mode field diameter is 8.6 μm at 1310 nm. The fiber loss is less than 0.4 dB/km at 1310 nm and less than 0.3 dB/km at 1550 nm. The mode field diameter and loss of the three fiber categories are detailed in Table 3.1. To facilitate the transmission of the optical signal with a good signal-to-noise ratio between the OLT and ONU, the link loss has to be guaranteed to be smaller than the power budget. The link loss of the

Table 2.3 Key parameters of uplink transmission

	PR20	PR30
Transmitter type	DML[*]	DML
Maximum launch power [dBm]	+4	+9
Minimum launch power [dBm]	−1	+4
Extinction ratio [dB]	6	6
Penalties due to transmitter and dispersion	3	3
Receiver type	APD	APD
Receiver sensitivity [dB m]	−28	−28
Power budget [dB]	27	32
Channel insertion loss [dB]	10~24	15~29
Distance [km]	20	20

*DML directly modulated laser.

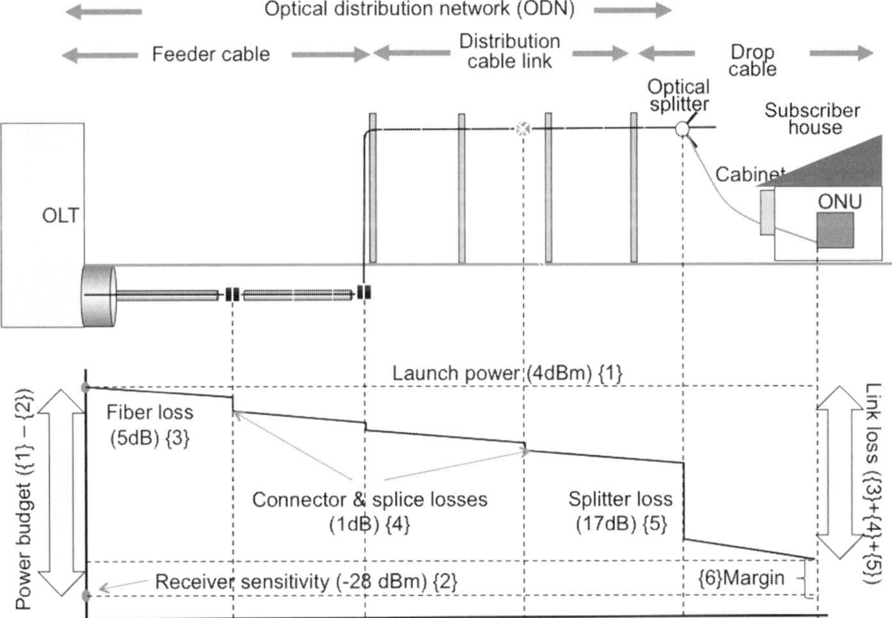

Figure 2.8 Power budget and link loss.

link between OLT and ONU in Fig. 2.8 is determined by the fiber cable loss, losses of connection and splicing, and the splitting loss with the margin as

$$\text{Link loss [dB]} = \{3\}\text{Fiber cable loss [dB/km]} \times \text{Distance [km]} + \{5\}\text{Splitter loss [dB]}$$
$$+ \{4\}\text{Connector and splice losses [dB][dB]}, \tag{2.1}$$

where the margin accounts for additional splices and/or connections. If the above link loss satisfies the power budget in Eq. (2.2), then proper operation of the PON system

Table 2.4 Maximum values of a conventional field plant of EPON and G-PON

	Maximum loss [dB]	Count
Splitter	17.0	1:32
Splice	0.067	6
Connection	0.15	5
Fiber cable loss (1310 nm)	0.35 [dB/km]	20 [km]
Fiber cable loss (1550 nm)	0.25 [dB/km]	20 [km]
Loss margin	1.0	
Link loss	26.2 [dB] (1310 nm), 24.2 [dB] (1550 nm)	

will be guaranteed:

$$\text{Power budget [dB]} = \{1\}\text{Launch power [dBm]} - \{2\}\text{Receiver sensitivity [dBm]}$$
$$\geq \text{Link loss [dB]} + \{6\}\text{Margin [dB]}. \tag{2.2}$$

In Table 2.4 the maximum loss values of a conventional field plant of the EPON and G-PON are summarized. The link losses are calculated to be 26.2 dB and 24.2 dB at 1310 nm and 1550 nm, respectively. If the values are applied to the power budget 32 dB (PR2D) in Table 2.3, the margins of 5.8 dB at 1310 nm and 7.8 dB at 1550 nm are obtained. The dispersion values at 1310 nm, 1490 nm, and 1550 nm are zero, 10 ps/nm/km, and 17 ps/nm/km. It is noted that the pulse broadening at 2.488 Gb/s due to the dispersion will cause little signal degradation for the 20 km physical reach.

Node consolidation enables the network operator to simplify the network structure and reduce the number of access sites. This is expected to enhance the overall cost efficiency of the network. NG-PON2 systems must support operator objectives concerning node consolidation by offering enhanced reach and capacity.

2.1.3.2 Ranging

The carrier sense multiple access with collision detection (CSMA/CD) protocol has been used in Ethernets to avoid collision when two or more terminals send out messages at the same time. This is because the terminals share a transmission medium in the Ethernet. CSMA/CD exhibits a low latency and high efficiency if the distance is within several hundred meters in a LAN environment. As the distance between the OLT and the ONUs extends to a few tens of kilometers in a PON, the efficiency of CSMA/CD significantly drops. This is because the round trip delay (RTD) becomes longer.

In current PONs the signal transmission timing of each ONU is adjusted according to this RTD so that the signals from individual ONUs arrive at different times and do not overlap each other, thus eliminating data re-transmission. The ranging is required to facilitate collisionless transmission. As the distance between the OLT and ONU normally differs from ONU to ONU, all the RTDs have to be determined accurately. According to ITU-T G.984.3 G-PON: Transmission convergence layer specification, ranging is conducted in the following manner. Here, the registration process of ONUs is omitted. The next step is the delay measurement as described in Fig. 2.9 [14].

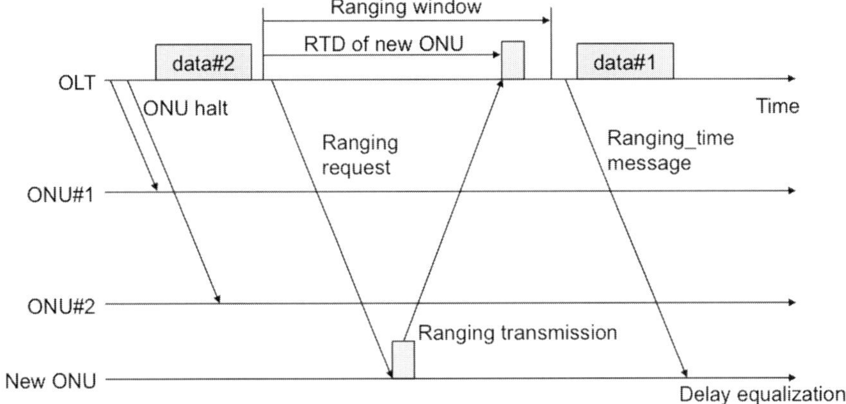

Figure 2.9 Ranging process of G-PON [14]. © Reprinted by permission of Academic Press.

- The OLT notifies all ONUs to halt uplink transmissions.
- The OLT sends out the ranging request to the target ONU, and upon receipt of the signal (ranging transmission) from the target ONU.
- The OLT measures the RTD according to the time when the signals for the delay measurement are received.
- The OLT notifies the ONU of the "equalization delay $(= T_{eqd} - T_{RTD})$" by sending a ranging message. T_{eqd} is a constant number and is set to be the maximum RTD value in the PON, for example, for an ONU 20 km from the OLT $T_{eqd} = 200$ μs, and T_{RTD} is the round trip time between the measured OLT and the ONU.
- The ONU memorizes the equalization delay and delays the subsequent uplink transmission timing by this value.

Through this ranging, all ranged ONUs possess an equivalent delay time T_{eqd}, given by the equalization delay plus its own T_{RTD}. In the case of EPON, all control messages have a *timestamp*, and the method is characterized by ranging and clock synchronization using the timestamp.

2.1.3.3 Dynamic bandwidth allocation

In the uplink, a time division multiple access (TDMA) mechanism, one of the multiple access protocols described in the next section, is used to avoid collision of optical signals that could occur at the splitter level. Each ONU has to restrict its transmission to predefined time-slots by operating in burst mode under the arbitration of the MAC protocol, which can allocate one or several time-slots to each ONU statically or dynamically. MAC provides addressing and channel access control mechanisms that make it possible for the OLT and a number of ONUs to communicate within a multiple access network that incorporates a shared medium.

All the ONUs, up to 32 in an EPON, have to share the uplink bandwidth. The point-to-multipoint connectivity is supported by MPCP described in Section 2.2.2, which uses standard Ethernet packets generated in the MAC control sublayer. Thus, each ONU can

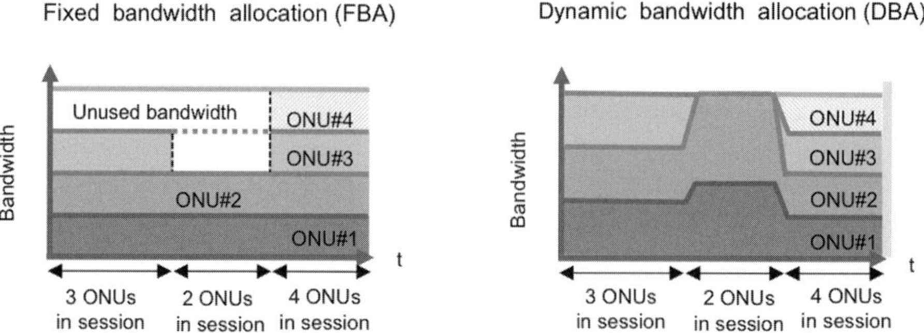

Figure 2.10 FBA and DBA [15]. © Reprinted by permission of NTT.

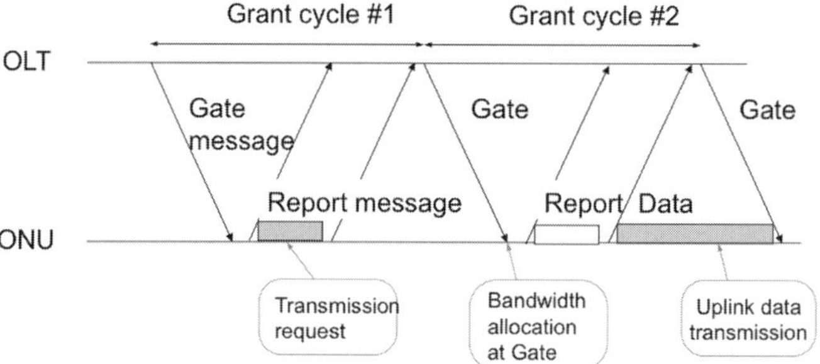

Figure 2.11 Procedure of DBA [15]. © Reprinted by permission of NTT.

communicate with the OLT without contention. There are static and dynamic control protocols. A static control protocol is fixed bandwidth allocation (FBA). FBA gives an equal bandwidth to each ONU, for example, 5 kB in order from ONU 1 to ONU 2, ... As shown in Fig. 2.10, there is no traffic from ONU 4 for a period, while ONU 1 and ONU 2 always send out traffic during their allocated time period. As a result, the uplink bandwidth cannot be fully utilized and a part of the bandwidth is unused and wasted [15]. The dynamic bandwidth allocation (DBA) protocol can solve this problem. The DBA protocol allocates the bandwidth only when the ONU has traffic to send, therefore, the uplink bandwidth can be fully utilized, and no bandwidth is wasted. The OLT controls the ONU's transmission window by the DBA protocol. The process of DBA shown in Fig. 2.11 is as follows.

- The ONU reports its queue status using REPORT messages when the data to transmit are generated and stored in the memory.
- The OLT then calculates the ONU transmission window length using DBA.
- The OLT notifies the ONU of the starting time and the allocated bandwidth by sending a GATE message.
- The ONU starts to transmit the data at the designated time. A REPORT message is sent together with the data for the next bandwidth allocation.

IEEE 802.3ah specifies the above framework in the MPCP, except for the calculation method in the second step. The DBA algorithm is left open to the operator of the PON service. Note that all time-driven events are synchronized to the PON clock, a 16 ns resolution counter that is carried in all MPCP messages. The ONU uses the received timestamp to lock to the OLT time base. The OLT uses returned timestamps to measure the ONU round-trip delay and schedule collision-free uplink transmissions.

The other functions GTC offers to configure are FEC and encryption. Due to the broadcast nature of the PON, this passive tree architecture may appear to lack privacy because each ONU has access to downlink data pertaining to other ONUs. To address this security concern, it is necessary to use signal encryption techniques. G-PON provides data scrambling and uses Advanced Encryption Standard (AES) to encrypt the downlink payload. The encryption key associated with each ONU can be sent in the clear uplink as no ONU can look at the uplink traffic coming from other ONUs in the same PON.

As the distance between the OLT and each ONU varies, each ONU needs to be synchronized in order to ensure collision-free transport. During initial activation and registration, the OLT uses a procedure called ranging, described in Section 2.1.3.2, in order to calculate the appropriate delay that each ONU must insert relative to the received downlink frame in order to synchronize the uplink phase of all ONUs on the PON. The GTC layer control is mainly operated via the PLOAM message protocol and some overhead fields referred to as embedded OAM. It includes management functions such as PMD management, GTC layer management, ONU activation, and encryption management. Voice service or Internet telephony is latency sensitive, while video broadcast and Internet access are bandwidth-demanding services.

2.2   Multiple access techniques

In Section 1.3, multiplexing techniques such as TDM and WDM were described. Multiplexing is introduced, which establishes simultaneously a number of channels in an optical fiber, while multiple access establishes simultaneously a number of connections. In a PON, simultaneous connections between subscribers and the central office can be set up via multiple access. Uplink data burst and downlink broadcasting data in PON are shown in Fig. 2.12. Without any pre-coordination, contention will occur among the uplink data bursts at the multiplexer if more than two subscribers send out data at the same time. Multiple access can avoid contention in time, wavelength, code space or subcarrier frequency (subcarrier frequency not shown), as shown in Fig. 1.7. Hybrid multiple access, combining two or three techniques among TDMA, WDMA, and OCDMA, could be a possible option.

2.2.1   Time division multiple access

The TDM-PON utilizing time division multiple access (TDMA) shown in Fig. 2.13 is of point-to-multipoint (P2MP) type. A star coupler located at a remote node splits the

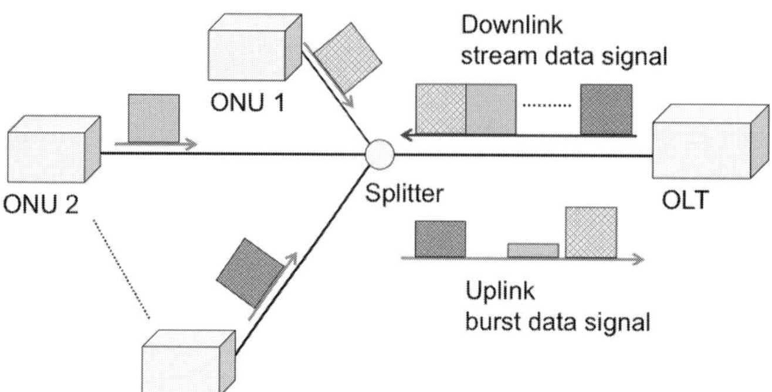

Figure 2.12 Uplink data burst and downlink broadcasting data in PON.

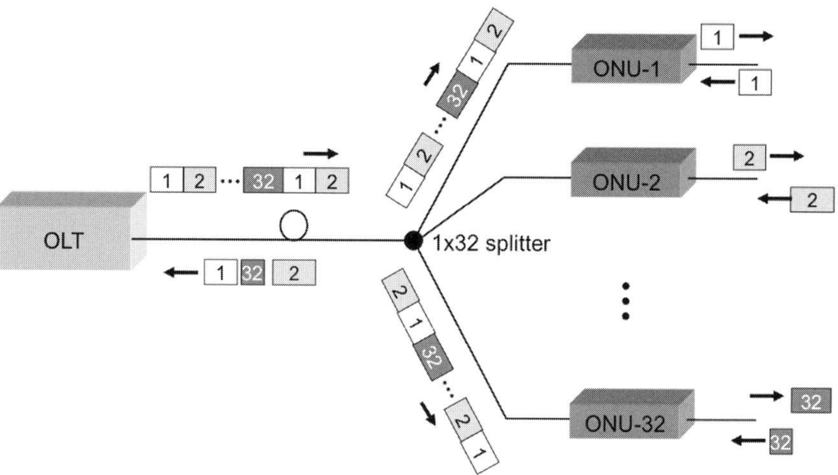

Figure 2.13 TDM-PON: point-to-multipoint (P2MP).

optical power of the downlink signal and combines the power of the uplink signal. It suffers inherently from loss of power according to the split count, and this will restrict the number of subscribers accommodated in a PON without optical amplification. Details of the star coupler are described in Section 5.1. Assume it accommodates 32 users in a CO. The TDM downlink data signal is broadcast to all ONUs, and each ONU can receive only the intended data by choosing the time slot every 32 slots. Timing synchronization between the OLT and all ONUs is needed, and this is facilitated by delivering the clock signal, for example an 8 kHz tone in G-PON, from the OLT to ONUs. For the uplink data burst, due to the nature of the P2MP architecture, the OLT coordinates the transmission so that TDMA is adopted to avoid contention. This is a task of the MAC layer in the data link layer described in Section 2.1.2. A straightforward way is to assign equal bandwidth to all ONUs. For example, the bandwidth allocated to each subscriber of G-PON is one 32nd of 1.24 Gb/s. DBA, shown in Fig. 2.10, is a more sophisticated bandwidth allocation which facilitates flexible bandwidth allocation to subscribers on demand.

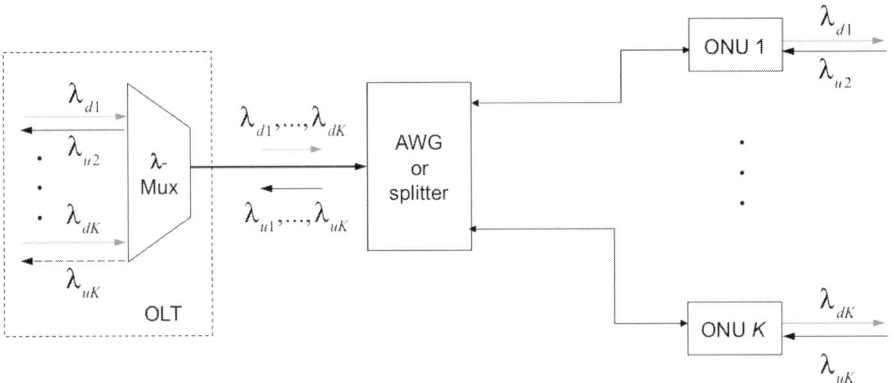

Figure 2.14 WDM-PON: point-to-point (P2P).

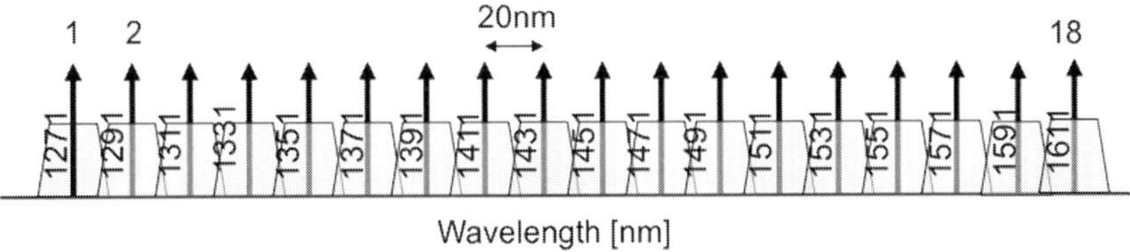

Figure 2.15 Coarse WDM (CWDM) PON.

2.2.2 Wavelength division multiple access

The WDM-PON utilizing wavelength division multiple access (WDMA) has a point-to-point (P2P) architecture as shown in Fig. 2.14. There is a logical point-to-point link between the OLT and every ONU since each ONU is assigned a different wavelength or wavelengths. Therefore, the bandwidth is symmetrical between the uplink and downlink, and in fact the bandwidth allocated to the subscriber can be unlimited because the bandwidth of a single wavelength can be fully occupied by the subscriber. This is distinct from the bandwidth-asymmetric TDMA. Either a star coupler or an arrayed waveguide grating (AWG) is located at a remote node depending on the architecture. The wavelength routing PON uses an AWG. Hence, no wavelength-selective receiver is required at the ONU for the downlink. When a star coupler is used instead of an AWG, each ONU has to select one of the broadcast wavelengths. For the uplink, the ONU uses another individual wavelength, and a wavelength-fixed light source with a different wavelength has to be placed at each ONU. This hinders maintenance as well as the low-cost implementation of ONUs, which benefits from mass production.

A main concern of WDM-PON is the scarcity of the available spectral region when a smooth migration from existing PONs such as gigabit PON and 10-gigabit PON is considered. There are currently two options for WDM-PON: coarse WDM (CWDM) PON and dense WDM (DWDM) PON. In CWDM, ITU G.694.2 defines 18 wavelength channels with 20 nm spacing in the wavelength range 1271 nm to 1611 nm as shown in Fig. 2.15. This might be a possible scenario for a green field deployment. Since

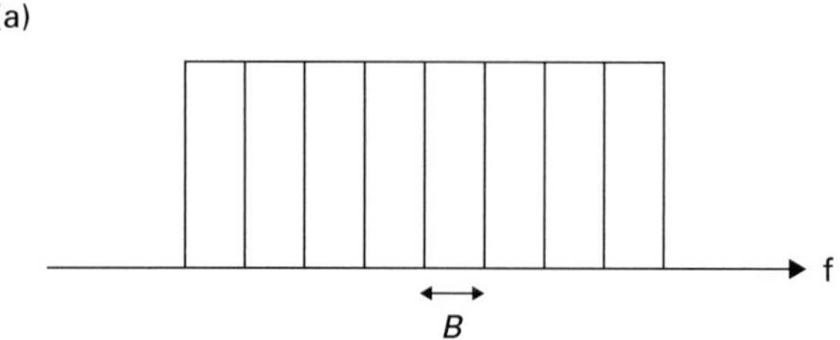

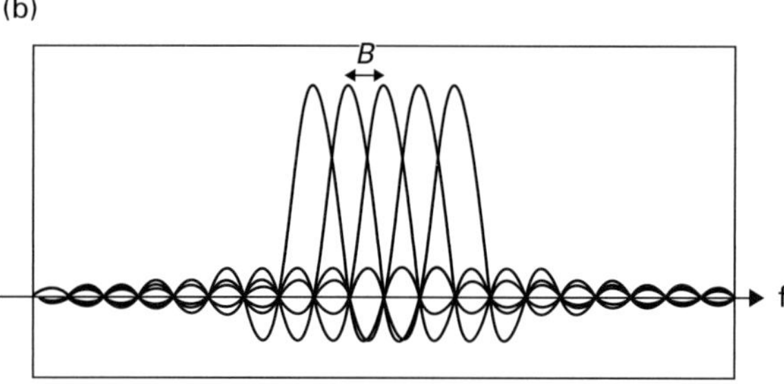

Figure 2.16 (a) Spectrum of Nyquist WDM and (b) spectrum of optical OFDM.

strict tuning of wavelengths is not needed for the CWDM-PON, a thermal control, called a thermo-electric cooler (TEC), is not required, making the cost less than that of DWDM-PON. Furthermore, the wavelength multiplexer is expected to have lower channel crosstalk than for DWDM-PON owing to the 20 nm channel spacing. However, a primary disadvantage of CWDM is that it lacks scalability due to the limited number of channels. Another disadvantage is that the shorter wavelength channels experience higher loss, thereby limiting the transmission distance or split ratio. A single-mode fiber without the water-peak around 1390 nm has to be used for this wide spectrum of transmission. A wide range of chromatic dispersion from zero to 20 ps/nm/km also makes a long reach difficult. The second option is conventional WDM. However, one can see from Fig. 2.4 that there is little available spectrum if WDM-PON needs to coexist with all the legacy PON systems. Only those technologies requiring a very narrow spectrum per wavelength and which are capable of tight control of the wavelength can coexist. From Fig. 2.4 the spectral region available for the WDM-PON would be about 30 nm around 1440 nm and about 30 nm around 1610 nm, except for the 1380 nm region where the absorption of the overtone of hydroxyl exists as shown in Fig. 3.1. A further factor that limits the available spectrum will be the legacy filter characteristics built in to deployed systems.

To realize spectrally efficient and high capacity transmission, novel optical frequency multiplexing techniques could be introduced in the access network in the future. There are two options: Nyquist WDM [16], shown in Fig. 2.16(a), and optical orthogonal frequency-division multiplexing (OFDM) [17], shown in Fig. 2.16(b). Hereafter, we refer to OFDM instead of "optical" OFDM. With these techniques, multiple optical carriers are generated and modulated at the transmitter, and coherent detection and digital signal processing (DSP) are employed for demultiplexing and demodulation at the receiver. In Nyquist WDM there should be no spectral overlap between the channels because each channel has a rectangular spectrum. The reason why this is referred to as Nyquist WDM is that the signal contains no frequencies higher than $B/2$ [Hz] where B denotes the symbol rate. The temporal waveform of the symbol pulse has a sinc-function profile with the zero-crossings at n/B ($n = \pm1, \pm2\ldots$). Therefore, there is neither crosstalk between the channels nor inter-symbol interference. In OFDM the signal is made up of the superchannel without guard interval, which is a set of very tightly spaced channels, that are typically termed "optical subcarriers." Crosstalk is apparently present in the spectrum but cancels out if proper filtering is performed at the receiver, and there is no inter-symbol interference (ISI) either. Interestingly, comparison of the capacity transmission performance between Nyquist WDM and no-guard-interval OFDM reveals that Nyquist WDM outperforms OFDM because OFDM suffers from a higher inter-carrier interference (ICI) than Nyquist WDM [18]. The ICI in the Nyquist-WDM superchannel is inherently low because the carriers have non-overlapping spectra, but the spectra of OFDM carriers overlap each other and the orthogonality between them can easily be violated by device imperfections in the transceiver, such as limited bandwidth and sampling rate. This is why the OFDM superchannel is prone to higher ICI. Details of OFDMA are described in Section 2.2.4.

An important issue of WDM-PON is that the transmitter at each ONU requires a laser with different wavelength, and this wavelength has to be controlled precisely. An approach to circumvent this problem is a colorless ONU technique. A colorless ONU is referred to as a non-wavelength-specific ONU. There are two types of colorless ONU: ONU with local emission and ONU with wavelength supply as illustrated in Fig. 2.17 [19]. Options for the local emission scheme include wavelength tuning using a coherent tunable laser diode (TLD) and spectrum slicing using an incoherent broadband light source (BLS). The TLD enables high-speed operation at 10 Gb/s and a higher bit rate but its cost will be a stumbling block in the provision of service at a reasonable rate. On the other hand a BLS, for example an amplified spontaneous emission (ASE) or a superluminescent diode (SLD), is cost effective and temperature-control free, but it requires spectrum-slicing, and the bit rate obviously does not exceed 1 Gb/s.

Another option is wavelength supply from the OLT to ONUs as illustrated in Fig. 2.18. The BLS can be configured with a mutually injection-locked Fabry–Pérot laser diode (MI-FPLD) [20]. In the MI-FPLD the output continuous wave from anti-reflection (AR)-coated FPLDa is injected into FPLDb, and vice versa. The configuration realizes an unpolarized BLS by polarization multiplexing. Each one of the longitudinal modes is filtered out by the AWG and delivered to each ONU. When the spectrum-sliced mode is injected into the individual FPLD, the oscillation wavelength of the FPLD is locked to the injected mode, yielding a high power output for the uplink. The injection locking has the capability of wavelength selectivity and amplification. This is the mechanism that the MI-FPLD together with individual FPLDs serves as the BLS. The MI-FPLD may be located at the OLT, while individual FPLDs are placed either at ONUs or at the OLT using a loop-back scheme.

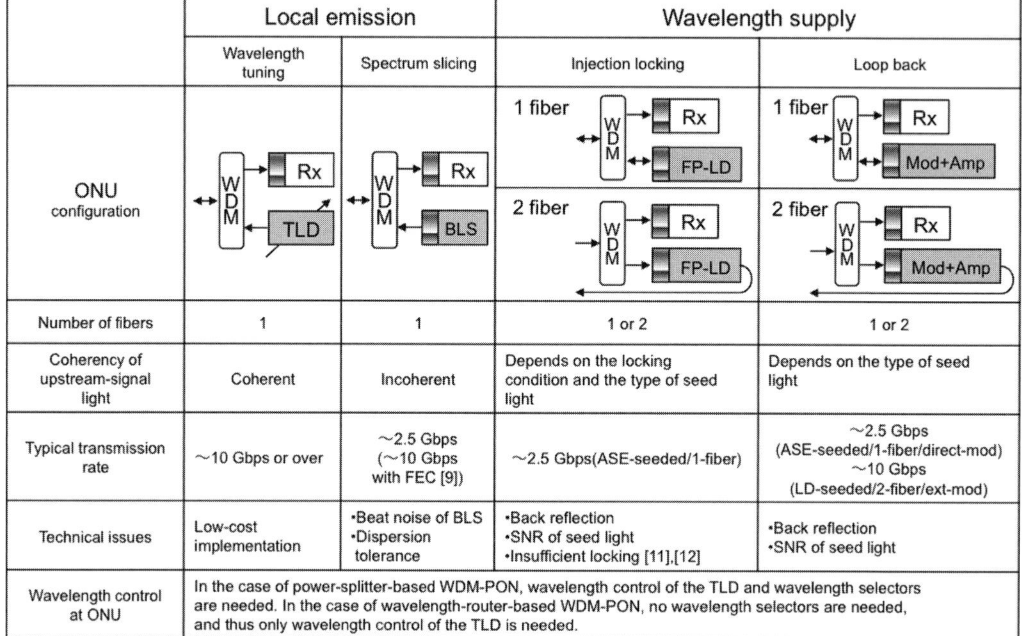

	Local emission		Wavelength supply	
	Wavelength tuning	Spectrum slicing	Injection locking	Loop back
ONU configuration				
Number of fibers	1	1	1 or 2	1 or 2
Coherency of upstream-signal light	Coherent	Incoherent	Depends on the locking condition and the type of seed light	Depends on the type of seed light
Typical transmission rate	~10 Gbps or over	~2.5 Gbps (~10 Gbps with FEC [9])	~2.5 Gbps(ASE-seeded/1-fiber)	~2.5 Gbps (ASE-seeded/1-fiber/direct-mod) ~10 Gbps (LD-seeded/2-fiber/ext-mod)
Technical issues	Low-cost implementation	•Beat noise of BLS •Dispersion tolerance	•Back reflection •SNR of seed light •Insufficient locking [11],[12]	•Back reflection •SNR of seed light
Wavelength control at ONU	In the case of power-splitter-based WDM-PON, wavelength control of the TLD and wavelength selectors are needed. In the case of wavelength-router-based WDM-PON, no wavelength selectors are needed, and thus only wavelength control of the TLD is needed.			

TLD:Tunable laser dio-de, BLS: Broadband light source,
Rx: Receiver, Mod: Modulator, Amp: Optical amplifier   : Wavelength selector (WS)

Figure 2.17 Colorless ONU: ONU with local emission and ONU with wavelength supply [19]. © Reprinted by permission of the Optical Society of America.

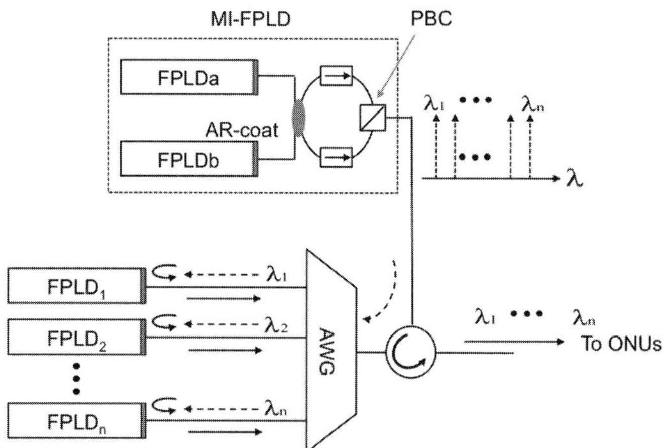

Figure 2.18 Colorless light source scheme using FPLD fed by mutual injection-locked Fabry–Pérot laser diodes (MI-FPLD) [21]. © Reprinted by permission of IEEE.

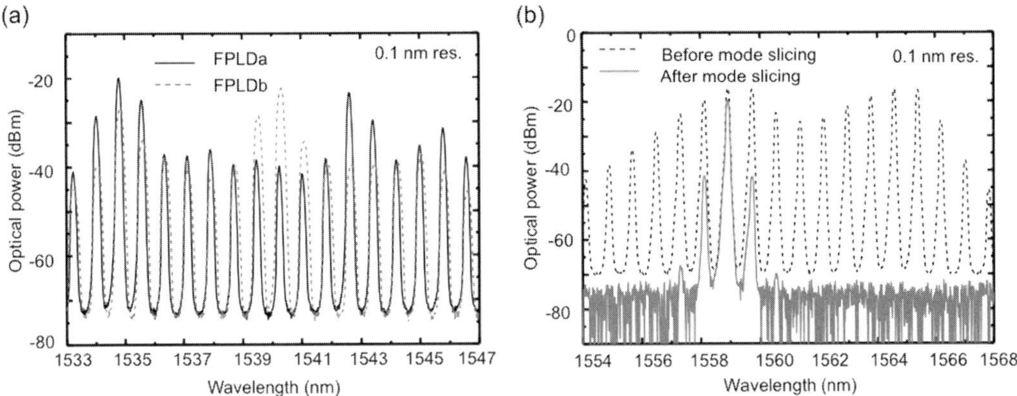

Figure 2.19 Measured optical spectra of (a) the two free-running FPLDs used in MI-FPLD and (b) output of MI-FPLD before and after mode slicing [21]. © Reprinted by permission of IEEE

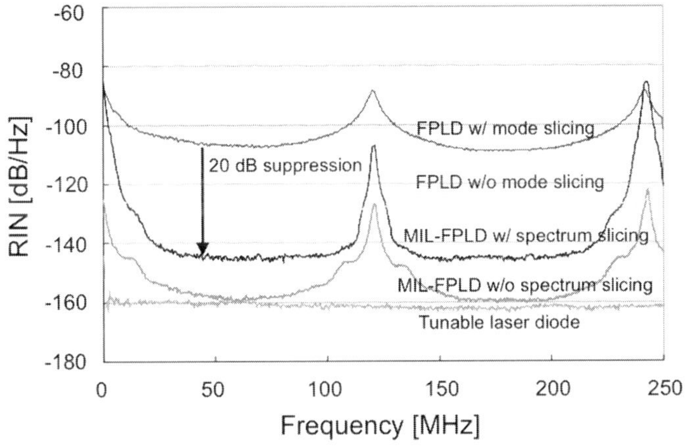

Figure 2.20 Measured RINs of the MI-FPLD in comparison with that of a tunable laser diode and a free-running FPLD [21]. © Reprinted by permission of IEEE.

In experiments [21], the output power of each FPLD was 3 dB m at a bias current of 60 mA. There is slight detuning in the longitudinal modes of the two emission spectra of FPLDa and FPLDb without mutual injection, and after mutual injection the spectra closely overlap each other as shown in Fig. 2.19(a). The 3 dB linewidth of each mode is measured to be 0.13 nm. After mutual injection, however, the output power of the MI-FPLD increases to 9 dBm and a very small change (0.01 nm) in linewidth is observed. The measured spectrum of the single mode at 1558.93 nm, which is sliced out from the BLS spectrum, exhibits excellent characteristics, with power as high as 3 dBm and side-mode suppression ratio of 21 dB, as shown in Fig. 2.19(b).

From the measured relative intensity noise (RIN) in Fig. 2.20, the RIN of the MI-FPLD is suppressed in comparison with individual lasers without injection-locking. The RIN represents the optical power fluctuation per hertz of a laser, and it is defined as the noise power within the measured bandwidth normalized by the average power. RIN originates from the beat between the electric fields of stimulated emission and spontaneous emission. The measured RIN of the free-running FPLD is higher than -130 dB/Hz, and after mode-slicing the RIN is increased to 106 dB/Hz. With mutual injection, the RIN is greatly reduced, for example, the RIN is measured to be 160 dB/Hz at 70 MHz. Strong RIN peaks spaced periodically at 122 MHz are observed. This is because the AR coating reduces the reflectivity of the front facet in each FPLD to $\sim 1\%$ but the residual reflection facilitates the formation of a new cavity between two rear facets of the FP-LDs. The distance between two FP-LDs is about 85 cm, which corresponds to a free spectral range of 122 MHz. Nevertheless, RIN of less than 140 dB/Hz is achieved even after spectrum slicing at wide frequency ranges of 25–100 MHz and 135–210 MHz. This can be eliminated or moved to a higher frequency out of interest.

2.2.3   Hybrid of wavelength and time division multiple access

Neither TDMA nor WDMA is suitable for accommodating both heavy users and light users at the same time. It will be difficult for TDMA-PON to provide a subscriber with sufficient bandwidth at any time, while WDMA-PON will have difficulty in providing a subscriber with a small bandwidth at a reasonable rate. This problem will be solved by combining heterogeneous PON systems in one optical line terminal (OLT) and ensuring flexible allocation of bandwidth to meet various users' needs. A hybrid WDM-TDMA-PON will be able to fulfill this requirement.

Figure 2.21 shows the configuration of WDM-TDMA-PON using wavelength-tunable laser diodes [22]. The OLT is equipped with a dynamic wavelength and bandwidth allocation (DWBA) controller and line cards (LCs) that transmit downlink signals and receive uplink signals at several wavelengths. Each ONU is assigned two designated wavelengths for the uplink and the downlink. Assume that the number of wavelengths is $2K$, K wavelengths each for uplink and downlink, and different wavelengths $\lambda_{u1} \sim \lambda_{uK}$ and $\lambda_{d1} \sim \lambda_{dK}$, respectively, are assigned to the uplink and downlink of PON branches $1 \sim K$. TDMA using M time-slots is adopted in a PON branch. Altogether $M \times K$ ONUs can be accommodated in the PON branch. An ONU is equipped with a wavelength-tunable burst-mode transmitter and a burst-mode receiver. At the OLT, M LCs have to be prepared, in which the burst-mode receiver is equipped with a tunable filter. The $1 \times K$ AWG can be replaced by an optical splitter if the wavelength count is small, and the insertion loss is comparable to that for the AWG. Uplink signals with different wavelengths from those of each PON branch are multiplexed by the $1 \times K$ AWG, divided by the $1 \times M$ coupler, and input into the LCs. A tunable filter (TF) in each LC tunes to the transmitting wavelength according to the wavelength-switching signal from the DWBA controller, and thus the uplink signal is received by the burst-mode receiver. An example of DWBA operation with two LCs and four PON branches is shown in the inset to Fig. 2.21. DWBA observes the requests from the ONUs and sets the wavelength and time matrices of LC1 and LC2. Suppose that there are light subscribers in PON branch 1 and heavy subscribers in PON branches $2 \sim 4$. After LC1 receives λ_{u1} from PON branch 1, in which the traffic demand is small, LC1 and LC2 receive λ_{u2}, λ_{u3}, and

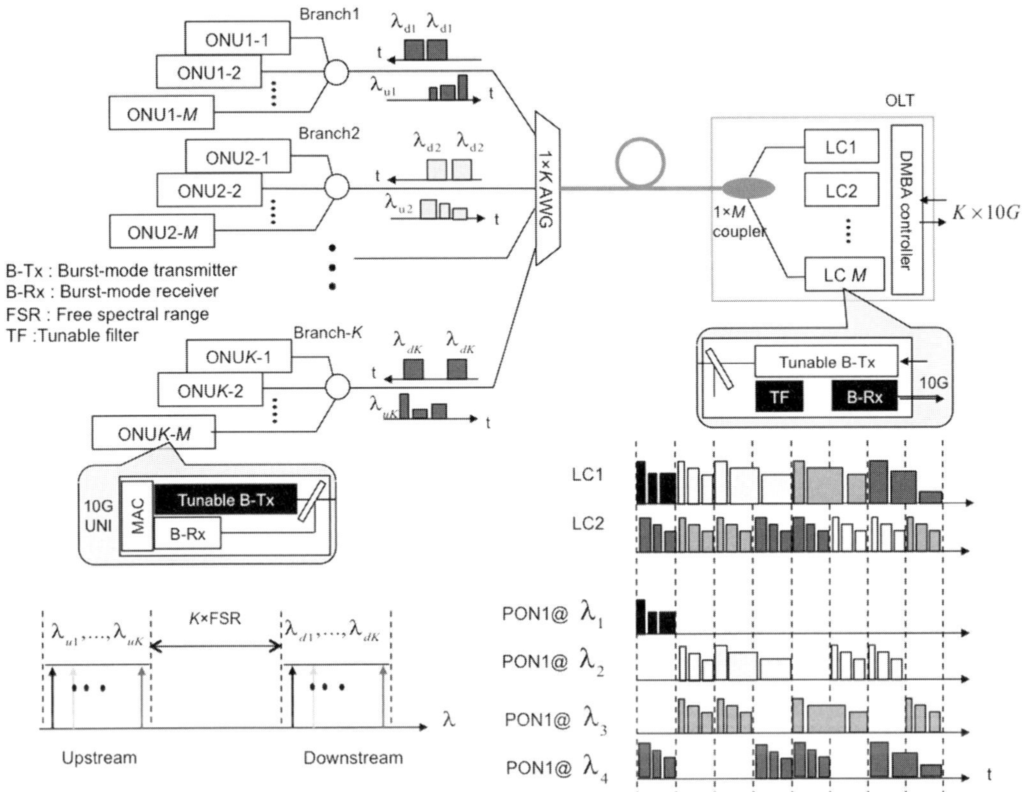

Figure 2.21 System configuration of WDM-TDMA-PON and the dynamic wavelength and bandwidth allocation (DWBA) [22]. © Reprinted by permission of the Optical Society of America (OFC2011).

λ_{u4} to meet the heavy data traffic demands from PON branches 2, 3, and 4 equally. This system can realize scalable and high capacity PON systems by increasing the number of accommodated ONUs and providing flexible load balance.

2.2.4 Orthogonal frequency division multiple access

Orthogonal frequency division multiple access (OFDMA) is emerging as an attractive candidate for the multiple access technique for NG-PON2 due to its high spectrum efficiency, flexibility in resource allocation, and resilience to fiber chromatic dispersion. The capability of elasticity in bandwidth allocation to subscribers, which OFDMA can offer, will be attractive for NG-PON, since the bandwidth required for an individual subscriber will range from 1 Gb/s to 100 Gb/s. Here, OFDM is reviewed briefly. The fundamental concept of OFDM was first formulated by Chang in a seminal paper in 1966 [25]. As very large scale integrated (VLSI) CMOS technology matured, OFDM was adopted as various standards in the mid-1990s. OFDM has been widely deployed in wireless communications such as wireless local area networks (WiFi, IEEE 802.11a/g),

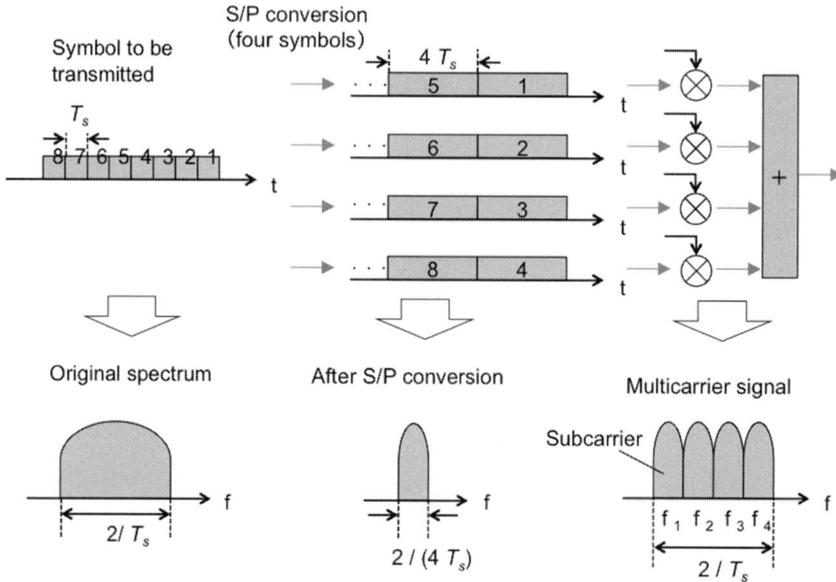

Figure 2.22 Schematic of multicarrier modulation. Modified after [26].

wireless metropolitan area networks (WiMAX, 802.16e), asymmetric digital subscriber lines (ADSL, ITU G.992.1), and fourth generation (4G) mobile communications technology, long-term evolution (LTE). For example, one of the LTE systems, which is allocated the bandwidth 20 MHz, can attain a bit rate of 480 Mb/s using 64QAM with the aid of a multiple-input multiple-output (MIMO) technique having 4×4 antennas. The spectral efficiency is 8 bit/s/Hz.

OFDM is a special class of multicarrier modulation, in which a single data link is transmitted over a number of lower rate subcarriers. The basic concept is the division of a high symbol rate data stream into several lower symbol rate data streams and modulation of several subcarriers using these substreams. The multicarrier modulation is illustrated in Fig. 2.22 [26]. The sequence of data symbols with symbol period T_s is serial-to-parallel converted and divided into L blocks of symbols ($L = 4$). The symbols $1\sim4$ and $5\sim8$ and the sinusoidal carrier wave of frequency f_i ($i = 1, 2, 3, 4$) are individually multiplied, combined into an OFDM signal, and transmitted. It is noteworthy that the spectrum width of the original symbol is $2/T_s$, while the spectrum of each subcarrier component is L compressed by a factor of L, that is, $1/(2T_s)$ ($L = 4$). Optical OFDM benefits from resilience to chromatic dispersion from this spectrum compression, compared with conventional single carrier binary OOK or PSK signals. In Fig. 2.23, the process used to generate an OFDM signal of symbols A, B, and C is shown schematically. For the purpose of illustration, the subcarrier sinusoidal wave is approximated as a bipolar function taking the values of $+1$ or -1, and $f_2(=2f_1)$, $f_3(=4f_1)$ are shown but $3f_1$ is omitted. Here, the time duration of each symbol is T_s. By taking the product of each data symbol and the subcarrier frequency $f_k(k = 1, 2, 3)$, followed by the summation of three subcarrier

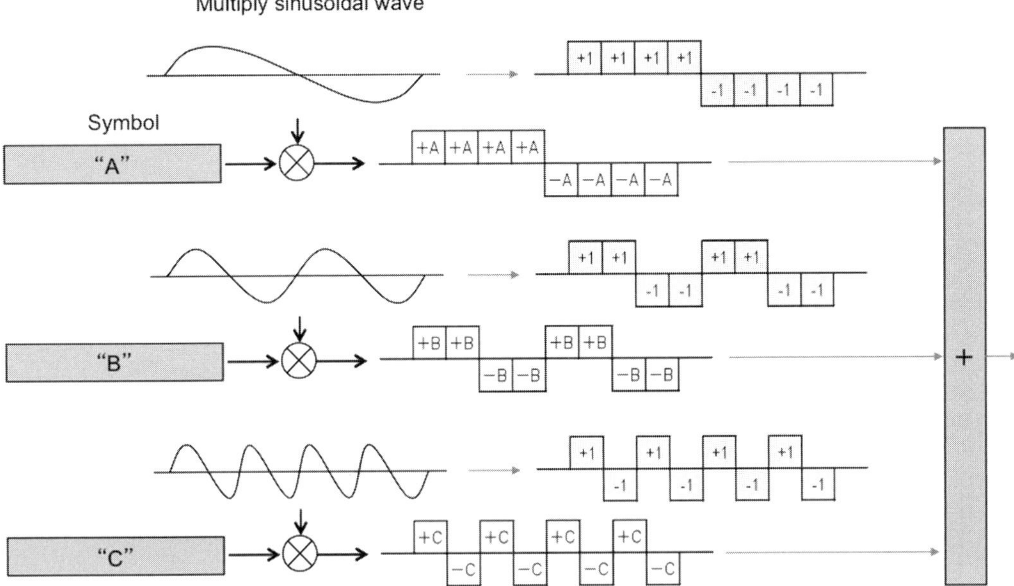

Figure 2.23 OFDM signal generation. Modified after [26].

components, the frequency spectrum of the OFDM signal is generated. This process is in the form of an inverse Fourier transform if we consider the limit of $T_s \to \infty$. At the receiver, the product of the transmitted signal and the subcarrier frequency, followed by the summation of four subcarrier components, each original symbol is decoded. For example, Fig. 2.24 shows the multiplication and summation carried out to demodulate symbol A. The subcarrier f_1 is multiplied with the received signal. Only symbol A is left demodulated, and the others, $\pm B$ and $\pm C$, are cancelled out. This is mathematically equivalent to the Fourier transform.

In wireless communication, in order to mitigate multi-path fading, a cyclic prefix (CP) is inserted in the transmitted signal. The CP is usually a copy of the tail part of the OFDM signal, and it is added as a precursor as shown in Fig. 2.25. Without the CP, several delayed signals, which take different paths during the propagation due to the fading, are received along with the desired signal, and these delayed signals cause interference with the desired signal. In Fig. 2.26, the desired signal and the delayed signal due to multi-path fading are illustrated. The received signal is the summation of the desired signal and the delayed signal. When the received signal is multiplied with the subcarrier frequency f_1, the symbol A is demodulated but symbols B and C also appear as interference. Let us see how the CP works. By multiplying the subcarrier frequency f_1 with the received signal, the summation of the desired signal and the delayed signal, symbol A is demodulated without any interference.

Figure 2.27 is a block diagram of a typical OFDM transmitter and receiver [27]. A binary data stream is converted from serial to parallel and after symbol mapping the

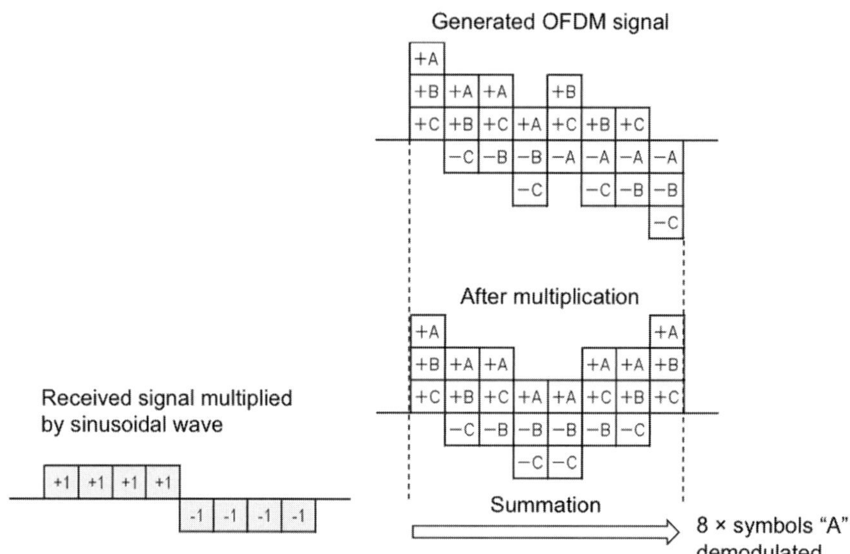

Figure 2.24 Demodulation of OFDM signal. Modified after [26].

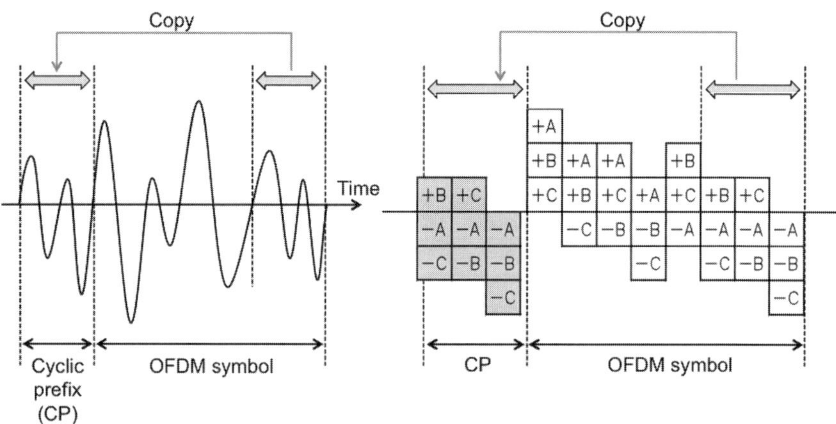

Figure 2.25 Insertion of cyclic prefix (CP) in the OFDM signal. Modified after [24].

low bit rate streams are simultaneously modulated onto orthogonal subcarriers. A very effective way of generating the OFDM subcarriers is by taking the IFFT. After the IFFT, the CP and training symbols (TS) are added and the signal is fed to the front end, followed by electrical-to-optical conversion, typically by an optical modulator. At the receiver, the front end consists of either coherent or direct detection. Subsequently, training symbols are used for symbol synchronization and compensation for impairments. The OFDM signal is then converted back to the frequency domain by taking the FFT. Finally, the binary data are recovered by demodulating the equalized signal.

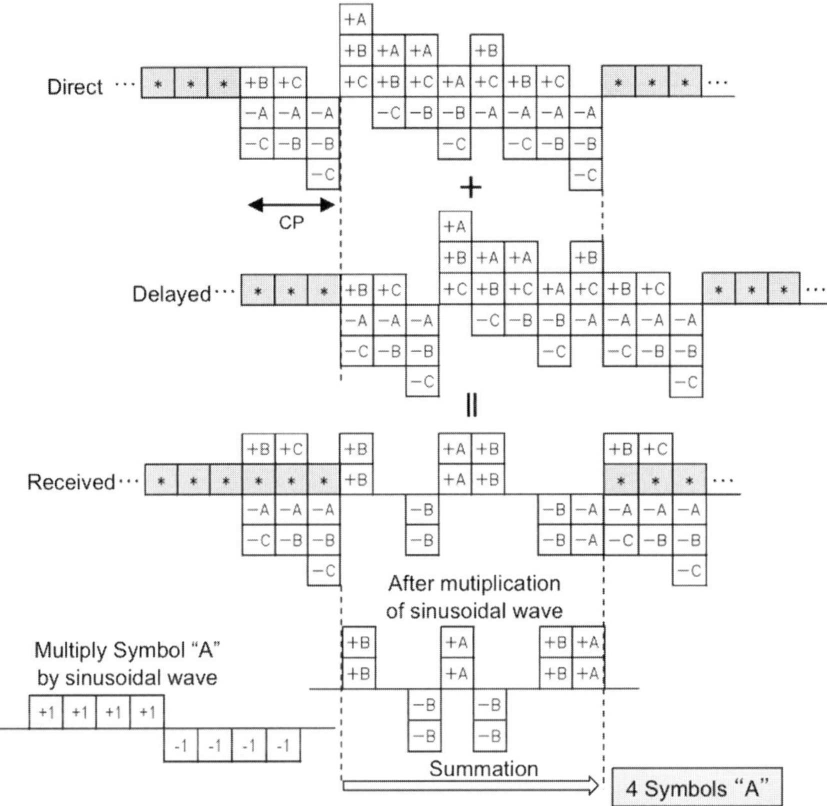

Figure 2.26 Demodulation of OFDM signal with cyclic prefix. Modified after [26].

The main drawback of OFDM is the inherently high peak-to-average power ratio (PAPR), which reduces the non-linear tolerance in optical fiber communications. Nevertheless, its unique capability of elastic bandwidth allocation is attractive not only for the long-haul transmission system but also for the access optical networks. Bandwidth elasticity will revolutionize the WDM network by introducing the concept of "flexible grid" as the standardization of ITU-T has already initiated in 2011, as shown in Fig. 1.26.

Let us turn our focus to OFDMA-PON. In Fig. 2.28 a conceptual schematic of an energy-efficient, elastic bandwidth coherent interleaved FDMA, so-called IFDMA-PON, is illustrated [28]. IFDMA is a special class of OFDMA. IFDMA is known for having the lowest PAPR of the modulated signal and for the simplicity of its subcarrier generation process without using the discrete Fourier transform (DFT). Hence, signal impairment due to fiber non-linearity is mitigated, and the power consumption of DSP circuits can be drastically reduced.

Elastic bandwidth provisioning can be realized by allocating different numbers of subcarriers to users. The more subcarriers assigned, the larger the bandwidth which can be provided. A fully bandwidth elastic PON based on the OFDMA concept might be

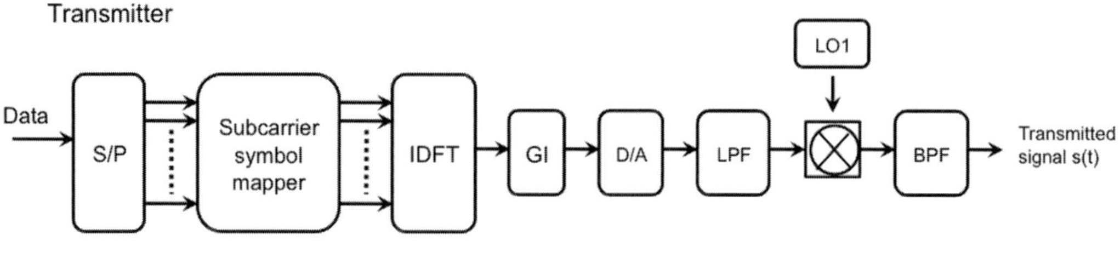

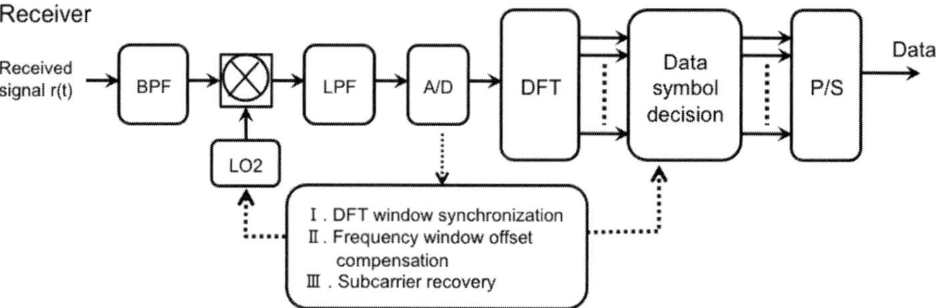

S/P: Serial-to-Parallel, GI: Guard Time Insertion, D/A: Digital-to-Analog,
(I)DFT: (Inverse) Discrete Fourier Transform, LPF: Low Pass Filter   BPF: Band Pass Filter

Figure 2.27 Block diagram of a typical OFDM transmitter and receiver [27]. © Reprinted by permission of Academic Press.

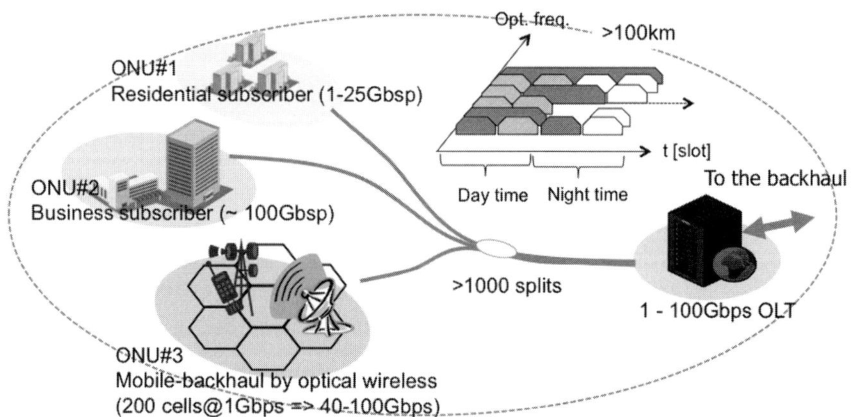

Figure 2.28 OFDMA-PON: elastic bandwidth provisioning [28]. © Reprinted by permission of the Optical Society of America (OFC2012).

realized with a coherent OFDMA-PON architecture where multiple optical in-phase and quadrature-phase (IQ) modulators at ONUs and a coherent receiver at the OLT enable the optical baseband OFDMA on a single wavelength. Owing to recent progress in digital coherent techniques for long-haul transmissions, it is becoming feasible to adopt coherent techniques in the access network. IM-DD OFDMA-PON uses cost-effective

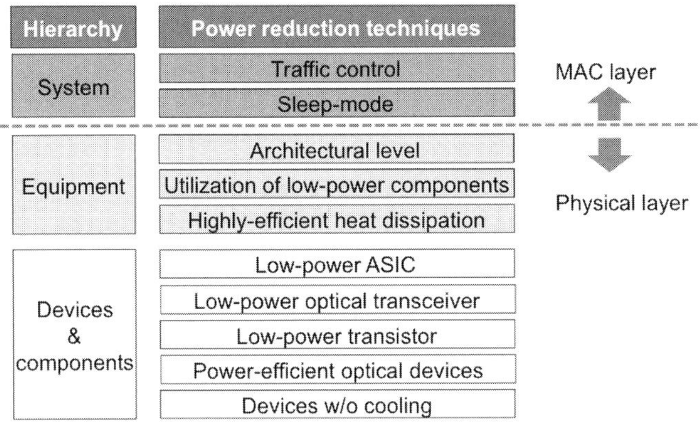

Figure 2.29 Hierarchical approach to power reduction of the PON system.

intensity modulation and direct detection (IM-DD), but each optical network unit (ONU) would require a different wavelength to avoid broadband beat noise at the optical line terminal (OLT) receiver. As a consequence, there is no longer a taboo in adopting coherent detection techniques to access networks, and coherent OFDMA has become a promising candidate for NG-PON.

2.3   Power saving operation

Access network equipment consumes a major share of the total network power, and power saving in telecommunication systems has become an increasingly important concern with respect to operators' expenditures (OPEX) and contributions to greenhouse gas emission. NG-PON2 systems must be designed in the most energy-efficient way also to reduce power consumption during normal operation without sacrificing service quality and user experience.

It is forecast that the electric power consumption of ICT will increase from 6% of the total amount in 2012 to 10% of the total amount in 2020 [29]. From the forecast of ICT power consumption in Japan in 2020 shown in Fig. 1.27, the power consumption of telecommunications and broadcasting will amount to 10%. It is noteworthy that the optical access network makes up 80% of the total electric power consumption of the network. This is why achieving low energy consumption in optical access networks has attracted broad research attention from both academia and industry. It should be remembered that the ONUs consume a large part of the power in access networks [30]. There are several approaches to reduction of the power consumption from different hierarchical levels, as shown in Fig. 2.29 [31]. ITU-T G.sup 45 Gigabit capable PON (GPON) power conservation standard categorizes the power saving states into three categories: power shedding, dozing, and sleeping [32]. Since the rectification of ITU-T G.sup 45, some ONU products of G-PON and EPON have added power saving modes as a feature.

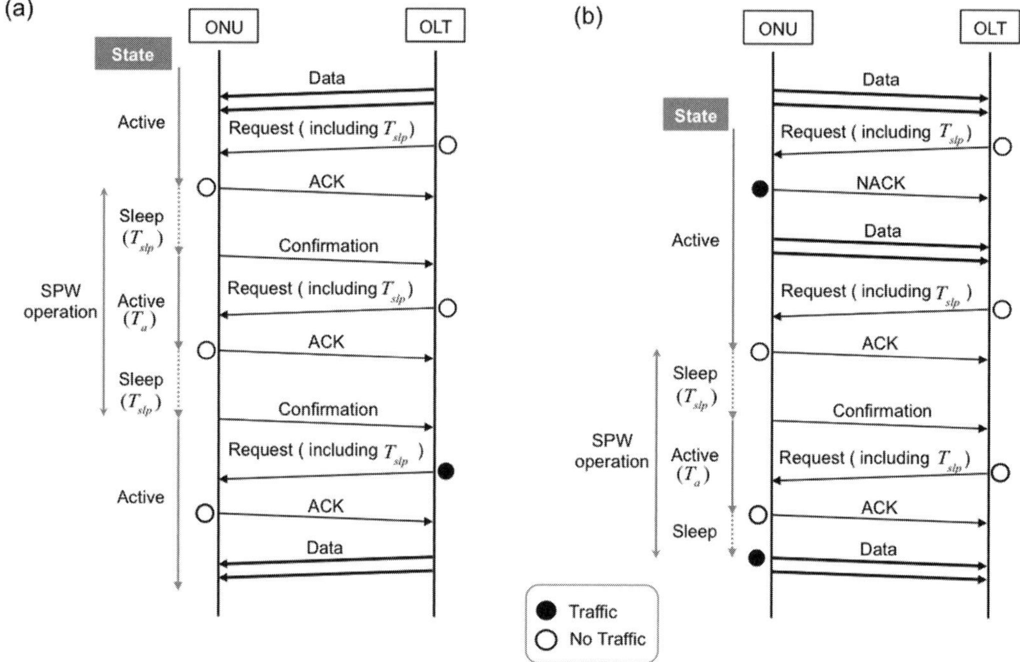

Figure 2.30 Operation of sleep and periodic wake-up (SPW) for (a) downlink traffic and (b) uplink traffic [31]. © Reprinted by permission of IEICE.

A common objective of power saving techniques of ONUs is to put the ONUs into lower power states. The ONU transmitter is already burst-mode capable, and it can turn on and off quickly during idle time slots. On the other hand, turning the ONU receiver on and off is more challenging because the downlink signals are broadcast to all ONUs. During lower power states, the ONU also faces the choice of selecting which functions and services to turn off. In the power shedding mode, used when the ONU operates under battery power, the ONU powers off or reduces the power to non-essential functions and services only. In the dozing mode, the ONU keeps all the downstream functions operational but turns off the transmitter and ignores OLT DBA bandwidth requests when the ONU does not have upstream traffic to send. In the sleeping mode, the ONU turns off virtually all functions and services to gain the greatest power saving potentials.

Here, one of the control mechanisms, SPW (sleep and periodic wake-up) is introduced. This uses the sleep mode and periodically exchanges messages between the OLT and ONU to determine whether the ONU should wake up, depending on its traffic [33]. The sequence of SPW operation is illustrated in Fig. 2.30. Let us look at the SPW downlink operation shown in Fig. 2.30(a). If there is no downlink traffic, the OLT sends the ONU a Request message about the sleep period T_{slp}. Upon receipt of the Request message, the ONU sends the OLT an ACK and enters into the sleep mode. When the sleep period expires, the sleeping ONU is activated, and a Confirmation message is sent to the OLT to check whether there is any downlink traffic for the ONU. Upon receipt of the Confirmation message, the OLT forwards to the ONU the frames from its buffer, which are stored during the sleep period. If there is no additional downlink traffic for the ONU, the OLT sends the ONU a Request message about the sleep period T_{slp}. If there is no additional downlink traffic, the OLT sends the ONU a Request message by setting $T_{slp} = 0$, in order to keep the ONU active and forward the downlink traffic. Note that the uplink bandwidth needed to send the control messages is consistently assigned by the DBA function of the OLT. An advantage is that it does not interfere with the existing DBA process. But there will be some detrimental effects on the delay performance. SPW operation for the uplink is illustrated in Fig. 2.30(b). Upon receiving a Request message, the ONU refuses to enter into the sleep mode when it is receiving uplink traffic and sends a NACK message to the OLT. If a sleeping ONU has uplink traffic from UNI, it is activated to forward the uplink traffic. Note that the uplink frame transfer is performed after the uplink bandwidth is allocated by the DBA.

In SPW operation, the number of ONUs which communicate with the OLT changes in each DBA cycle, since a sleeping ONU remains registered even if the ONU does not communicate with the OLT, and the OLT consistently assigns the minimum uplink bandwidth to each registered ONU. Otherwise, the sleeping ONU cannot send a control message in every DBA grant cycle.

A technique which would put the ONU in sleep mode through energy-conscious MAC-layer control and scheduling such as SPW will be a powerful solution to reduce power consumption. The OLT and ONUs have to be operated for 24 hours, in contrast to an energy efficient light bulb, for example, a domestic LED light bulb which is not likely to be switched on for 24 hours, and hence energy saving is limited. The overall energy consumption per bit of a PON, E_{PON}, is expressed as the summation of the energy of the OLT, E_{OLT}, and that of the ONU, E_{ONU}, as

$$E_{PON} = E_{OLT} + N_{ONU} \cdot (1.0 - \eta_{sleep}) \cdot E_{ONU} \tag{2.3}$$

where η_{sleep} $(0 < \eta_{sleep} < 1)$ is the sleep mode rate coefficient. $\eta_{sleep} = 0$ denotes operation without the sleep mode. For example, we consider an EPON providing symmetric 1 Gb/s connection and 32 customers sharing a connection to an OLT. The OLT equipment shelf is capable of supporting 32 EPON lines ($N_{ONU} = 32$) and consumes about 100 W. The splitter is unpowered. Each ONU consumes 5 W, corresponding to $E_{ONU} = 5$ ~ 40 nJ/bit by taking into account that the average bit rate per customer is one eighth of the maximum line rate of 1 Gb/s. As the overall energy consumption of a PON is dominated by the ONUs, the sleep mode can be a powerful technique for energy reduction. It should be remembered that the power consumption of current network equipment does not typically scale with utilization, resulting in very low efficiencies at low utilization. It is also noted that the power consumption stays constant at a bit rate below 1 Gb/s per customer but it scales above 1 Gb/s. If we look at the breakdown of power consumption of an ONU, the electronics is the largest portion: 70%, 10%, and 20% for the electronics, optical components, and power conversion loss, respectively. Hence, improvements in complementary metal oxide semiconductor (CMOS) and optical technology should lead to improvements in energy efficiency in future generations of network equipment. The per annum "business as usual" improvement rates for these subsystems are electronics 26% and optical components 5% [30].

Problems

2.1 Discuss the differences between G-PON and EPON.

2.2 By comparing DBA and FBA, discuss their advantages and disadvantages.

2.3 Explain why ranging is necessary in PON.

2.4 By comparing TDMA and WDMA, discuss their advantages and disadvantages.

2.5 Explain what the sleep mode in PON is. Why is it necessary?

3   Light propagation in optical fibers

3.1   Loss

Glass was made by the ancient Egyptians in 1 BC, as described in Section 1.2. It was sufficiently transparent in the region of visible wavelengths, but was opaque in the infrared region. In 1966 Dr. Kao, 2009 Nobel Prize laureate for Physics, first proposed the concept of cladded glass fibers as a "new form of communication medium" and predicted that the dielectric fiber loss could be reduced to as low as 20 dB/km in his seminal paper [1]. Since then, there has been a dramatic improvement in silica-based optical loss achieved as a function of time in silica-based optical fibers as shown in Fig. 1.5. The lowest fiber loss coefficient achieved in a pure silica core fiber is 0.1484 dB/km [2]. It is conjectured that a loss of 0.03 dB/km might be achievable with a hollow-core fiber [3]. Further challenges in the long run include the quest for exotic materials which drastically reduce attenuation. If further drastic loss reduction could be achieved in the future, the economic impact would be enormous, as a result of extension of the repeater span.

A typical loss spectrum of silica fiber is shown in Fig. 3.1 [4]. There are several loss mechanisms in silica optical fiber: one is the intrinsic absorption of silica glass, and others include the extrinsic loss due to the absorption of impurities and loss due to Rayleigh scattering. The intrinsic absorption α_{UV} of pure silica in the ultraviolet (UV) region of the spectrum in amorphous media is associated with electronic transitions in the band gap and shows "Urbach behavior." The absorption edge decays exponentially with increased wavelength and is expressed by

$$\alpha_{UV} \propto e^{a/\lambda}. \tag{3.1}$$

Another intrinsic absorption loss α_{IR} is in the infrared band, and its tail extends into the near infrared as

$$\alpha_{IR} \propto e^{-b/\lambda}. \tag{3.2}$$

However, these intrinsic absorption losses in the transmission windows of interest around 1550 nm and 1300 nm are negligible. The extrinsic impurity ion absorption is caused by the hydroxyl ion OH^- and metallic ions. Residual hydrogen and moisture in the fiber after the manufacturing process result in SiOH, and this affects the optical properties due to the fundamental absorption band at 2730 nm and its overtones at 950 nm, 1240 nm, and 1380 nm. The OH absorption at 1380 nm is called the "water peak," and fiber almost

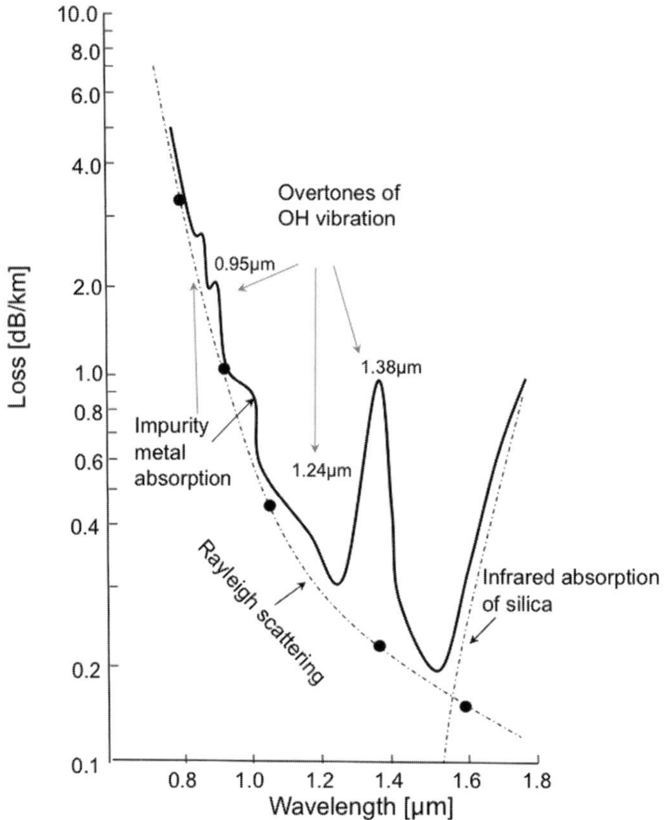

Figure 3.1 Intrinsic losses of silica fiber [4]. © Reprinted by permission of Pearson Education, Inc.

free from this absorption is categorized as having a low water peak, as summarized in Table 3.1. More details will be described in Section 3.3. Rayleigh scattering loss $\alpha_{Rayleigh}$ is caused by inhomogeneities such as minute changes in refractive index and density fluctuations, which are small compared with the wavelength of light, and is expressed by

$$\alpha_{Rayleigh} \propto A\lambda^{-4},$$ (3.3)

where A is a material constant which depends both on composition and on the glass preparation method. Superimposed on the above intrinsic absorption and Rayleigh scattering, there is a variety of extrinsic loss mechanisms which can limit the achievable level of optical loss. As a consequence, the most transparent spectral region exists around 1550 nm, and the minimum loss is typically 0.20 dB/km.

The fiber loss is defined as the ratio of the output power P_{out} to the input power P_{in} and is often expressed by α_0 dB/km:

$$\alpha_0 = -10 \log_{10}\left(\frac{P_{out}}{P_{in}}\right) \ [\text{dB/km}].$$ (3.4)

Table 3.1 Categories and characteristics of single-mode fibers, G652, G656, and G.657

	G.652				G.656	G.657	
	G.652A	G.652B	G.652C	G.652D		G.657A	G.657B
Mode field diameter at 1310 nm	8.6~9.5 μm	8.6~9.5 μm	8.6~9.5 μm	8.6~9.5 μm	7.0~11.0 μm	8.6~9.5 μm	8.6~9.5 μm
Cutoff wavelength	≤1260 nm	≤1260 nm	≤1260 nm	≤1260 nm	≤1450 nm	≤1260 nm	≤1260 nm
Bending loss at 1625 nm (R30 mm, 100 turns)		≤0.1 dB	≤0.1 dB	≤0.1 dB		≤0.5 dB	≤0.1 dB (10 turns)
Bending loss at 1550 nm (R30 mm, 100 turns)	≤0.1 dB						≤0.03 dB (10 turns)
Bending loss at 1625 nm (R15 mm)						≤1.0 dB (10 turns)	≤0.1 dB (10 turns)
Bending loss at 1550 nm (R15 mm)						≤0.25 dB (10 turns)	≤0.03 dB (10 turns)
Attenuation at 1625 nm		≤0.4 dB/km	≤0.4 dB/km	≤0.4 dB/km	≤0.4 dB/km		≤0.4 dB/km
Attenuation at 1550 nm	≤0.4 dB/km	≤0.35 dB/km	≤0.3 dB/km	≤0.3 dB/km	≤0.35 dB/km	≤0.3 dB/km	≤0.3 dB/km
Attenuation at 1383 nm			≤1310 nm value	≤1310 nm value	≤0.4 dB/km at 1460 nm	≤1310 nm value	
Attenuation 1310 nm	≤0.5 dB/km	≤0.4 dB/km	≤0.4 dB/km	≤0.4 dB/km		≤0.4 dB/km	≤0.5 dB/km

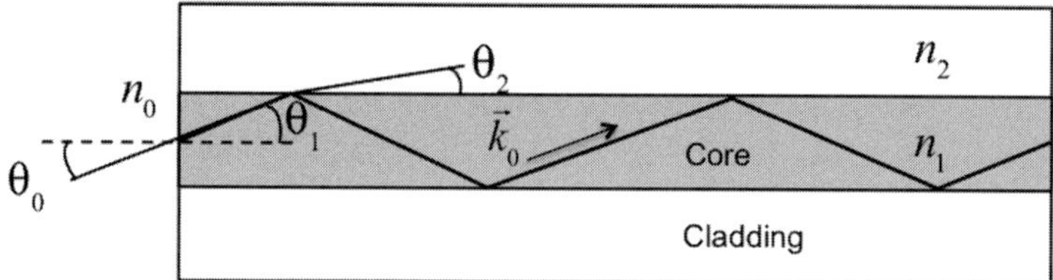

Figure 3.2 Ray trace propagating in an optical fiber.

On the other hand, the attenuation coefficient α is expressed as

$$\alpha = -\ln\left(\frac{P_{out}}{P_{in}}\right) \ [1/\text{km}]. \tag{3.5}$$

The relation between α_0 in dB/km and α in km^{-1} is given by

$$\alpha = -\ln\left(10^{-\alpha_0/10}\right). \tag{3.6}$$

3.2   Waveguide mode

Geometric optics or ray theory will give an intuition of how a ray propagates in a waveguide. The ray is defined as the line that crosses the surface of the constant phase of the electric field of light at right angles. This approach is only valid for a narrow beam of light, and this is the case with an optical fiber with core radius much larger than the operation wavelength and/or a sufficient difference in refractive index between the core and the cladding. In a homogeneous medium the ray takes a straight path as is shown by the wave vector in Fig. 3.2. Snell's law at the input face governs the relation of the incident and the refractive angles, and is expressed as

$$n_1 \sin\theta_1 = n_0 \sin\theta_0, \tag{3.7}$$

and at the boundary between the core and cladding

$$n_1 \cos\theta_1 = n_2 \cos\theta_2. \tag{3.8}$$

Since $n_1 > n_2$, only $\theta_2 < \theta_1$ satisfies Snell's law. For $\theta_2 = 0$ the angle θ_1 of the incident ray to the boundary is maximized and is defined as the critical angle θ_c given by

$$\cos\theta_c = \frac{n_2}{n_1} \cong 1 - \Delta, \tag{3.9}$$

where the relative difference in refractive index Δ is defined as

$$\Delta = \frac{n_1^2 - n_2^2}{2n_1^2} \cong \frac{n_1 - n_2}{n_1}. \tag{3.10}$$

Therefore, only a ray with incident angle $\theta_1 < \theta_c$ is allowed to propagate by taking a zig-zag path along the waveguide. This is the phenomenon of "total internal reflection."

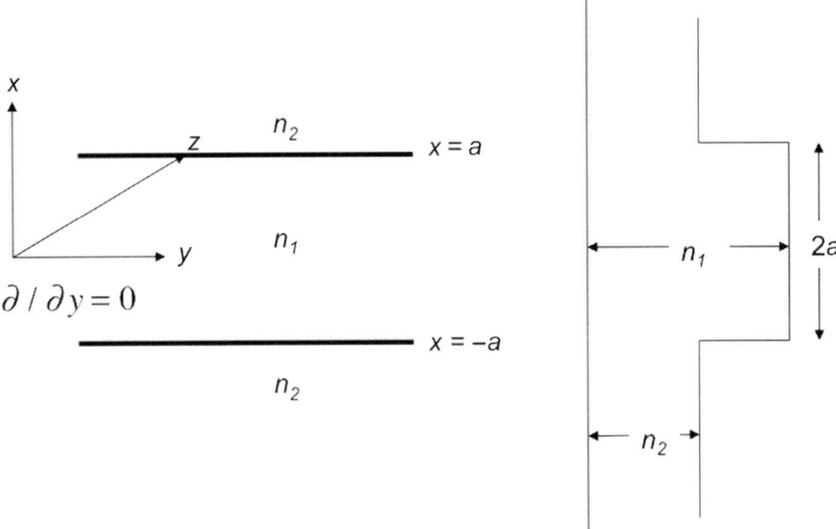

Figure 3.3 Two-dimensional slab waveguide.

The critical angle determines the maximum incident angle from air to the fiber end face $\theta_0^{\max}$:

$$\theta_0^{\max} = \sin^{-1}\frac{\sqrt{n_1^2 - n_2^2}}{n_0}.\tag{3.11}$$

If $\theta_1 > \theta_c$, the ray penetrates the cladding to the outside, and is no longer totally reflected back to the core at the boundary. Then, a question arises. Does the ray that is allowed to propagate along the waveguide hit the core–cladding boundary at an arbitrary angle? The answer is no. The standing wave has to be preserved in the cross-section to the waveguide axis, and the incident angle θ_1 has to satisfy the following condition:

$$2(2a)k_0 n_1 \sin\theta_1 + 2\Phi = 2p\pi \, (p = 1, 2, \ldots),\tag{3.12}$$

where the ray undergoes a phase shift Φ at the core–cladding boundary, and k_0 is the wave number in vacuum. The phase shift is called the Goos–Hanschen shift and is given by

$$\Phi = -2\tan^{-1}\sqrt{\frac{2\Delta}{\sin^2\theta_1} - 1}.\tag{3.13}$$

This requires θ_1 and the propagation constant along the waveguide axis to be discrete.

Before we go into detail of guided modes of a circular waveguide, for simplicity let us consider the two-dimensional slab waveguide in the yz plane shown in Fig. 3.3. Without loss of generality, the slab waveguide modes can be a good model for a circular waveguide such as an optical fiber. There are two types of modes in such a slab waveguide: transversal electric (TE) and transversal magnetic (TM) modes. The TE mode has only three field components: E_y, H_x, H_z as shown in Fig. 3.4. Assume that the slab is infinitely extended in the yz plane and there is no variation in the y direction, that

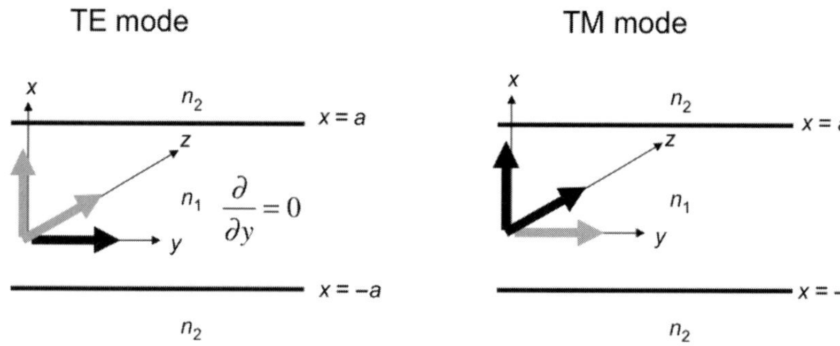

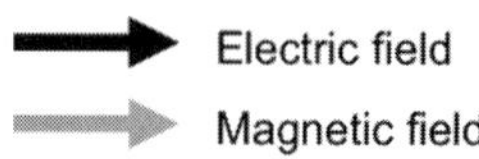

Figure 3.4 Field components of TE and TM modes.

is $\partial/\partial y = 0$. We will consider the time harmonic complex electric field vector of a monochromatic wave expressed by

$$\mathbf{E} = \mathbf{E}(x)\exp[j(\omega t - \beta z)], \tag{3.14}$$

where ω is the angular frequency of the wave, and β is the propagation constant along the z-axis in a medium. Here, the z-axis is taken to be along the fiber axis. $\mathbf{E}(x)$ is the transversal vector electric field. Maxwell's equations are expressed as

$$\nabla \times \mathbf{E} = -\mu_0 \frac{\partial \mathbf{H}}{\partial t}$$

$$\nabla \times \mathbf{H} = \varepsilon_0 n^2 \frac{\partial \mathbf{E}}{\partial t}. \tag{3.15}$$

where ε_0 and μ_0 are the permittivity and permeability in vacuum, respectively. Given a vector $\mathbf{E}$ that is twice differentiable, the wave equation results when $\nabla \cdot \mathbf{E} = 0$:

$$\nabla^2 \mathbf{E} - n^2 \varepsilon_0 \mu_0 \frac{\partial^2 \mathbf{E}}{\partial t^2} = 0. \tag{3.16}$$

For the TE mode having electromagnetic components (E_y, H_x, H_z), Eq. (3.16) is reduced to

$$\frac{\partial^2 E_y}{\partial x^2} + \left(k_0^2 n^2 - \beta^2\right) \quad E_y = 0 \; \left(k_0^2 = \omega^2 \mu_0 \varepsilon_0\right) \tag{3.17}$$

By substituting the electric field of the TE mode

$$\mathbf{E} = \mathbf{e}_y E_y(x) e^{-j\beta z} \tag{3.18}$$

into Eq. (3.15), we obtain

$$H_x = -(\beta/\omega \, \mu_0) E_y$$

$$H_z = (j/\omega \, \mu_0) \partial E_y / \partial x. \tag{3.19}$$

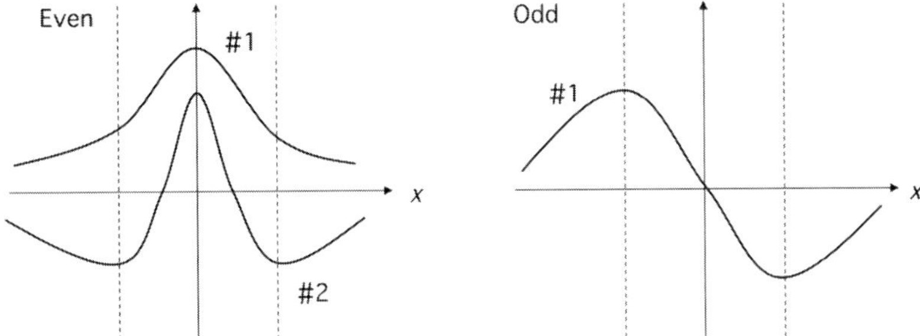

Figure 3.5 Field distributions of guided modes of a slab waveguide.

By substituting Eq. (3.19) into Eq. (3.17), the wave equation is rewritten as

$$\frac{\partial^2 E_y}{\partial x^2} + \kappa^2 E_y = 0, \quad (|x| \le a)$$

$$\frac{\partial^2 E_y}{\partial x^2} - \gamma^2 E_y = 0 \quad \text{elsewhere} \tag{3.20}$$

where

$$\kappa = \left(n_1^2 k_0^2 - \beta^2\right)^{1/2}$$

$$\gamma = \left(\beta^2 - n_2^2 k_0^2\right)^{1/2}. \tag{3.21}$$

There are two types of mode field, one is an even mode having symmetric distribution with respect to $x = 0$, and the other is an odd mode having asymmetric distribution as shown in Fig. 3.5. By taking into account that the field must vanish at $x = \pm\infty$, the field is expressed as

$$E_y = A \left\{ \begin{array}{c} \cos \kappa x \\ \sin \kappa x \end{array} \right\}, \quad |x| \le a$$

$$= B e^{-\gamma x}, \quad x \ge a$$

$$= B e^{\gamma x}, \quad x \le a$$

$$H_z = (-j\kappa/\omega\mu_0) A \left\{ \begin{array}{c} \sin \kappa x \\ \cos \kappa x \end{array} \right\}, \quad |x| \le a$$

$$= (-j\gamma/\omega\mu_0) B e^{-\gamma x}, \quad x \ge a$$

$$= (j\gamma/\omega\mu_0) B e^{\gamma x}, \quad x \le a. \tag{3.22}$$

The field components of the TE mode must satisfy the boundary conditions, because the tangential electric and magnetic fields are continuous, at the two dielectric interfaces at $x = a$ and $x = -a$:

$$A \left\{ \begin{array}{c} \cos \kappa a \\ \sin \kappa a \end{array} \right\} = B e^{-\gamma a}$$

$$-A \frac{jk}{\omega\mu_0} \left\{ \begin{array}{c} \sin \kappa a \\ -\cos \kappa a \end{array} \right\} = -\frac{j\gamma}{\omega\mu_0} B e^{-\gamma a}, \tag{3.23}$$

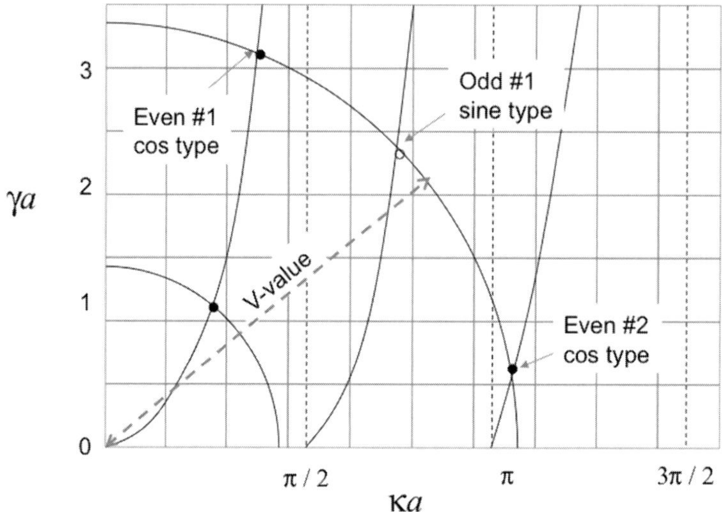

Figure 3.6 Graphical solution of the eigenvalue equation [5]. © Reprinted by permission of Shokodo.

The continuity of E_y and H_z at $x = a$ derives the eigenvalue equation for the propagation constant β:

$$ka\frac{\sin \kappa a}{\cos \kappa a} = ka \tan \kappa a = \gamma a \text{ (Even mode)}$$

$$ka\frac{\cos \kappa a}{\sin \kappa a} = ka \cot \kappa a = -\gamma a \text{ (Odd mode)}, \tag{3.24}$$

and

$$\frac{A}{B} = \frac{e^{-\gamma a}}{\left\{ \begin{array}{c} \cos \kappa a \\ \sin \kappa a \end{array} \right\}}. \tag{3.25}$$

We introduce the normalized frequency V-value defined as

$$v^2 \equiv (\kappa a)^2 + (\gamma a)^2 = (k_0 a)^2 (n_1^2 - n_2^2)$$
$$v = n_1 k_0 a \sqrt{2\Delta}. \tag{3.26}$$

The propagation constant β is obtained graphically as the eigenvalue of Eq. (3.24) as shown in Fig. 3.6 [5]. As κa decreases to π with decreasing V-value, γa approaches zero, and then $\tan \kappa a$ becomes zero. This leads to $\beta = n_2 k_0$ from Eq. (3.21), and the wave propagates at the speed of the cladding region. In the regime $V \ll \pi$', even-mode 2 does not exist and becomes *cut off*. But this is not the case with even-mode 1. When V approaches zero, there is always an intersection between the V-value circle and $\tan \kappa a$. This means there is no cutoff for even-mode 1. The *single mode operation* is maintained in the region $V < \pi/2$. The odd-mode 1 is cut off in the region $V < \pi/2$. On the other hand, as the V-value goes to infinity, γ eventually approaches the V-value, resulting in $k = 0$. This means that β becomes equal to $n_1 k_0$ from the definition of V-value in Eq. (3.26).

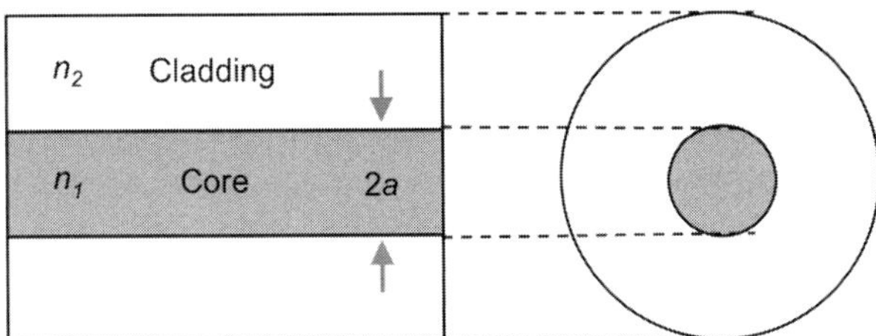

Figure 3.7 Geometry of step-index fiber.

3.3 Single-mode fiber

We will extend the waveguide mode analysis to an optical fiber with a step index profile, as shown in Fig. 3.7. It has a cylindrical core of radius a with refractive index n_1, and cladding with refractive index n_2 which extends to infinity. As for the slab waveguide described in Section 3.2, the propagation constant of the waveguide mode is limited within the region $n_2 k_0 \leq \beta \leq n_1 k_0$. The weakly guiding approximation $\Delta \ll 1$ leads to the simplest but practically useful description of the modes [6]. These modes are the linear polarized LP mode. The longitudinal field component E_z is the solution of the wave equation in cylindrical coordinates

$$\frac{\partial^2 E_z}{\partial r^2} + \frac{1}{r}\frac{\partial E_z}{\partial r} + \frac{1}{r}\frac{\partial^2 E_z}{\partial \theta^2} + \left(\omega^2 \mu_0 \varepsilon - \beta_0^2\right) E_z = 0, \tag{3.27}$$

where ε is a function of r. We assume the transverse electric field component $E_z(r)$ has periodicity in ϕ, $\exp(-jl\phi)$ and z dependence of the form $\exp(j\beta z)$,

$$E_z(r, \phi, z) = E_z(r)\exp(-jl\phi)\exp(i\beta_0 z) \quad (l = 0, \pm 1, \pm 2, \ldots). \tag{3.28}$$

In the core the wave equation is rewritten as

$$\frac{\partial^2 E_z(r)}{\partial r^2} + \frac{1}{r}\frac{\partial E_z(r)}{\partial r} + \left(\kappa^2 - \frac{l^2}{r^2}\right) E_z(r) = 0, \tag{3.29}$$

and in the cladding it is reduced to the form of

$$\frac{\partial^2 E_z(r)}{\partial r^2} + \frac{1}{r}\frac{\partial E_z(r)}{\partial r} - \left(\gamma^2 + \frac{l^2}{r^2}\right) E_z(r) = 0. \tag{3.30}$$

By taking into account that the field must vanish at $r = \pm\infty$, the field components in the core will take the form of the first kind Bessel functions along the radius, and the modified Bessel functions of the second kind is applied in the cladding. These Bessel functions are plotted in Fig. 3.8:

$$E_z(r, \phi) = \begin{cases} A_l J_l(\kappa r)\cos l\phi & r < a \text{ (core)} \\ A_l \frac{J_l(\kappa a)}{K_l(\gamma a)} K_l(\gamma r)\cos l\phi & r > a \text{ (cladding)} \end{cases}$$

$$H_z(r, \phi) = \begin{cases} B_l J_l(\kappa r)\cos l\phi & r < a \text{ (core)} \\ B_l \frac{J_l(\kappa a)}{K_l(\gamma a)} K_l(\gamma r)\cos l\,\phi & r > a \text{ (cladding)}. \end{cases} \tag{3.31}$$

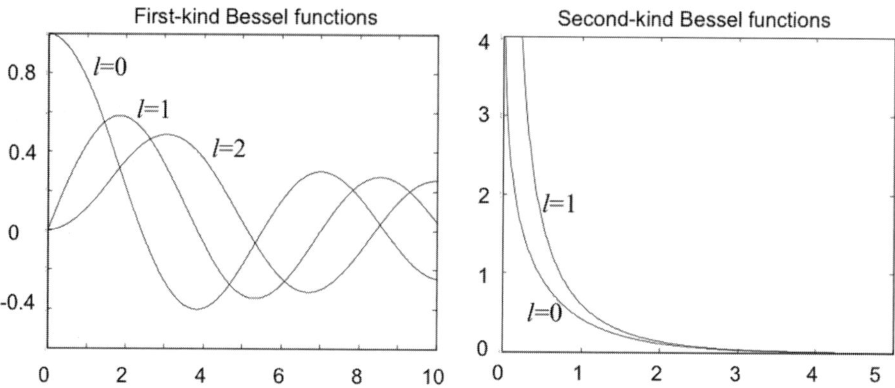

Figure 3.8 Bessel functions of the first kind j_l and the modified Bessel function of the first kind.

The transversal field components are expressed using E_z and H_z as

$$E_r = \frac{-j}{\omega^2 \varepsilon \mu_0 - \beta_0^2}\left(\beta_0 \frac{\partial E_z}{\partial r} + \omega\mu_0 \frac{1}{r}\frac{\partial H_z}{\partial \phi}\right)$$

$$E_\phi = \frac{-j}{\omega^2 \varepsilon \mu_0 - \beta_0^2}\left(\beta_0 \frac{1}{r}\frac{\partial E_z}{\partial \phi} - \omega\mu_0 \frac{\partial H_z}{\partial r}\right), \tag{3.32}$$

$$H_r = \frac{-j}{\omega^2 \varepsilon \mu_0 - \beta_0^2}\left(-\omega\varepsilon \frac{1}{r}\frac{\partial E_z}{\partial \phi} + \beta \frac{\partial H_z}{\partial r}\right)$$

$$H_\phi = \frac{-j}{\omega^2 \varepsilon \mu_0 - \beta_0^2}\left(\omega\varepsilon \frac{\partial E_z}{\partial r} + \beta \frac{1}{r}\frac{\partial H_z}{\partial \phi}\right). \tag{3.33}$$

By imposing the boundary condition that all tangential field components match at the interface between the core and the cladding, the propagation constant β is determined as the eigenvalue of the characteristic equation. Since the weakly guiding approximation $n_1 \cong n_2$ is valid, we obtain the characteristic equation of the linearly polarized LP_{ml} mode using the recurrence relations for J_m and K_m,

$$\frac{\kappa J_{m-1}(\kappa a)}{J_m(\kappa a)} = -\frac{\gamma K_{m-1}(\gamma a)}{K_m(\gamma a)}. \tag{3.34}$$

Setting $\gamma = 0$, the cutoff value of β from $J_{m-1}(u)$ is obtained. The cutoff of the first-order or fundamental linear polarized mode LP_{11} mode is obtained from

$$J_0(\kappa) = 0. \tag{3.35}$$

Here, we introduce the normalized propagation constant b defined as

$$b = \frac{(\beta/k_0)^2 - n_2^2}{n_1^2 - n_2^2} = \frac{(\gamma a)^2}{(\kappa a)^2 + (\gamma a)^2} \quad (0 < b < 1). \tag{3.36}$$

In Fig. 3.9 the normalized propagation constant calculated from Eq. (3.34) is plotted as a function of V-value. Here the hybrid mode represents the rigorous solution of the wave equation. The cutoff V-value of the LP_{01} mode is $v_{c1} = 2.405$, and below $v < 2.405$ is the single-mode operation region in which only the fundamental mode

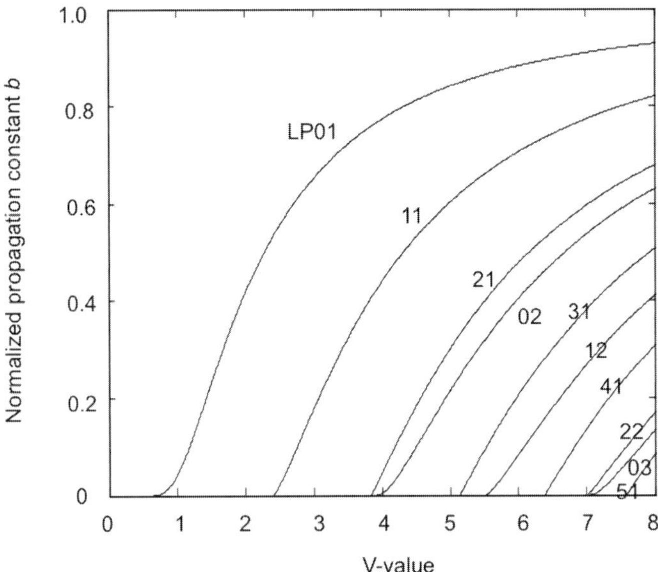

Figure 3.9 Normalized propagation constant b versus V-value.

LP$_{01}$ can propagate. For example, a typical value of the relative refractive index is $\Delta = 0.3\%$ at the operation wavelength of $\lambda = 1550$ nm with core radius $a = 5$ μm. The electric fields and the intensity patterns of the three low-order modes are shown in Fig. 3.10. It is interesting to see how much of the power of the mode field is confined in the core. The amount of power that is contained inside the core is given by

$$P_{core} = \frac{1}{2} \int_0^{2\pi} \int_0^a r(E_x H_y^* - E_y H_x^*) \, dr \, d\phi. \tag{3.37}$$

For the fundamental LP$_{01}$ mode, the power outside the core is approximated as

$$\frac{P_{clad}}{P} = \left(\frac{1}{v}\right)^2 \left\{ (\kappa a)^2 + \frac{(\gamma a)^2 J_m^2(\kappa a)}{J_{m-1}(\kappa a) J_{m+1}(\kappa a)} \right\}. \tag{3.38}$$

In Fig. 3.11 the ratio is plotted as a function of V-value. It is noted that 20% of the total power is guided in the cladding at around $v = 2.4$.

There are several categories of single-mode fiber, depending on the losses, dispersions etc. ITU-T has defined specifications G652 ~ G557:

- G.652 (Standard SMF)
- G.653 (Dispersion-shifted SMF, DSF)
- G.654 (Cutoff shifted SMF)
- G.655 (Non-zero dispersion-shifted SMF, NZDSF)
- G.656 (Non-zero dispersion for wideband optical transport)
- G.657 (Bending insensitive SMF).

Table 3.1 lists characteristics of single-mode fibers specified by G.652, G.655, and G.657, mainly focusing on attenuation and bending loss [7–9]. The main differences

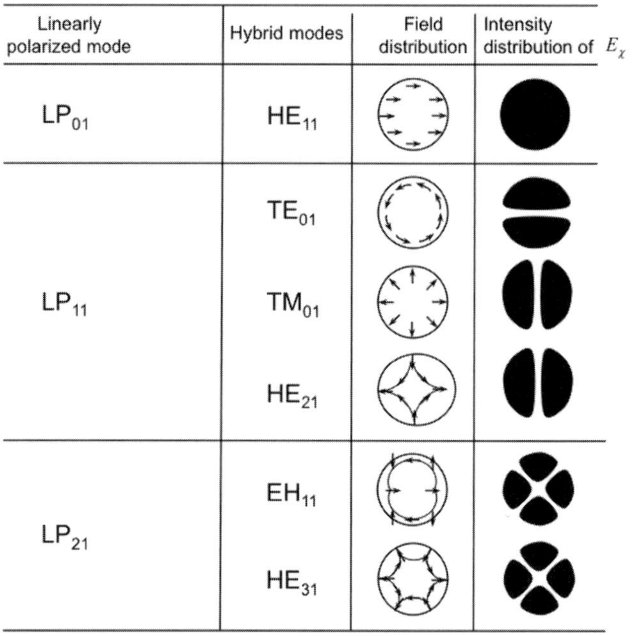

Linearly polarized mode	Hybrid modes	Field distribution	Intensity distribution of E_χ
LP_{01}	HE_{11}		
LP_{11}	TE_{01}		
	TM_{01}		
	HE_{21}		
LP_{21}	EH_{11}		
	HE_{31}		

Figure 3.10 Electric fields and intensity patterns of the three low-order modes.

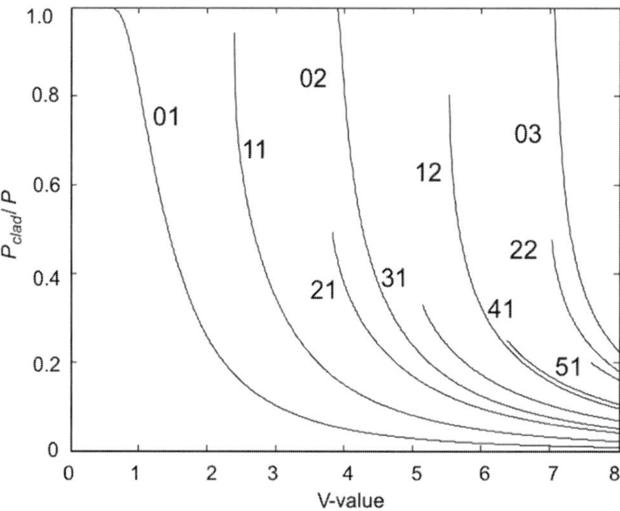

Figure 3.11 Power ratio inside the core versus V-value.

among the G.652 family are in the loss and the polarization mode dispersion (PMD). G.652.A and G.652.B are conventional, having the OH absorption peak at 1380 nm, while G.652.C and G.652.D are almost free of the water peak. The upper limit of the L-band is set at 1625 nm, except for G.652.A.

In addition to the intrinsic loss of optical fibers described in Section 3.1, the overall loss of optical fiber cable deployed in the plant includes losses caused by bending and

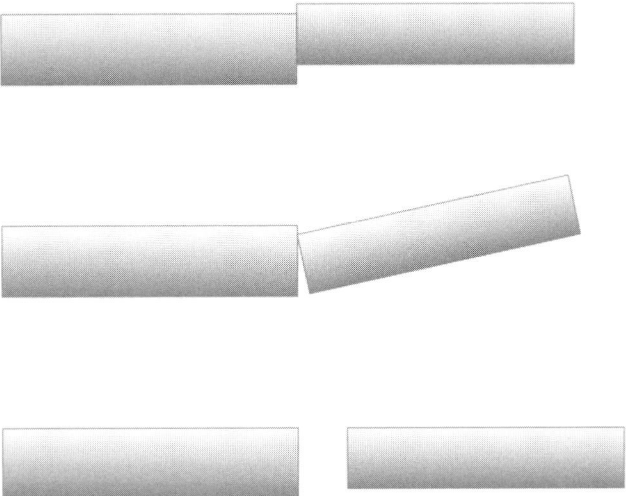

Figure 3.12 Possible types of geometrical imperfections at the splice point: offset of fiber axis, tilt, and longitudinal separation.

geometrical imperfections at the point of fiber splicing. In designing PON systems, the lower the splice loss and bending loss, obviously the less tight the power budget will become. Within the weakly guiding approximation, the guided mode is very nearly transverse, and the field of the fundamental mode LP_{01} can be approximated as Gaussian:

$$E_r \propto \exp\left(-\frac{r^2}{w^2}\right), \tag{3.39}$$

where w represents the mode field radius, defined by $1/e$ field radius. In fact the field distribution of the LP_{01} mode matches the Gaussian field almost perfectly at $v = 2.4$, and the best match is obtained at $v = 2.8$ [10]. The mode field radius normalized by the core radius of a step-index fiber is expressed as a function of V-value in the form

$$\frac{w}{a} = 0.65 + \frac{1.619}{v^{3/2}} + \frac{2.879}{v^6}. \tag{3.40}$$

According to ITU-T G.652, the nominal value of mode field diameter of a single-mode fiber is set $8.6 \sim 9.5$ μm with a tolerance of ± 0.7 μm at 1310 nm [8].

We will consider the splice loss of single-mode fiber using the Gaussian approximation. There are three possible types of geometrical imperfections: offset of fiber axis, tilt, and longitudinal separation, shown in Fig. 3.12. Here, we focus on the splice loss due to offset, which is of practical importance. Assume that the mode field radii of two fibers are w_1 and w_2, the transmission coefficient in the presence of the offset d is expressed by [10]

$$T_{offset} = \left(\frac{2w_1 w_2}{w_1^2 + w_2^2}\right)^2 \exp\left(-\frac{2d^2}{w_1^2 + w_2^2}\right). \tag{3.41}$$

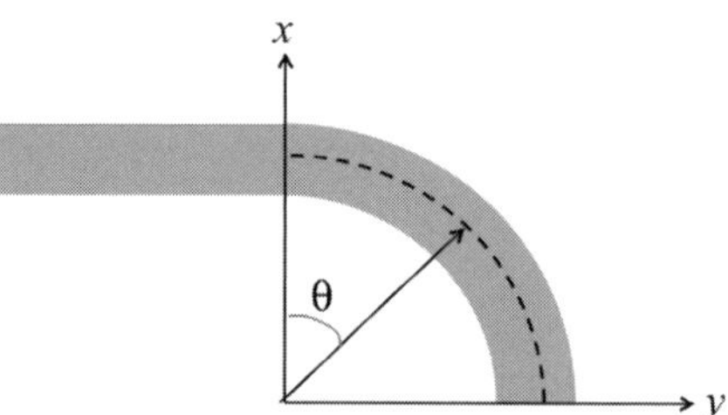

Figure 3.13 Geometry of a bent fiber.

With the offset $d = 0$ the loss due to the field mismatch is obtained as

$$T_{offset} = \left(\frac{2w_1 w_2}{w_1^2 + w_2^2} \right)^2.$$
(3.42)

For the case with identical fibers $w_2 = w_1$, the transmission coefficient is reduced to

$$T_{offset} = \exp \left(-\frac{d^2}{w_1^2} \right).$$
(3.43)

For a given power budget for a PON system, see Section 2.1.3.1, the target value for link loss is set by determining the target values for all loss factors, including fiber cable loss, splitter loss and connector and splice loss. For example, in order to satisfy a splice loss of 0.067 dB, the offset value between identical fibers has to be smaller than $d = 0.5$ μm for the mode field radius $w = 8$ μm when there is assumed to be no longitudinal separation and tilt.

Next, we consider bending loss. The geometry of a bent fiber is illustrated in Fig. 3.13, in which a straight section leads into a bend of uniform radius of curvature r_c. The effective refractive index profile $n_{eff}(r)$ for the fundamental mode in the weakly guiding approximation and assuming $r_c \gg r$ is given by [11]

$$n_{eff}^2(r) = n^2(r) + 2n_1^2 \left(\frac{r}{r_c} \right) \cos \phi.$$
(3.44)

This implies that the mode field shifts toward the outer side of the bend due to the bend, and the maximum shift occurs at $\phi = 0$, that is in the plane of the curvature. As the radius of the bend becomes smaller, the shift becomes larger. A physical interpretation of the bending loss mechanism is as follows. Since the speed of the mode field decreases toward the outer side along the direction of the radius of curvature, the mode field at the inner position travels faster so that the mode field at the outer side cannot catch up with it, resulting in it being broken off, i.e., radiation loss. Therefore, a smaller radius bend results in a larger shift, and larger bend loss occurs.

ITU-T G.652 recommends the allowable bending loss for a bend radius of 37.5 mm. This may sometimes be too large for fiber installation, and hence a more stringent requirement is desirable to ease fiber installation, particularly in-house wiring, and at the same time to downsize the closure and outlet box. A large tolerable bend radius will also ease the cable manufacturing process because microbending loss is inevitable. There is a category of bend insensitive single-mode fiber, standardized in ITU-T G.657 as summarized in Table 3.1. Its specifications are much tighter than those of ITU-T G.652,

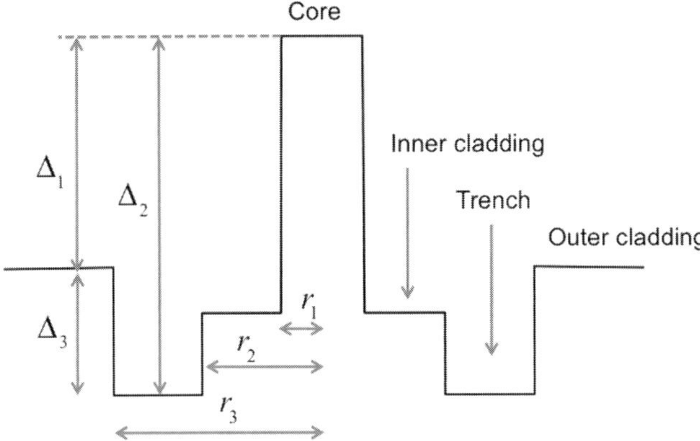

Figure 3.14 Refractive index profile of an optical fiber with trench.

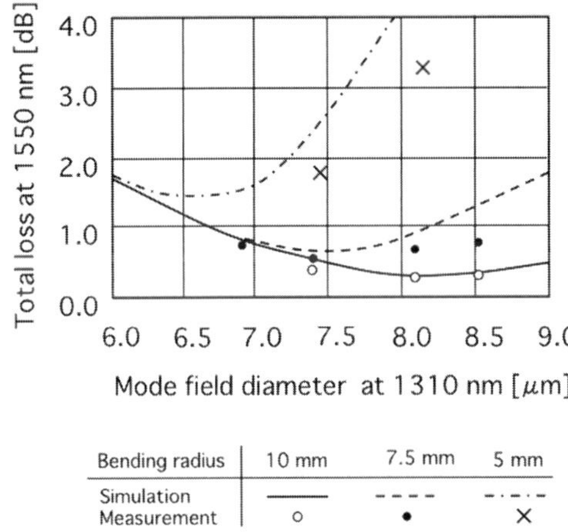

Bending radius	10 mm	7.5 mm	5 mm
Simulation	——	- - - -	- · - · -
Measurement	○	●	×

Figure 3.15 Total loss including the splice loss of two points and the bending loss plotted as a function of the mode field diameter [12]. © Reprinted by permission of IEICE.

including a bending loss ≤ 0.05 dB for one turn with a bend radius of 15 mm at a wavelength of 1550 nm [12]. A bend insensitive single-mode fiber has a trench in its refractive index profile as illustrated in Fig. 3.14, which is the same profile as that of the dispersion-shifted fiber (DSF) in ITU-T G.653. Typical values of the parameters are $\Delta_1 = 0.35 \sim 0.70\%$, $\Delta_2 = 0$, $\Delta_3 = -0.25\%$, $r_2/r_1 = 3.5$, and $r_3/r_1 = 5.5$. There is a tradeoff between bend loss and splice loss. For smaller mode field, the splice loss due to the offset increases according to Eq. (3.43) while, on the other hand, the bend loss decreases. This is because the mode field is confined more tightly. In Fig. 3.15 the total loss of test single-mode fibers having the trench, including the splice loss of two points and the bending loss, is plotted as a function of the mode field diameter. The points indicate measured values of test fibers. The mode fields of the fibers range from 6.90 μm to 8.52 μm. The bend radii are 10 mm, 7.5 mm, and 5 mm, and the number of turns is ten. These results confirm that the trench profile can reduce bend loss, compared to a step-index single-mode fiber. Note that the total loss without trench is around 2 dB. It is also seen that there is an optimum parameter of trench profile for given conditions of curvature of bending and the splice.

3.4 Two-mode fiber

In between the multimode fiber and the single-mode fiber there is a special class of optical fiber, the so-called two-mode fiber or few-mode fiber. This classification is based upon the number of guided modes. Here, we focus on the two-mode fiber having two guided modes, LP_{01} and LP_{11}. If the two polarizations and the two circular dependences are taken into account, the mode count is six, including LP_{01}^x, LP_{01}^y, $LP_{11\|}^x$, $LP_{11\perp}^x$, $LP_{11\|}^y$, and $LP_{11\perp}^y$. The two-mode fiber was invented in 1978 at two NTT laboratories independently [13]. One of the inventors continued preparing two-mode fibers and conducted the proof-of-concept experiments [14, 15]. This research was initiated in the late 1970s when single-mode fiber attracted lots of attention, instead of multimode fiber, for transmission lines of 400 Mb/s and above, but it was thought that the low-loss splice of single-mode fibers was a big challenge. Therefore, the motivation for the two-mode fiber was enlargement of the core radius to ease the difficulty in splicing found with the single-mode fiber. Soon after the invention of the two-mode fiber, the fusion splice loss went down rapidly, in only two to three years, to 0.1 dB/km[1] and hence unfortunately the two-mode fiber has never seen the light of day since its invention. The revisit to the two-mode fiber is motivated by the capacity crunch of the optical fiber transmission system, arising as an emerging issue since 2010, because the transmission capacity per fiber has almost hit the limit, at a bit rate of 100 Tb/s. A major laboratory experiment at 100 Tb/s in a single core fiber over 165 km has already been reported using state-of-the-art fibers, transmitters, receivers, and digital signal processing algorithms [16]. There is growing interest in space division multiplexing (SDM) for a possible solution to this capacity crunch. There are two approaches: one is via a multicore fiber having a number of cores embedded in cladding of the fiber (see Fig. 1.10(a)), and the other is mode division multiplexing (MDM) via a multimode fiber (see Fig. 1.10(b)), which supports more than one waveguide mode so the modes can be exploited as independent channels. For this reason exotic fibers such as two-mode fibers and multicore fibers have been attracting attention as an ultra-high capacity transmission medium. Since the rapid advance in digital signal processing techniques for wireless communications, such as the fourth generation long-term evolution (4G LTE) which adopts orthogonal frequency division multiple access (OFDMA), mode multiplexing using a few modes, combined with multiple-input multiple-output (MIMO) in Fig. 1.11, is thought to be a possible solution to increasing the transmission capacity.

Obviously, the price one has to pay for MDM is the penalty due to group velocity dispersion (GVD) between the fundamental LP_{01} mode and the higher-order modes. MIMO signal processing is used to resolve the mode couplings between the guided modes, but

---

[1] The fusion splicing of single-mode fibers requires fine alignment between the two cores, and this was performed manually by monitoring the power at the opposite end of the fiber to be spliced while a continuous wave laser was launched from the other end. Nowadays, fusion splicing can be performed automatically without manual positioning of fibers (see Fig. 1.12).

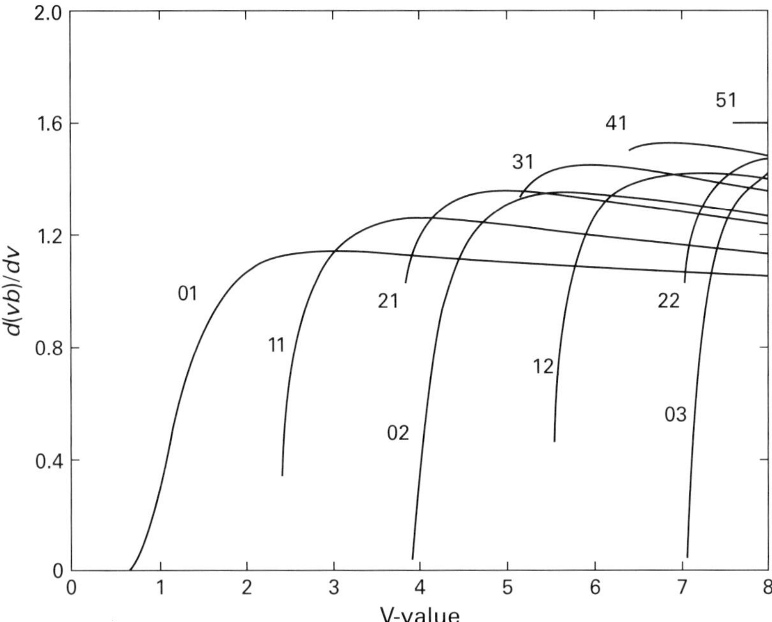

Figure 3.16 Normalized group delay of linearly polarized (LP) modes versus the V-value of step-index fiber.

its computational load grows exponentially as the GVD increases [17]. There are also issues of mode excitation and mode conversion caused at the fiber splice. One will notice from Fig. 3.10 that the mode field distribution of the LP_{01} mode ($l = 0$) has a peak at the core center $r = 0$ while, on the other hand, the field of the LP_{11} mode ($l = 1$) has a ring shape with zero intensity at the core center. This difference makes it difficult to stably excite the two modes equally, and thus the transmission characteristic depends on the mode excitation. If there is any tilt and axis offset between fibers to be spliced, mode conversion between two modes will occur. The mode dependent loss will make this problem more complicated. Another issue is the difficulty in mode-multiplexing and demultiplexing of LP_{01} and LP_{11} modes. However, these issues might not be a serious problem in the near future when the MIMO technique is applied to MDM, and as DSP chip becomes much more powerful the computational load will not be too heavy for real-time DSP.

The normalized group delays of low-order linearly polarized modes in a step-index fiber are plotted as a function of V-value. As seen from Fig. 3.16, there is a crossover between LP_{01} and LP_{11} modes at $V_0 \simeq 3.0$. If the operation wavelength is set properly at V_0 there is no GVD between the two LP_{01} and LP_{11} modes. In reality there will be deviations from the ideal settings in the operation wavelength and the fiber parameters such as the radius and refractive index of the core. These deviations, even if they are small, will cause a large GVD since the two dispersion curves in Fig. 3.16 cross at a relatively steep angle. One approach is to make the two dispersion curves cross at as small an angle as possible under the condition that the V-value of crossing is smaller or equal to the cutoff V-value V_{c2} of the next higher-order mode, the LP_{21} mode. In a

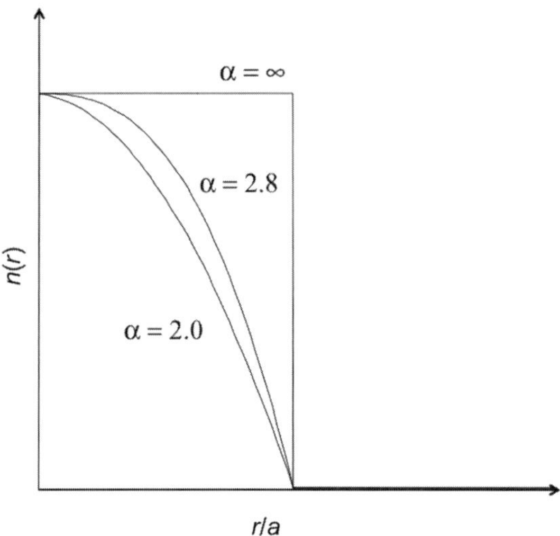

Figure 3.17 Refractive index profile of α power-law.

step-index fiber the V_0 value of around 3.0 is much smaller than the cutoff V-value of the LP_{21} mode, $V_{c2} = 3.83$. The index profile of α power-law with $\rho = 1$ in Fig. 3.17 is expressed as

$$
n(r) = \begin{cases} n_1 \left\{ 1 - 2\rho\Delta \left(\frac{r}{a}\right)^{\alpha} \right\}^{1/2} & (0 \leq r \leq a) \\ n_2 \left(= n_1\sqrt{1 - 2\Delta} \right) & (r > a) , \end{cases}
\tag{3.45}
$$

where α denotes the index profile parameter, Δ is the difference in relative index, ρ is the parameter representing the refractive-index step at the core–cladding interface, and a is the core radius. Note that $\alpha = \infty$ corresponds to a step-index profile. Assuming that Δ is small enough to adopt the weakly guiding approximation, the propagation constants and the group delays of LP_{01} and LP_{11} modes for the α power-law are calculated numerically. The eigenvalue equation of the propagation constant of the LP mode is given in Appendix 3.1 [18]. In Fig. 3.18 the normalized group delays of LP_{01} and LP_{11} modes for $\alpha = 2.8$, and $\rho = 1$ are plotted as a function of V-value. Group delays of the two modes match at $V = 5.1$ which corresponds to V_{c2}. Therefore, this α value will be optimum. This is an example of the optimum design. The differential modal group delay (DMD) between LP_{01} and LP_{11} modes, $\Delta\tau$, caused by deviation of the values of α and V is of practical importance. The tolerable ranges of α and V become much larger, compared with the step-index fiber [14]. In Fig. 3.19 the DMD $\delta\tau$ is plotted as functions of V-value and α. The LP_{21} mode can propagate in the region $V > 5.1$ with $\alpha = 2.8$. Of practical significance is the result that $\delta\tau$ caused by the deviation in V-value from the target V-value of $\delta\tau = 0$ is relatively small. For the WDM two-mode fiber transmission links, a small and flat DMD over a wide spectral range is desirable. It is a requisite, in particular, for MIMO MDM transmission in order to reduce the computational complexity of MIMO. An approach to realize the DMD-flattened two-mode fiber link in a wide spectral range is to use the DMD compensation technique [18]. Recently, DMD compensation

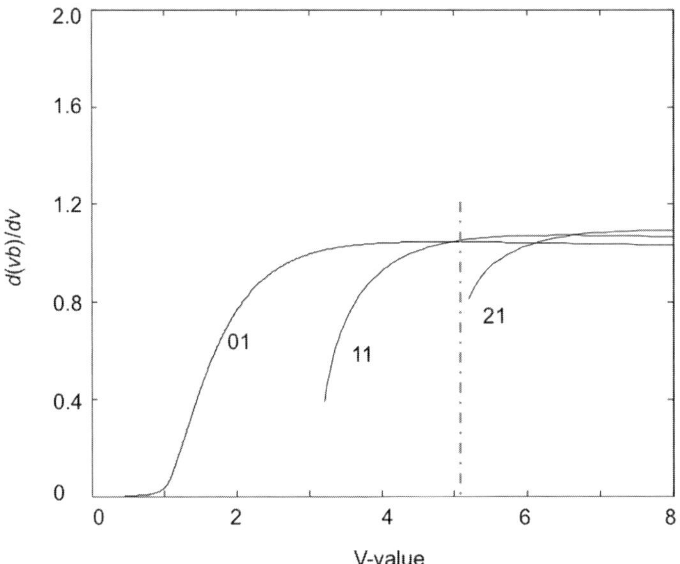

Figure 3.18 Normalized group delays of LP_{01} and LP_{11} modes versus V-value for $\alpha = 3.08$, $y = 0$, and $\rho = 1$.

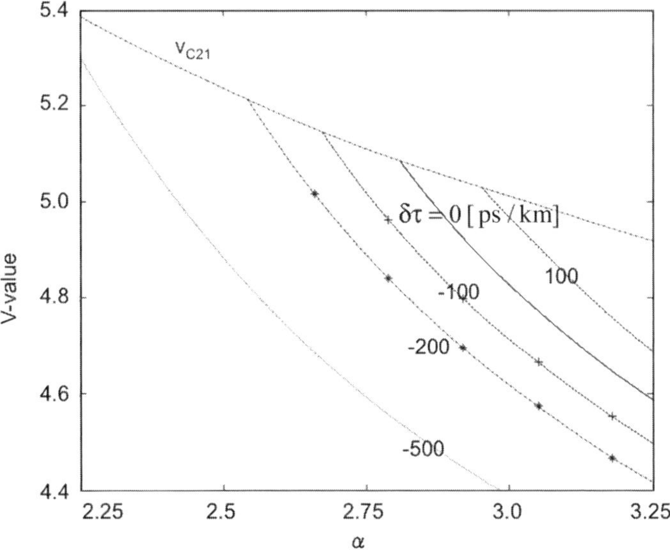

Figure 3.19 V-value plotted as a function of α for various values of group delay difference (DMD) between the two modes $\Delta\tau$.

using two-mode fibers having positive and negative chromatic dispersions between the LP_{01} and LP_{11} modes has been demonstrated experimentally [19]. Two sorts of test two-mode fibers are prepared, having positive and negative DMD. The fiber parameters are designed so that the signs of DMD are opposite in the entire C-band, and the

DMD slope becomes small. The refractive-index profile is an α power-law expressed by Eq. (3.45) with $\alpha = 1.6 \sim 1.9$, with trench $\rho \cong 1.3$.

3.5   Coupled-mode theory

In practice the coupled-mode theory serves as a basis of the key optical devices used in PON systems such as the 3-dB coupler, the star coupler, fiber Bragg grating filter, and distributed feedback (DFB) laser etc. The coupled-mode equation is derived from perturbation theory. The magnitude of the perturbation has an impact in such a way that the propagation constants of the waveguide mode of individual waveguides are perturbed to some extent but their fields remain unchanged. Let us consider the simplest couple mode system consisting of two closely adjacent waveguides. Hence, the field E of a coupled waveguide is expressed by the linear combination of the transversal mode fields $E_i(\mathbf{r})$ $(i = 1, 2)$ of individual waveguides 1 and 2

$$E(\mathbf{r}, z) = a_1(z)E_1(x\mathbf{r}) + a_2(z)E_2(\mathbf{r}). \tag{3.46}$$

Hereafter, we eliminate (x, y) from the mode field. The amplitudes a_1 and a_2 of modes 1 and 2, in the absence of coupling, have the propagation constants β_1 and β_2 of individual waveguide modes,

$$\frac{da_1}{dz} = -j\beta_1 a_1$$

$$\frac{da_2}{dz} = \mp j\beta_2 a_2 \tag{3.47}$$

where the signs $(-)$ and $(+)$ denote forward and backward waves, respectively. Suppose that the two waves are weakly coupled by some means, a_1 is affected by a_2, and vice versa, then the coupled-mode equations are expressed as

$$\frac{da_1}{dz} = -j\beta_1 a_1 + \kappa_{12} a_2$$

$$\frac{da_2}{dz} = \mp j\beta_2 a_2 + \kappa_{21} a_1 \tag{3.48}$$

where the coupling coefficients are constrained within the weak coupling regime

$$|\kappa_{12}|, |\kappa_{21}| \ll \beta_1, \beta_2. \tag{3.49}$$

Assume that the coupled waveguide is lossless. We normalize a_1 and a_2 so that the power of the modes becomes $|a_1|^2$ and $|a_2|^2$. As we consider the lossless case, the total net power P has to be conserved

$$P = |a_1|^2 \pm |a_2|^2 = \text{constant} \tag{3.50}$$

The power conservation law requires that the power be independent of distance z:

$$\frac{dP}{dz} = \frac{d\,|a_1|^2}{dz} \pm \frac{d\,|a_2|^2}{dz} = \frac{d}{dz}(a_1 a_1^* \pm a_2 a_2^*) = 0 \tag{3.51}$$

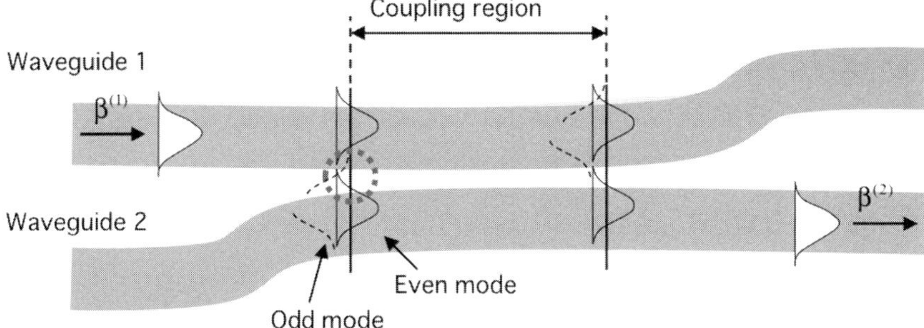

Figure 3.20 Geometry of a co-directional coupler.

from which it follows that

$$(\kappa_{12}^* \pm \kappa_{21})a_1 a_2^* + (\kappa_{12} \pm \kappa_{21}^*)a_1^* a_2 = 0. \tag{3.52}$$

The requirement for the above equation to hold regardless of the amplitude is

$$\kappa_{12} \pm \kappa_{21}^* = 0. \tag{3.53}$$

For waves carrying power in the same direction,

$$\kappa_{12}\kappa_{21} = -|\kappa_{12}|^2, \kappa_{21} = -\kappa_{12}^*. \tag{3.54}$$

For waves carrying power in the opposite direction,

$$\kappa_{12}\kappa_{21} = |\kappa_{12}|^2, \kappa_{21} = \kappa_{12}^*. \tag{3.55}$$

3.5.1 Co-directional coupling

Let us start with the coupling between two waveguides close to each other with the modes propagating in the same direction, as shown in Fig. 3.20. By adopting the constraint on the coupling coefficient in Eq. (3.54), the coupled-mode equation is given by

$$\frac{da_1}{dz} = -j\beta_1 a_1 + \kappa_{12}a_2$$

$$\frac{da_2}{dz} = -j\beta_2 a_2 + \kappa_{21}a_1. \tag{3.56}$$

The solution in the form of a matrix $[S_{ij}]$ $(i, j = 1, 2)$ is expressed as

$$\begin{bmatrix} a_1(z) \\ a_2(z) \end{bmatrix} = e^{-j\beta_m z} \begin{bmatrix} \cos \beta_0 z - j\frac{\Delta}{\beta_0}\sin \beta_0 z & \frac{\kappa_{12}}{\beta_0}\sin \beta_0 z \\ \frac{\kappa_{21}}{\beta_0}\sin \beta_0 z & \cos \beta_0 z + j\frac{\Delta}{\beta_0}\sin \beta_0 z \end{bmatrix} \begin{bmatrix} a_1(0) \\ a_2(0) \end{bmatrix}$$

or

$$\begin{bmatrix} a_1(z) \\ a_2(z) \end{bmatrix} = e^{-j\beta_m z} \begin{bmatrix} \cos \beta_0 z - j\frac{\Delta}{\beta_0}\sin \beta_0 z & \frac{\kappa_{12}}{\beta_0}\sin \beta_0 z \\ -\frac{\kappa_{12}^*}{\beta_0}\sin \beta_0 z & \cos \beta_0 z + j\frac{\Delta}{\beta_0}\sin \beta_0 z \end{bmatrix} \begin{bmatrix} a_1(0) \\ a_2(0) \end{bmatrix} \tag{3.57}$$

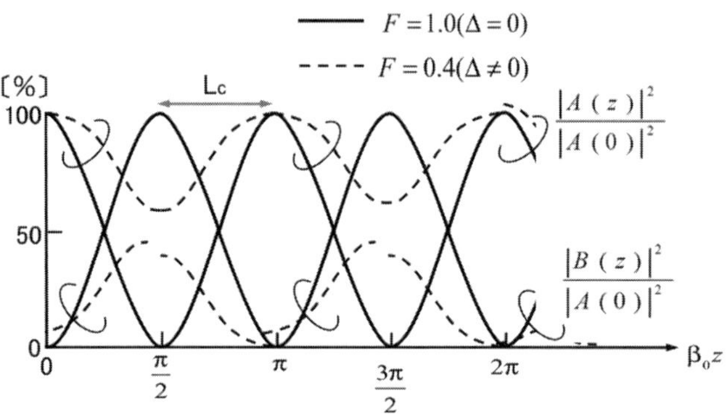

Figure 3.21 Power exchange as a function of $\beta_0 z$ for $F = 0.4$ and $F = 1.0$.

where the propagation constant of the coupled mode is given by $\beta_m \pm \beta_0$ where $\beta_m = \frac{\beta_1 + \beta_2}{2}$ and

$$\beta_0 = \sqrt{\Delta^2 + |\kappa_{12}|^2}, \quad \Delta = \frac{\beta_1 - \beta_2}{2}. \tag{3.58}$$

Details of the derivation of the solution are described in Appendix 3.3. It is noteworthy that the above transfer matrix $[S_{ij}]$ $(i, j = 1, 2)$ is unitary, satisfying $[S_{ij}^+][S_{ij}] = I$ where the operation $+$ indicates complex conjugate of the transpose, and I is the identity matrix. This operation leads to

$$\begin{aligned}
|S_{11}|^2 + S_{12}^* S_{21} &= |S_{11}|^2 + |S_{12}|^2 = 1 \\
S_{11}^* S_{12} + S_{12}^* S_{22} &= 0 \\
S_{21}^* S_{11} + S_{22}^* S_{21} &= 0 \\
S_{21}^* S_{12} + |S_{22}|^2 &= |S_{21}|^2 + |S_{22}|^2 = 1.
\end{aligned} \tag{3.59}$$

When $\kappa_{21} = -\kappa_{12}^*$ in Eq. (3.54) is taken into account, the reciprocity can be confirmed. The fields of coupled modes in Fig. 3.20 take the symmetric and asymmetric distributions, according to the signs $(+)$ and $(-)$, respectively. Our interest is in the power exchange between two waveguides. For simplicity, let us consider the case with only the input to waveguide 1, and $a_2(0) = 0$,

$$\left| \frac{a_1(z)}{a_1(0)} \right|^2 = \cos^2 \beta_0 z + \left(\frac{\Delta}{\beta_0} \right)^2 \sin^2 \beta_0 z = 1 - F \sin^2 \beta_0 z,$$

$$\left| \frac{a_2(z)}{a_1(0)} \right|^2 = F \sin^2 \beta_0 z,$$

$$F \equiv \left(\frac{|\kappa_{12}|}{\beta_0} \right)^2 = \frac{1}{1 + \left(\frac{\beta_1 - \beta_2}{2|\kappa_{12}|} \right)^2} < 1. \tag{3.60}$$

For $F = 1.0$, full power transfer occurs between two waveguides as shown in Fig. 3.21, and this is only the case with $\Delta = 0$, that is, of the degenerate propagation constant $\beta_1 = \beta_2$ between the two waveguides. For $F = 0.4$ only part of the power is exchanged between the waveguides. The coupling period L_c in Fig. 3.21 is defined as a half period of the power exchange as

$$L_c = \frac{\pi}{2\beta_0}.$$ (3.61)

Using the coupled-mode theory, practical optical components such as the 3-dB coupler, optical switch and tunable filter can be designed. A 3-dB coupler can be designed with $\Delta = 0$ and $\beta_1 = \beta_2 = \beta$. The transfer matrix is written as

$$\begin{bmatrix} a_1(z) \\ a_2(z) \end{bmatrix} = e^{-j\beta_m L} \begin{bmatrix} \cos \beta_0 L & \frac{\kappa_{12}}{|\kappa_{12}|} \sin \beta_0 L \\ -\frac{\kappa_{12}^*}{|\kappa_{12}|} \sin \beta_0 L & \cos \beta_0 L \end{bmatrix} \begin{bmatrix} a_1(0) \\ a_2(0) \end{bmatrix}.$$ (3.62)

By setting the coupling length $\beta_0 L = \pi/4$, Eq. (3.62) is rewritten as

$$\begin{bmatrix} a_1(z) \\ a_2(z) \end{bmatrix} = \frac{1}{\sqrt{2}} e^{-j\beta_m L} \begin{bmatrix} 1 & \frac{\kappa_{12}}{|\kappa_{12}|} \\ \frac{-\kappa_{12}^*}{|\kappa_{12}|} & 1 \end{bmatrix} \begin{bmatrix} a_1(0) \\ a_2(0) \end{bmatrix},$$ (3.63)

and its expression in the intensity shows the operation of the 3-dB coupler:

$$\begin{bmatrix} |a_1(z)|^2 \\ |a_2(z)|^2 \end{bmatrix} = \frac{1}{2} \begin{bmatrix} 1 & 1 \\ 1 & 1 \end{bmatrix} \begin{bmatrix} |a_1(0)|^2 \\ |a_2(0)|^2 \end{bmatrix}.$$

2×2 switching between two waveguides occurs by changing $\beta_0 L$ from $\pi/2$ to π by any means, for example using the electro-optic effect described in Section 4.2.3. A tunable filter can also be designed by coupling between non-identical waveguides.

3.5.2　Contra-directional coupling

Contra-directional coupling will provide a basic mechanism for the fiber Bragg grating (FBG), which plays an important role in optical encoders and decoders described in Section 7.3.1. Contra-directional coupling is defined as the coupling between two lightwaves having opposite group velocities. What we learn from co-directional coupling tells us that there is a good chance of coupling between two lightwaves propagating nearly at the same velocity. However, this is not the case when the lightwaves propagate in the opposite direction. How can lightwaves propagating in opposite directions couple with each other? Intuitively, a perturbation $\beta_{perturb}$ will be required in the waveguide to satisfy $\beta \simeq -\beta + \beta_{perturb}$. A grating formed by a periodically corrugated reflecting surface, as shown in Fig. 3.22, serves as the perturbation. The period of the grating is Λ, and n is the refractive index of the medium. Contra-directional coupling is considered to be a special class of Bragg deflection caused by the grating having a periodically corrugated reflecting surface. This is the case with the reflection angle of $-90°$ shown in Fig. 3.23.

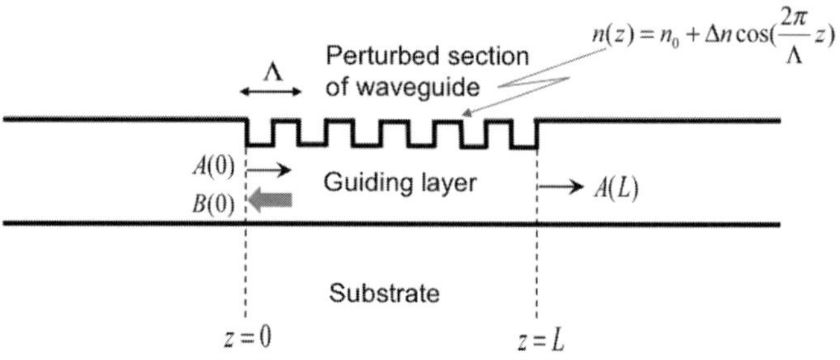

Figure 3.22 Grating formed by a periodically corrugated reflecting surface.

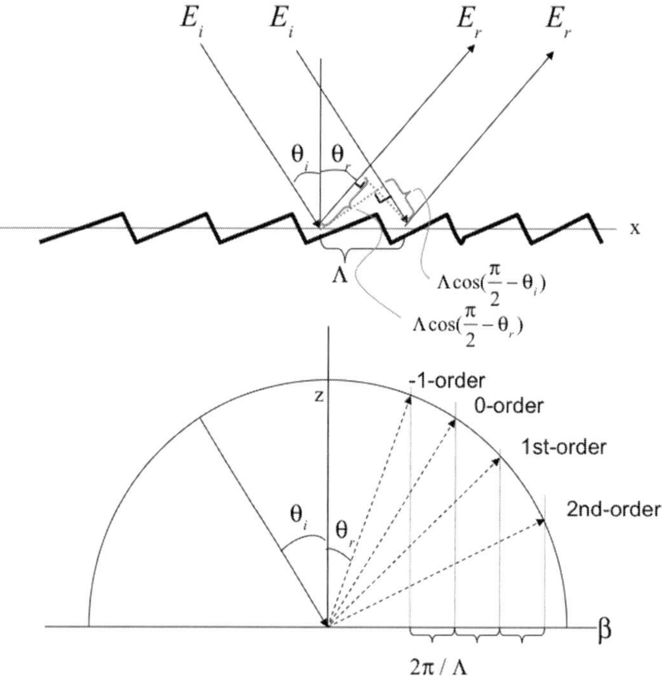

Figure 3.23 Bragg deflection at a reflection grating.

Reflected waves are excited so as to satisfy the boundary condition of zero tangential electric fields E_i and E_r on the surface of the grating

$$\Lambda \cos\left(\frac{\pi}{2} - \theta_r\right) - \Lambda \cos\left(\frac{\pi}{2} - \theta_i\right) = l\lambda, \ (l = 0, \pm 1, \pm 2, \ldots). \qquad (3.64)$$

This leads to

$$kn \sin\theta_\gamma = kn \sin\theta_i + \frac{2\pi}{\Lambda} l, \qquad (3.65)$$

where the reflection grating corresponds to -1-order deflection. By setting $\theta_i = -\theta_r = \pi/2$ and $l = -1$, the relation between the period of the grating and the Bragg wavelength is determined as

$$\Lambda = \frac{\lambda_B}{2n}.\tag{3.66}$$

We will begin with the coupled-mode equation. We consider the constraint on the coupling coefficients $\kappa_{21} = \kappa_{12}^*$ in Eq. (3.53). Let a and b, respectively, be the amplitudes of the forward and backward waves, then the coupled-mode equation is written as

$$\frac{da}{dz} = -j\beta a + \kappa_{ab}b$$
$$\frac{db}{dz} = j\beta b + \kappa_{ab}^* a.\tag{3.67}$$

Suppose that a periodic perturbation of the guiding structure is introduced:

$$n(z) = n_0 + \Delta n \cos\left(\frac{2\pi}{\Lambda}z\right).\tag{3.68}$$

In the presence of a cosinusoidal perturbation, the lightwave propagates along the periodic structure and acquires modulation with the spatial dependence $\cos(2\pi/\Lambda)z$, and the sideband components will be generated

$$\exp(-j\beta z)\cos\frac{2\pi}{\Lambda}z = \frac{1}{2}\exp\left\{-j\left(\beta - \frac{2\pi}{\Lambda}\right)z\right\} + \frac{1}{2}\exp\left\{-j\left(\beta + \frac{2\pi}{\Lambda}\right)z\right\}$$
$$\tag{3.69}$$

where $(\beta - 2\pi/\Lambda)$ in the first term on the right-hand side can be close to $-\beta$, and hence this causes the coupling between the forward and backward propagating lightwaves. On the other hand, there is no chance for $(\beta + 2\pi/\Lambda)$ in the second term to become close to $-\beta$. By incorporating the spatial dependence into the coupling coefficient, the coupled-mode equations are rewritten as

$$\frac{da}{dz} = -j\beta a + \kappa e^{-j(2\pi/\Lambda)z}b$$
$$\frac{db}{dz} = j\beta b + \left\{\kappa e^{-j(2\pi/\Lambda)z}\right\}^* a = j\beta b - \kappa^* e^{+j(2\pi/\Lambda)z}a.\tag{3.70}$$

where $\kappa_{ab} = \kappa$. This equation can be reduced to a coupled-mode equation with space-independent coefficients by introducing new variables $a = Ae^{-j\frac{\pi}{\Lambda}z}$, $b = Be^{j\frac{\pi}{\Lambda}z}$. By introducing the detuning parameter $\delta = \beta - \pi/\Lambda$, the coupled-mode equations can be simplified:

$$\frac{dA}{dz} = -j\delta A + \kappa B$$
$$\frac{dB}{dz} = j\delta B - \kappa^* A.\tag{3.71}$$

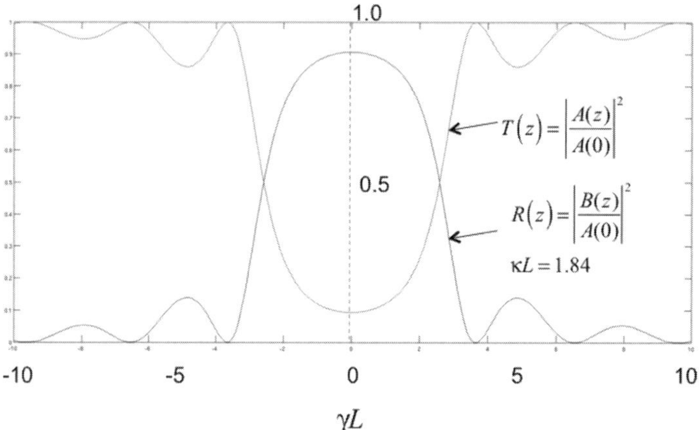

Figure 3.24 Numerical reflection coefficient plotted as a function of detuning parameter δ.

The solutions are given by

$$B(z) = \frac{\kappa^* A(0) \sinh \gamma (z - L)}{\gamma \cosh \gamma L + j\delta \sinh \gamma L} \tag{3.72}$$

$$A(z) = \frac{\gamma \cosh \gamma (z - L) - j\delta \sinh \gamma (z - L)}{\gamma \cosh \gamma L + j\delta \sinh \gamma L} A(0) \tag{3.73}$$

where the propagation constants of the coupled-modes γ are given by

$$\gamma = \pm\sqrt{|\kappa|^2 - \delta^2}. \tag{3.74}$$

Details of the derivation of the solution are described in Appendix 3.4. The transmission and reflection coefficients in the amplitude at $z = 0$ are

$$t(z) = \frac{A(z)}{A(0)} = \frac{\gamma \cosh \gamma (z - L) - j\delta \sinh \gamma (z - L)}{\gamma \cosh \gamma L + j\delta \sinh \gamma L},$$

$$r(z) = \frac{B(z)}{A(0)} = \frac{\kappa^* \sinh \gamma (z - L)}{\gamma \cosh \gamma L + j\delta \sinh \gamma L}. \tag{3.75}$$

Then, the power reflection and transmission coefficients are obtained

$$T(z) = \left|\frac{A(z)}{A(0)}\right|^2 = \frac{\gamma^2 + |\kappa|^2 \sinh^2 \gamma (z - L)}{\gamma^2 + |\kappa|^2 \sinh^2 \gamma L},$$

$$R(z) = \left|\frac{B(z)}{A(0)}\right|^2 = \frac{|\kappa|^2 \sinh^2 \gamma (z - L)}{\gamma^2 + |\kappa|^2 \sinh^2 \gamma L}. \tag{3.76}$$

The numerical reflection and transmission coefficients are plotted as a function of δ in Fig. 3.24. It is interesting to check that the power conservation law holds at any point on the z-axis:

$$T(z) - R(z) = \text{constant} \tag{3.77}$$

and

$$T(L) + R(0) = T(0) + R(L) = 1. \tag{3.78}$$

At Bragg's wavelength λ_B of practical interest, $\delta = 0(\gamma = \pm|\kappa|)$, and Eqs. (3.76) are rewritten as

$$T(z) = \left|\frac{A(z)}{A(0)}\right|^2 = \frac{\cosh^2 \gamma(z-L)}{\cosh^2 \gamma L},$$

$$R(z) = \left|\frac{B(z)}{A(0)}\right|^2 = \frac{\sinh^2 \kappa(z-L)}{\cosh^2 \kappa L}. \tag{3.79}$$

The power reflection coefficient at the input $z = 0$ is given by

$$R(0) = \tanh^2 \kappa L. \tag{3.80}$$

3.5.3 Distributed Bragg reflection laser diode

From the above obtained reflection coefficient $r(z)$ in Eq. (3.75), the phase shift Θ is given by

$$\Theta = \frac{\pi}{2} \tan^{-1} \left[\frac{\delta}{\gamma} \tanh(\gamma L) \right]. \tag{3.81}$$

The phase shift which the lightwave undergoes after the round trip along the grating between $z = 0$ and L is

$$2\Theta = \pi + 2 \tan^{-1} \left[\frac{\delta}{\gamma} \tanh(\gamma L) \right] \cong \pi + 2\delta L. \tag{3.82}$$

From Eq. (3.82) the forward and backward lightwaves interfere destructively at Bragg's wavelength λ_B, that is $\delta = 0$. As a consequence, no laser oscillation can be expected at $\lambda = \lambda_B$. To make the forward and backward lightwaves interfere constructively, an intentional phase shift of $\pi/2$ has to be added so that the phase shift in the round trip 2Θ satisfies

$$2\Theta = \pi + 2\delta L = \pm 2m\pi \quad (m = 0, 1, 2, \ldots). \tag{3.83}$$

From Eq. (3.66)

$$\delta = \beta - \frac{\pi}{\Lambda} = \frac{2\pi n_{eff}}{\lambda} - \frac{2\pi n_{eff}}{\lambda_B} \cong \frac{2\pi n_{eff}(\lambda_B - \lambda)}{\lambda_B^2}, \tag{3.84}$$

where n_{eff} denotes the effective refractive index in the medium. Suppose that the gain bandwidth covers the spectral region of $m = -1, 0$, only two longitudinal modes are allowed to lase at wavelengths on both sides of the Bragg wavelength λ_B:

$$\lambda = \lambda_B \pm \frac{\lambda_B^2}{4n_{eff}L}. \tag{3.85}$$

In Fig. 3.25(a) and (b), spectra of both the two-mode and single-mode lasings of a distributed feedback laser, respectively, are shown [19].

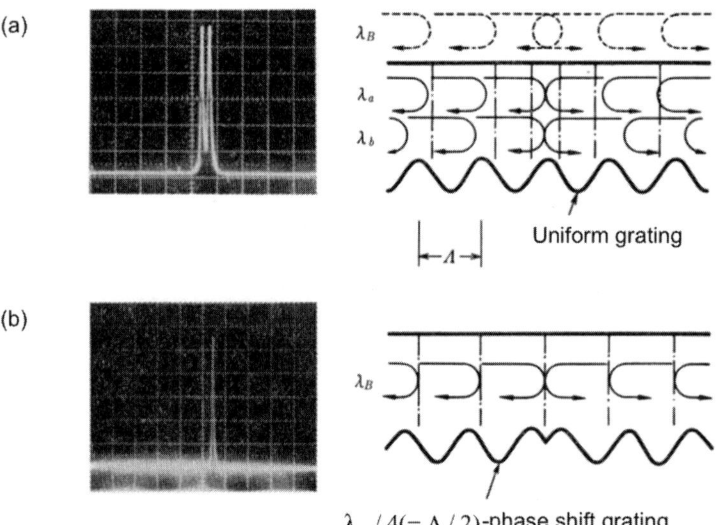

Figure 3.25 Lasing spectra of two-mode and single-mode distributed feedback laser [19]. © Reprinted by permission of Ohmsha.

3.6   Dispersion

A complex electric field vector of a monochromatic wave is expressed as

$$\mathbf{E} = \mathbf{E}(\mathbf{r}) \exp[j(\omega_0 t - \beta_0 z)] \tag{3.86}$$

where β_0 is the propagation constant at $\omega = \omega_0$ in a medium. Here, the z-axis is taken to be along the fiber axis. The amplitude is given by

$$|\mathbf{E}| = |\mathbf{E}(\mathbf{r})| \cos(\omega_0 t - \beta_0 z). \tag{3.87}$$

The phase velocity is the speed of the wave front and is obtained by taking the derivative of the phase plane $\beta_0 z - \omega_0 t = $ constant with respect to z,

$$v_p \equiv \frac{dz}{dt} = \frac{\omega_0}{\beta_0} = \frac{c}{n} \quad \left(c = \frac{1}{\sqrt{\varepsilon_0 \mu_0}} \right). \tag{3.88}$$

Note that the phase velocity is the speed of the phase front of the wave, and this is different from the speed of the optical pulse. The optical pulse is the envelope of the wave packet having different frequency components. For simplicity, let us consider two monochromatic waves having slightly different frequencies. The amplitude of the electric field is written as the sum of two waves:

$$|\mathbf{E}| = |\mathbf{E}(\mathbf{r})| \, [\cos\{\beta_0(\omega_0 + \Delta\omega)z - (\omega_0 + \Delta\omega)t\}$$
$$+ \cos\{\beta_0(\omega_0 - \Delta\omega)z - (\omega_0 - \Delta\omega)t\}] \tag{3.89}$$

Assume that the pulse has narrow spectral width, and hence $\Delta\omega \ll \omega_0$ so that the propagation constant is approximated in a Taylor expansion

$$\beta(\omega_0 \pm \Delta\omega) \approx \beta_0 \pm \beta_1 \Delta\omega$$

$$\beta_n \equiv \left.\frac{d^n\beta}{d\omega^n}\right|_{\omega=\omega_0} \tag{3.90}$$

$$|\mathbf{E}| \cong 2|\mathbf{E}(\mathbf{r})|\cos\Delta\omega(\beta_1 z - t)\cos(\beta_0 z - \omega_0 t). \tag{3.91}$$

The group velocity v_g, the speed of the envelope, is obtained by taking the derivative of the phase $\beta_1 z - t = \text{constant}$ with respect to z

$$v_g = \frac{dz}{dt} = \frac{1}{\beta_1}. \tag{3.92}$$

The group delay is obtained as

$$\tau_g = \frac{1}{v_g} = \beta_1. \tag{3.93}$$

The second-order derivative $\beta_2 = d^2\beta/dt^2$ is called the group velocity dispersion or GVD parameter. The optical fiber is a dispersive medium because Kramers–Kronig relations govern the relationship between the real part of the refractive index and the absorption, described by the imaginary part of the refractive index, resulting in the wavelength dependence of both the refractive index and the absorption. Because of this phenomenon, the GVD is called chromatic dispersion.

When we consider the plane wave in a lossless medium with refractive index n for simplicity, the propagation constant β_0 is expressed as

$$\beta_0 = nk_0. \tag{3.94}$$

By taking the derivative of the propagation constant β_0 with respect to the angular frequency ω, the group delay time and the group velocity are obtained as

$$\tau_g = \frac{d\beta_0}{d\omega} = \frac{d(kn)}{d\omega} = n + k\frac{dn}{d\omega} = \frac{n}{c} - \frac{\lambda}{c}\frac{dn}{d\lambda} \equiv \frac{N}{c}, \tag{3.95}$$

$$v_g = \frac{c}{N}, \tag{3.96}$$

where N is the group index defined as

$$N \equiv n - \lambda\frac{dn}{d\lambda}. \tag{3.97}$$

The GVD or chromatic dispersion D is given by

$$D = \frac{1}{c}\frac{dN}{d\lambda} = -\frac{\lambda}{c}\frac{d^2n}{d\lambda^2}. \tag{3.98}$$

The chromatic dispersion D is related to the second derivative of the propagation constant defined in Eq. (3.90) as

$$\beta_2 = -\frac{\lambda^2}{2\pi c}D. \tag{3.99}$$

Table 3.2 Coefficients of the Sellmeier equation [20]

	λ_l [μm]	B_i
1	0.0684043	0.6961663
2	0.1162414	0.4079426
3	9.896161	0.8974794

It is convenient to remember that the value of phase velocity in a fused-silica optical fiber is roughly 5 ns/m or 5 ps/mm.

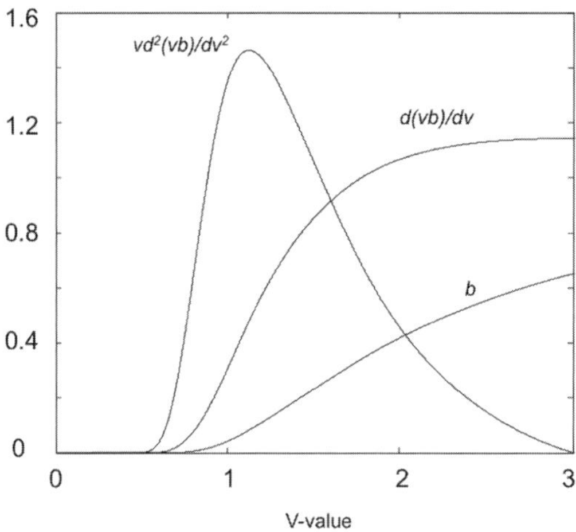

Figure 3.26 Waveguide parameters $\beta, b,\ d(vb)/dv,\ vd^2(vB)/dv^2$ plotted as a function of V-value.

In a waveguide, the GVD consists of two kinds of dispersion: material dispersion and waveguide dispersion. By taking the derivative of the normalized propagation constant b in Eq. (3.36) with respect to the angular frequency ω, we have

$$\beta_1 = \frac{d\beta}{d\omega} = c\left\{ N_2 + (N_1 - N_2)\frac{d(vb)}{dv} \right\}. \tag{3.100}$$

By taking the derivative $d\beta/d\omega$ with respect to ω once again, the GVD is obtained as

$$\beta_2 = \frac{1}{c}\left\{ \frac{dN_2}{d\omega} + \frac{d}{d\omega}(N_1 - N_2)\frac{d}{dv}(vb) + \frac{1}{\omega}\frac{N_1^2 - N_2^2}{n_1 + n_2}v\frac{d^2}{dv^2}(vb) \right\}. \tag{3.101}$$

Obviously the first and the second terms, respectively, represent the material dispersion and the waveguide dispersion, and the second term is the cross-term of material and waveguide dispersions. In Fig. 3.26 the waveguide dispersion parameters in Eqs. (3.36), (3.100), and (3.101) are plotted as a function of V-value. The Sellmeier equation is an empirical relationship between refractive index and wavelength and is given by

$$n^2(\lambda) = 1 + \sum_{j=1}^{3} \frac{B_j\lambda^2}{\lambda^2 - \lambda_j^2}. \tag{3.102}$$

Note that the wavelength is in units of μm. For convenience, the coefficients of typical fused silica fiber are provided in Table 3.2.

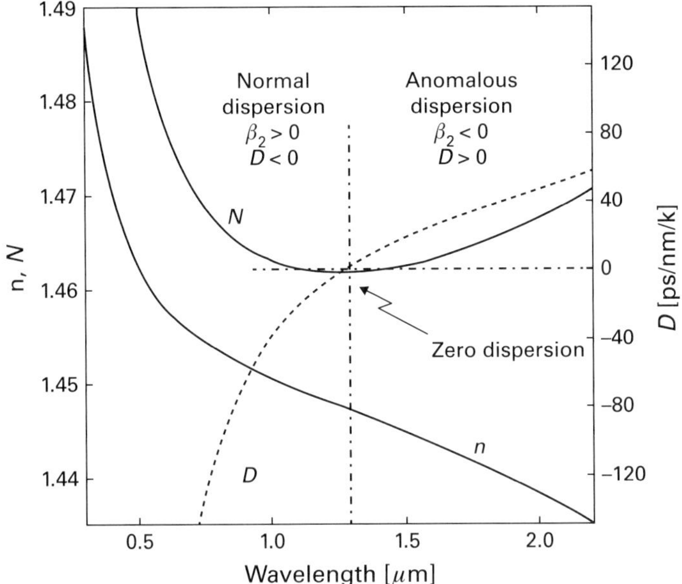

Figure 3.27 Refractive index n, group index N, group delay τ_g, and the chromatic dispersion D of fused silica plotted as a function of wavelength.

The Sellmeier equation is commonly used to calculate the material dispersion. It should be remembered that in nominal silica optical fibers, germanium (Ge), phosphorus (P), and fluorine (F) are doped to increase or decrease the refractive index of the core and cladding, and hence their Sellmeier coefficients differ slightly from those of the above pure silica. In Fig. 3.27 the refractive index n, the group delay N, and the chromatic dispersion D of fused silica are plotted as a function of wavelength. The pulse propagates fastest at around the wavelength of 1300 nm where the group delay τ_g becomes minimum, dividing into two wavelength regions of *normal* dispersion and *anomalous* dispersion. In the normal dispersion region, the longer wavelength component propagates faster than the shorter wavelength component, or the lower frequency component propagates faster than the higher frequency component. In the anomalous dispersion region the opposite is the case. Accordingly, the dispersion D crosses zero at the wavelength around 1300 nm. In the early age of optics, only the visible spectral region was developed and there was little knowledge of the infrared region. Therefore, the refractive index was thought to have the tendency that as the wavelength becomes shorter the refractive index becomes larger. In the infrared region the opposite was thought – as the wavelength becomes shorter the refractive index becomes smaller – although this is not the case. This is why the infrared region beyond 1300 nm is called the anomalous region. In Table 3.3 the dispersion, dispersion slope, and polarization dispersion of the single-mode fibers, G.652, G.655, and G.656 are summarized [7–9, 21]. The main differences among the G.655 family are in their dispersion coefficients. There are two values in the polarization mode dispersion (PMD) of the G.652 family. The PMD values of G.652A and C are 0.5 ps/km$^{1/2}$, and the value is 0.2 ps/km$^{1/2}$ in G.652B.

Table 3.3 Dispersion characteristics of single-mode fibers, G.652, G.655, and G.656

	G.655					G.652	
	G.655A	G.655C	G.655D	G.655E	G.656	G.652A and C	G.652B
Zero dispersion wavelength						1300–1340 nm	1300–1340 nm
Dispersion coefficient [ps/nm/km] (1530–1565 nm)	Min 0.1 Max 6.0	Min 1.0 Max 10.0					
Dispersion coefficient [ps/nm/km]			1460–1550 nm Min $\frac{700}{90}(\lambda - 1460) - 4.20$ Max $\frac{2.91}{90}(\lambda - 1460) + 3.29$ 1550–1625 nm Min $\frac{2.97}{75}(\lambda - 1550) + 2.80$ Max $\frac{5.06}{75}(\lambda - 1550) + 6.20$	1460–1550 nm Min $\frac{5.42}{90}(\lambda - 1460) + 0.64$ Max $\frac{4.65}{90}(\lambda - 1460) + 4.66$ 1550–1625 nm Min $\frac{3.30}{75}(\lambda - 1550) + 6.06$ Max $\frac{4.12}{75}(\lambda - 1550) + 9.31$	1460–1550 nm Min $\frac{2.60}{90}(\lambda - 1460) + 1.00$ Max $\frac{4.68}{90}(\lambda - 1460) + 4.60$ 1550–1625 nm Min $\frac{0.98}{75}(\lambda - 1550) + 3.60$ Max $\frac{4.72}{75}(\lambda - 1550) + 9.28$		
Dispersion slope [ps/nm^2/km]	+ or −	+ or −				0.092	0.092
Polarization dispersion [ps/$\sqrt{\text{km}}$]	0.5	0.20	0.20	0.20	0.20	0.5	0.20

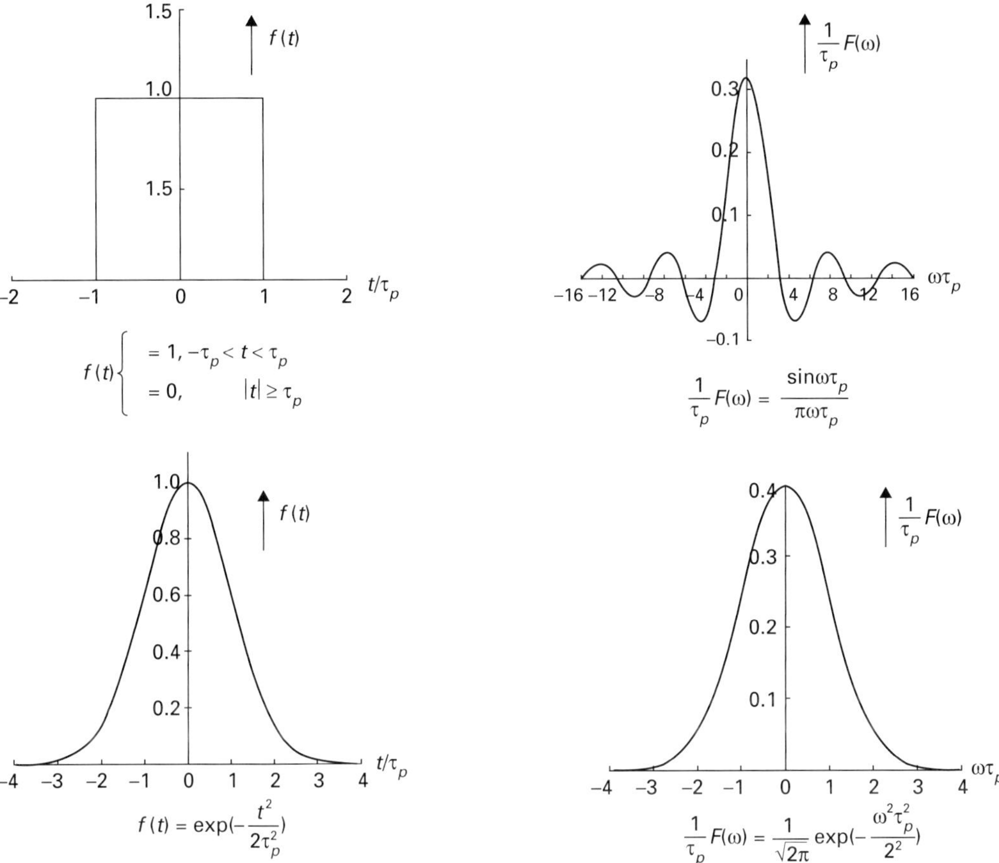

Figure 3.28 Examples of Fourier transforms.

3.7   Pulse broadening

Let us consider pulse propagation in an optical fiber. The examples of Fourier transforms in Fig. 3.28 show that a temporal pulse waveform contains different frequency components. The frequency spectrum becomes discrete for the case of a pulse train at a designated repetition rate. On the other hand, it is continuous for a single-shot pulse. Since optical fibers are dispersive media, different frequency components propagate at different group velocities, resulting in the distortion of the pulse waveform. Let us consider a pulse propagating along the z-axis. The electric field in circular coordinates is written as

$$E(r, \phi, z) = \frac{1}{2} R(r, \phi) A(z, t) \exp\left[j(\beta_0 z - \omega_0 t)\right], \tag{3.103}$$

where $A(z, t)$ represents the envelope of the pulse, $R(r, \phi)$ the distribution of the electric field in the fiber cross-section in circular coordinates, β_0 the propagation constant at $\omega = \omega_0$, and ω_0 the center angular frequency. The propagation of the pulse envelope is governed by the linear Schrödinger equation (LSE) as

$$j\frac{\partial A}{\partial z} - \frac{1}{2}\beta_2\frac{\partial^2 A}{\partial T^2} + \frac{j}{2}\alpha A = 0, \tag{3.104}$$

where α denotes the attenuation coefficient of the fiber, and T is measured in a frame of reference moving with the pulse at the group velocity v_g ($T = t - z/v_g$). We will consider the case in which the carrier wavelength is far away from the zero-dispersion wavelength so that the contribution of the higher-order derivative β_3 term is negligible. To eliminate α we will introduce $U(z, T)$,

$$A(z, T) = \sqrt{P_0}\exp(-\alpha z/2)U(z, T), \tag{3.105}$$

where P_0 denotes the peak power of the pulse. Then, Eq. (3.84) is rewritten as

$$j\frac{\partial U}{\partial z} - \frac{1}{2}\beta_2\frac{\partial^2 U}{\partial T^2} = 0. \tag{3.106}$$

For the case with zero dispersion $\beta_2 = 0$, the pulse propagates without any change in the waveform except for attenuation. We will use the Fourier transform to solve the LSE. $\tilde{U}(z, \omega)$ is the Fourier transform of $U(z, T)$ given by

$$U(z, T) = \frac{1}{2\pi}\int_0^\infty \tilde{U}(z, \omega)e^{-j\omega T}d\omega. \tag{3.107}$$

The LSE in the time domain is reduced to the LSE in the frequency domain:

$$j\frac{\partial \tilde{U}}{\partial z} + \frac{1}{2}\beta_2\omega^2\tilde{U} = 0. \tag{3.108}$$

Here, we use the Fourier exchange rule $-j\omega \rightarrow \partial/\partial t$, $-\omega^2 \rightarrow \partial^2/\partial t^2$, the solution of Eq. (3.108) is obtained as

$$\tilde{U}(z, \omega) = \tilde{U}(0, \omega)\exp\left(\frac{j}{2}\beta_2\omega^2 z\right). \tag{3.109}$$

By performing an inverse Fourier transform, the temporal waveform is obtained as

$$U(z, T) = \frac{1}{2\pi}\int_0^\infty \tilde{U}(0, \omega)\exp\left(\frac{j}{2}\beta_2\omega^2 z\right)\exp\left(-j\omega T\right)d\omega. \tag{3.110}$$

The solution is valid for an input pulse having an arbitrary profile. For a simple but practical example, we assume that the input pulse at $z = 0$ has a Gaussian profile

$$U(0, T) = A_0\exp\left[-\frac{1}{2}\left(\frac{T}{T_0}\right)^2\right], \tag{3.111}$$

where A_0 is the peak amplitude, and the parameter T_0 represents the half-width at $1/e$ intensity of the pulse envelope. Before we solve the LSE, an interesting property of a Gaussian pulse is reviewed. The full width at half-maximum T_{FWHM} of the intensity is obtained as

$$A_0^2\exp\left\{-\frac{(T_{FWHM}/2)^2}{T_0^2}\right\} = \frac{A_0^2}{2}$$

$$T_{FWHM} = 2\sqrt{\ln 2}\,T_0 \approx 1.665 T_0. \tag{3.112}$$

By Fourier transforming $U(0, T)$,

$$\tilde{U}(0, \omega) = \int_{-\infty}^{\infty} U(0, T)e^{j\omega T} dT = \left(2\pi T_0^2\right)^{1/2} A_0 \exp\left[-\frac{\omega^2 T_0^2}{2}\right]. \quad (3.113)$$

The full width at half-maximum f_{FWHM} of the spectrum in the intensity is obtained as

$$f_{FWHM} = \frac{\sqrt{\ln 2}}{\pi T_0}. \quad (3.114)$$

The product of the full width at half-maximum temporal and spectral widths becomes constant:

$$f_{FWHM} \times T_{FWHM} = \frac{\sqrt{\ln 2}}{\pi T_0} \times 2\sqrt{\ln 2}\,T_0 = \frac{2 \ln 2}{\pi} = 0.441. \quad (3.115)$$

The pulse that satisfies the above relation has the narrowest spectrum and is called the "Fourier-transform-limited" Gaussian pulse.

The solution for the Gaussian input pulse is obtained from Eq. (3.108) as [22]

$$U(z, T) = \frac{T_0^2}{\sqrt{T_0^2 - i\beta_2 z}} \exp\left\{-\frac{T^2}{2(T_0^2 - i\beta_2 z)}\right\}$$

$$= \frac{T_0^2}{\sqrt{T_0^2 - i\beta_2 z}} \exp\{-j\phi(T)\} \times \exp\left\{-\frac{1}{2}\left(\frac{T}{T_1}\right)^2\right\}. \quad (3.116)$$

The pulse width is given by

$$\frac{T_1}{T_0} = \left(1 + \frac{|\beta_2|^2}{T_0^4}z^2\right)^{1/2} = \left[1 + \left(\frac{z}{L_D}\right)^2\right]^{1/2}, \quad (3.117)$$

and the phase of the pulse is given by

$$\phi(T) = -\frac{\text{sgn}(\beta_2)(z/L_D)}{1 + (z/L_D)^2}\frac{T^2}{T_0^2} + \frac{1}{2}\tan^{-1}\left(\frac{z}{L_D}\right),$$

$$\text{sign}(x) = \begin{cases} 1, & x > 0 \\ -1, & x < 0 \end{cases} \quad (3.118)$$

where L_D is the dispersion length defined as

$$L_D = \frac{T_0^2}{|\beta_2|}. \quad (3.119)$$

In Fig. 3.29 the dispersion-induced broadening of a Gaussian pulse is plotted as a function of varying propagation length. The Gaussian pulse maintains its shape after propagation, except for the phase change, and its width broadens by a factor of $\sqrt{2}$ at $z = L_D$. As the value of dispersion $|\beta_2|$ becomes larger, the dispersion length L_D becomes shorter. Since L_D goes to infinity for zero dispersion, this results in $z/L_D = 0$.

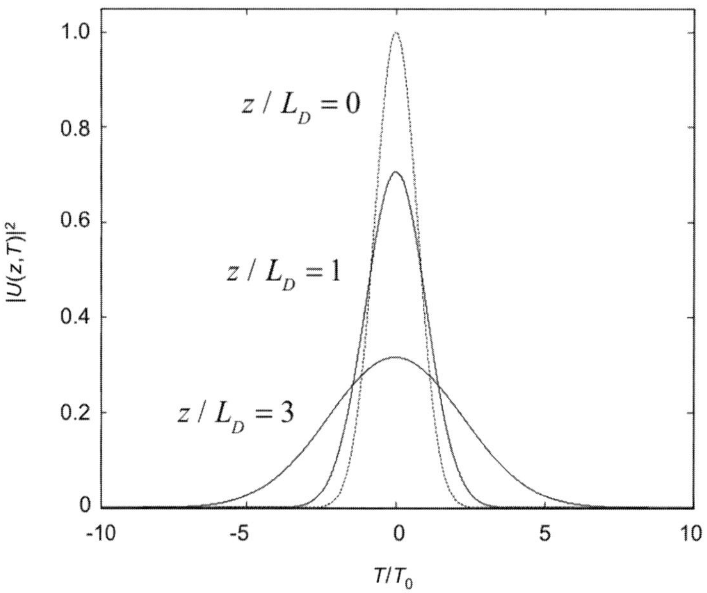

Figure 3.29 Dispersion-induced broadening of a Gaussian pulse.

Therefore, no pulse broadening occurs. The instantaneous frequency $\delta\omega(T)$, which is obtained from the time derivative of the phase in Eq. (3.118), is given by

$$\delta\omega(T) = -\frac{\partial\phi}{\partial T} = \frac{\mathrm{sgn}(\beta_2)(2z/L_D)}{1 + (z/L_D)^2} \frac{T}{T_0^2}. \tag{3.120}$$

For the normal dispersion region $\beta_2 > 0$, the frequency increases linearly from its precursor to the tail and changes sign from minus to plus when it crosses ω_0 at the center of the pulse, $T = 0$. The opposite is the case for the anomalous dispersion region $\beta_2 < 0$. This phenomenon is called frequency chirp. According to the nomenclature of laser optics, the positive and negative chirps, respectively, are referred to as "blue shift" and "red shift." It should be remembered that as the product $\beta_2 \cdot \delta\omega > 0$ regardless of the sign of β_2, the precursor of the pulse goes faster than the tail, taking into account the group delay characteristic in Fig. 3.27. As a consequence, the dispersion induces pulse broadening.

For complicated pulse shapes, the FWHM cannot always be a true measure of the pulse width. The width of such pulses is more accurately described by the root-mean-square (rms) σ width of the pulse,

$$\sigma = \left\{ \langle T^2 \rangle - \langle T \rangle^2 \right\}^{1/2} \tag{3.121}$$

where

$$\langle T^m \rangle = \frac{\int\limits_{-\infty}^{\infty} T^m \, |U(z, T)|^2 \, dT}{\int\limits_{-\infty}^{\infty} |U(z, T)|^2 \, dT}. \tag{3.122}$$

From Eq. (3.117) the initial rms width of the Gaussian pulse is given by

$$\sigma = \sigma_0 \left[1 + \left(\beta_2 L 2\sigma_0^2\right)^2\right]^{1/2} = \sigma_0 \left\{1 + \left(\frac{L}{L_D}\right)^2\right\}^{1/2},$$

$$\sigma_0 = \frac{T_0}{\sqrt{2}}. \tag{3.123}$$

Here, it is assumed that the spectral width of the light source is much narrower than the spectral width of the input pulse σ_0. Pulse broadening caused by the non-monochromatic light source is neglected. The dispersion-induced broadening can be minimized by choosing the optimum value of σ_0. For a given transmission distance, the minimum value of σ occurs for $\sigma_0 = (|\beta_2| L/2)^{1/2}$, and the minimum rms width is given by

$$\sigma_{\min} = \sqrt{|\beta_2| L}. \tag{3.124}$$

The pulse broadening induces an unwanted intersymbol interference (ISI) in which a symbol interferes with subsequent symbols, eventually leading to errors in the decision at the receiver. Therefore, the bit rate of optical transmission systems is limited by the ISI. The limiting bit rate can be obtained in a typical system as $\varepsilon = 0.491$ for the power penalty of 2 dB caused by ISI. This means that the pulse cannot occupy more than ε in one bit time duration T_B:

$$\sigma_{\min} < \varepsilon T_B \quad (0 < \varepsilon < 1)$$

$$\sqrt{|\beta_2| L} \leq \varepsilon T_B. \tag{3.125}$$

It is noted that the bit rate scales as $1/\sqrt{L}$ rather than $1/L$. For example, the chromatic dispersion of SMF at the wavelength of 1550 nm is $D = 17$ ps/nm/km, and the transmission distance is 112 km at the bit rate of 10 Gb/s; it is reduced one sixteenth at the bit rate of 40 Gb/s.

3.8 Dispersion compensation

Optical components for dispersion compensation are commercially available, and there is a wide variation including single-mode dispersion compensation fibers (DCFs), fiber Bragg gratings (FBGs), and thin film etalons. According to a thorough review of dispersion compensating fibers [23], their use was proposed in 1980 and first demonstrated in 1992. Dispersion-managed fibers were proposed in 1997, and have proven to be a suitable choice for high capacity long-distance transatlantic and transpacific transmission. A dispersion-managed fiber is normally composed of a positive dispersion fiber with positive dispersion slope ($D > 0$) and a large effective area, in combination with a negative dispersion fiber which has a negative dispersion slope ($D < 0$), referred to as an inverse dispersion fiber (IDF).

Suppose that the pulse broadening after propagation of distance L_1 in fiber 1 with the dispersion β_{21} is compensated for by using dispersion compensation fiber 2 with

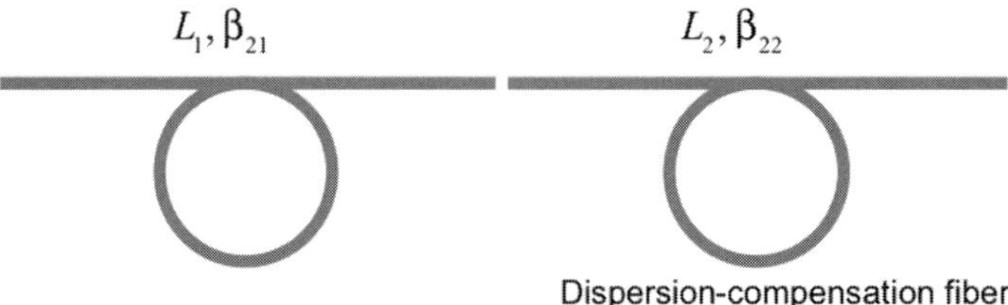

Figure 3.30 Schematic diagram of dispersion compensation.

the dispersion β_{22} at the distance of $L_1 + L_2$ as shown in Fig. 3.30. The overall pulse broadening is described using Eq. (3.110),

$$U(L_1 + L_2, T) = \frac{1}{2\pi} \int_0^\infty \tilde{U}(0, \omega) \exp\left\{ \frac{j}{2}\omega^2 \left(\beta_2^{(1)} L_1 + \beta_2^{(2)} L_2 \right) - j\omega T \right\} d\omega \tag{3.126}$$

where $\beta_2^{(i)}$ and L_i $(i = 1, 2)$ denote the GVD parameters and the lengths of fibers 1 and 2, respectively. In order to cancel the pulse broadening, the output pulse has to be same as the input pulse,

$$U(L_1 + L_2, T) = U(0, T). \tag{3.127}$$

This leads to the requirement

$$\beta_2^{(1)} L_1 + \beta_2^{(2)} L_2 = 0. \tag{3.128}$$

One has to keep in mind that this compensation scheme only considers the GVD parameter, but the dispersion compensation scheme which takes into account the dispersion slope, the third-order dispersion β_3, is of practical use.

An alternative to the optical dispersion compensation technique is electronic dispersion compensation using a digital coherent technique, coherent detection with the aid of digital signal processing (DSP). As the phase-shift-keying modulation format has become a desirable choice for DWDM long-haul transmission, rather than the conventional OOK signal, the digital coherent technique has an edge over optical dispersion compensation. The real and imaginary parts of the detected signal are sampled and converted to digital signals by analog-to-digital converters (ADCs). A real-time optical channel recovery has been demonstrated at the bit rate of 100 Gb/s using a digital coherent signal processor as shown in Fig. 1.15 [24]. The specifications of a 127 Gb/s DSP chip are summarized in Table 3.4. The frequency domain equalizer (FDE) compensates for chromatic dispersion up to 40,000 ps/nm, corresponding to approximately 2000 km transmission in a standard single-mode fiber at 1550 nm. An adaptive equalizer fulfills the roles of polarization demultiplexing, polarization mode dispersion (PMD) compensation, frequency offset compensation, carrier phase recovery, and sampling clock recovery.

However, 10 Gb/s transmission systems based upon the direct-detection OOK format rely on optical techniques for compensation because of their reasonable cost and better performance such as low latency and low insertion loss. Optical compensation still has an edge over the electronic counterpart in terms of less complexity, lower latency, less power consumption, and smaller noise accumulation.

Table 3.4 Specifications of 100 Gb/s DSP chip

Item	Specification
Modulation format	PDM-QPSK
Line side signal bit rate	127.156 Gb/s
Oversampling ratio	2.0
Chromatic dispersion compensation	$\pm 40{,}000$ ps/nm
Polarization mode dispersion compensation	100 ps
Polarization tracking speed	50 kHz
Frequency offset compensation	± 5 GHz
FEC net coding gain	10.8 dB at BER $= 10^{-15}$
Recovery time	<50 ms
Process	40 nm CMOS

Appendix 3.1

Given a vector $\mathbf{A}$ that is twice differentiable, the identity holds:

$$\nabla \times (\nabla \times \mathbf{A}) = \nabla (\nabla \cdot \mathbf{A}) - \nabla^2 \mathbf{A} \tag{A3.1.1}$$

where

$$\nabla = \mathbf{u_x}\frac{\partial}{\partial x} + \mathbf{u}_y\frac{\partial}{\partial y} + \mathbf{u}_z\frac{\partial}{\partial z}. \tag{A3.1.2}$$

The field components of the TE mode must satisfy the boundary conditions, the tangential $\mathbf{E}$ and $\mathbf{H}$ fields being continuous, at the two dielectric interfaces at $x=a$ and $x=-a$. Solutions must vanish at $x = \pm\infty$. The continuity of E_y and H_z at $x=a$ gives the eigenvalue equations

$$A \begin{Bmatrix} \cos\kappa a \\ \sin\kappa a \end{Bmatrix} = Be^{-\gamma a}$$

$$-A\frac{jk}{\omega\mu_0} \begin{Bmatrix} \sin\kappa a \\ -\cos\kappa a \end{Bmatrix} = -\frac{j\gamma}{\omega\mu_0}Be^{-\gamma a}. \tag{A3.1.3}$$

The eigenvalue for non-zero values of constants, A and B must satisfy

$$\begin{vmatrix} \begin{Bmatrix} \cos\kappa a \\ \sin\kappa a \end{Bmatrix} & -e^{-\gamma a} \\ k\begin{Bmatrix} \sin\kappa a \\ -\cos\kappa a \end{Bmatrix} & -\gamma e^{-\gamma a} \end{vmatrix} = 0. \tag{A3.1.4}$$

The characteristic equations of TE mode are derived:

$$\gamma \begin{Bmatrix} \cos\kappa a \\ \sin\kappa a \end{Bmatrix} = k \begin{Bmatrix} \sin\kappa a \\ -\cos\kappa a \end{Bmatrix}$$

$$ka\frac{\sin\kappa a}{\cos\kappa a} = ka\tan\kappa a = \gamma a \quad \text{(Even mode)}$$

$$ka\frac{\cos\kappa a}{\sin\kappa a} = ka\cot\kappa a = -\gamma a \quad \text{(Odd mode)} \tag{A3.1.5}$$

Appendix 3.2

$$
-\frac{\gamma a K_{m-1}(\gamma a)}{K_m(\gamma a)} - \frac{\{\rho - 1/\zeta_m(\gamma a)\}}{(\alpha + 2)}(\gamma a)^2
$$

$$
= \frac{\kappa a J_{m-1}(\kappa a)}{J_m(\kappa a)} + \frac{(\rho - 1)^2}{\alpha}(\kappa a)^2
\tag{A3.2.1}
$$

where ζ_m is defined as

$$
\zeta_m(w) = \frac{K_m(w)^2}{K_{m-1}(w)K_{m+1}(w)}.
\tag{A3.2.2}
$$

In the case with $\alpha = \infty$ and $\rho = 1$, Eq. (A3.2.1) is simplified to

$$
-\frac{\gamma K_{m-1}(\gamma a)}{K_m(\gamma a)} = \frac{\kappa J_{m-1}(\kappa a)}{J_m(\kappa a)}.
\tag{A3.2.3}
$$

This is the characteristic equation of the step-index fiber given in Eq. (3.34).

Appendix 3.3

Assume $\exp(-j\beta z)$ dependence, the amplitudes a_1 and a_2 take the forms

$$
a_1(z) = A_1(z)e^{-j\beta_m z}
$$

$$
a_2(z) = A_2(z)e^{-j\beta_m z}
\tag{A3.3.1}
$$

where

$$
\beta_m = \frac{\beta_1 + \beta_2}{2}.
\tag{A3.3.2}
$$

The field E of the coupled waveguide is written as

$$
E = A_1(z)e^{-j\beta_m z}E_1 + A_2(z)e^{-j\beta_m z}E_2.
\tag{A3.3.3}
$$

The general solutions with arbitrary constants are

$$
A_1(z) = A_{10}e^{\gamma z}, \quad A_2(z) = A_{20}e^{\gamma z}
$$

$$
a_1(z) = A_{10}e^{(\gamma - j\beta_m)z}, \quad a_2(z) = A_{20}e^{(\gamma - j\beta_m)z}.
\tag{A3.3.4}
$$

By substituting in Eq. (3.31), the determinantal equation is derived:

$$
\begin{vmatrix} \gamma + j\Delta & -\kappa_{12} \\ -\kappa_{21} & \gamma - j\Delta \end{vmatrix} = 0
\tag{A3.3.5}
$$

where

$$
\Delta = \frac{\beta_1 - \beta_2}{2}.
\tag{A3.3.6}
$$

The solution γ is obtained as the eigenvalue of the determinantal equation

$$
\gamma^{\pm} = \pm j\sqrt{\Delta^2 + |\kappa_{12}|^2} = \pm j\beta_0
$$

$$
\beta_0 = \sqrt{\Delta^2 + |\kappa_{12}|^2}
\tag{A3.3.7}
$$

The eigenvector is given by

$$\left(\frac{A_{10}}{A_{20}}\right)^{\pm} = \frac{\pm\beta_0 - \Delta}{\kappa_{21}} = \frac{\pm\beta_0 - \Delta}{-\kappa_{12}^*}. \qquad (A3.3.8)$$

The product of two eigenvalues always becomes negative. This means that two coupled-mode fields are symmetric and asymmetric:

$$\left(\frac{A_{10}}{A_{20}}\right)^{+} \cdot \left(\frac{A_{10}}{A_{20}}\right)^{-} = -1 < 0. \qquad (A3.3.9)$$

The values of the eigenvectors A_{10} and A_{20} can be determined by the initial condition at $z = 0$. By substituting Eqs. (A3.3.7) and (A3.3.8) in Eq. (A3.3.4), the solution in the form of transfer matrix Eq. (3.40) is obtained.

Appendix 3.4

The general solutions with arbitrary constants of Eq. (3.53) are

$$A(z) = A_0 e^{-\gamma z}, \quad B(z) = B_0 e^{-\gamma z}$$
$$a(z) = A_0 e^{-(\gamma + j\frac{\pi}{\Lambda})z}, \quad b(z) = B_0 e^{(-\gamma + j\frac{\pi}{\Lambda}z)z}. \qquad (A3.4.1)$$

The solutions are rewritten as

$$\begin{aligned}
A(z) &= A_0^+ e^{-\gamma z} + A_0^- e^{+\gamma z} \\
&= (A_0^+ + A_0^-)\cosh \gamma z - (A_0^+ - A_0^-)\sinh \gamma z, \\
B(z) &= B_0^+ e^{-\gamma z} + B_0^- e^{+\gamma z} \\
&= (B_0^+ + B_0^-)\cosh \gamma z - (B_0^+ - B_0^-)\sinh \gamma z. \qquad (A.3.4.2)
\end{aligned}$$

The arbitrary constants A_0 and B_0 can be determined by the initial conditions at $z = 0$, $A(0) \neq 0$ and $B(L) = 0$.

From the coupled-mode equation, the determinantal equation is derived:

$$\begin{vmatrix} \gamma - j\delta & \kappa \\ \kappa^* & \gamma + j\delta \end{vmatrix} = 0. \qquad (A3.4.3)$$

The eigenvalues are obtained as

$$\gamma = \pm\sqrt{|\kappa|^2 - \delta^2}. \qquad (A3.4.4)$$

For the *stop band* of $\delta < \kappa$, the solutions are of the form $\exp(\pm\gamma z)$ which grow and decay exponentially. In the range of $\delta > \kappa$ they are periodic functions.

Problems

3.1 Loss and attenuation constant. Convert $\alpha_0 = 0.3\,[\mathrm{dB/km}]$ to $\alpha\,[km^{-1}]$.

3.2 Geometric optics of slab waveguide. Calculate the critical angle θ_c and the maximum incident angle $\theta_0^{\max}$ to the fiber endface for the case $\Delta = 0.01$, $n_0 = 1.0$, $n_1 = 1.45$.

3.3 Show that the transfer matrix $[S_{ij}]$ $(i, j = 1, 2)$ of the 2×2 coupled-mode system in Eq. (3.57) is unitary which satisfies Eq. (3.60).

3.4 Two couplers having non-degenerate modes in tandem $\Delta = (\beta_1 - \beta_2)/2 \neq 0$.

$$
\begin{bmatrix} a_1(z) \\ a_2(z) \end{bmatrix} = e^{-j\beta_m z}
\begin{bmatrix}
\cos \beta_0 z - j\frac{\Delta}{\beta_0} \sin \beta_0 z & \frac{\kappa_{12}^*}{\beta_0} \sin \beta_0 z \\
-\frac{\kappa_{12}^*}{\beta_0} \sin \beta_0 z & \cos \beta_0 z + j\frac{\Delta}{\beta_0} \sin \beta_0 z
\end{bmatrix}
\begin{bmatrix} a_1(0) \\ a_2(0) \end{bmatrix}.
$$

Consider the two couplers connected in a flip-flop configuration. For simplicity, it is assumed that $L_1 = L_2 = L$.

(1) Set the value of κ_{12}/β_0 so that $a_1(L) = a_2(L)$ at $\beta_0 L = \pi/2$.

(2) Show 100% power transfer from Input 1 to Output 2 occurs when the difference in the propagation constants $\Delta = (\beta_1 - \beta_2)$ is reversed in the middle of the coupling region at $z = L$.

3.5 Fiber loop mirror. Consider the fiber loop mirror. In the loop, the input lightwave is split by the 3-dB coupler and guided into two arms. The clockwise and counter-clockwise waves are coupled again via the 3-dB coupler. This is considered to be two 2×2 3-dB couplers ($\Delta = 0$) which are connected in tandem. The configuration is different from the non-linear loop mirror in Fig. 5.9 in that there is no arm to launch the control pulse.

(1) Derive the transfer matrix of the field for the fiber loop mirror.

(2) For a single input light from the west arm, calculate the output.

3.6 Bragg's wavelength λ_B. Calculate the power reflection coefficient $R(0)$ for the case with $\delta = 0$, $\kappa L = 1.84$.

3.7 DFB laser. Calculate the wavelength difference between two longitudinal modes for the case with $L = 300$ μm, $n_{eff} = 3.5$, $\lambda_B = 1550$ nm.

3.8 Dispersion compensation. Consider the dispersion compensation scheme after propagating the fiber transmission line of the length L_1 and the chromatic dispersion β_{21} at $\lambda = \lambda_0$. Use another fiber of length L_2 and dispersion β_{22}.

(1) Derive the requirement for dispersion of the compensation fiber.

(2) Calculate the dispersion D_2 of the compensation fiber for the case with $L_2 = 1$ km, $D_1 = 17$ [ps/km/nm] at $\lambda_0 = 1\,550$ nm and $L_1 = 50$ km.

4 Fundamentals of transmission systems

4.1 Detection theory

In communication systems, as shown in Fig. 4.1, the information sources include speech waveforms, image waveforms, and text files. The source encoder converts the source output to a binary sequence, and the channel encoder (often called a modulator) processes the binary sequence for transmission over the channel. The channel decoder (demodulator) regenerates the incoming binary sequence, and the source decoder recreates the original source information. In a binary digital optical transmission system, bit error of the transmitted signal occurs when the receiver decides whether the signal is either a mark or a space, that is, logical data "1" or "0." At the receiver, the optical signal is converted to an electrical signal by the photodetector, and the electrical signal is equalized through the low-pass filter (LPF) and amplified so as to keep its amplitude constant by automatic gain control (AGC). In the decision circuit the threshold is set: the signal is decoded as "1" when the signal level is higher than the threshold, while the signal is decoded as "0" when the signal level is lower than the threshold. This decision has to be made with proper timing. The clock signal is recovered from the received random bit stream having jitter. This is the so-called 3R function of the repeater: reshaping including equalization and amplification, retiming, and regenerating, as illustrated in Fig. 4.2(a). Figure 4.2(b) shows the detected signal at the receiver (on the top), the original clean signal without noise (in the middle), and the clock signal recovered from the detected signal [1]. The threshold has to be set so as to minimize the error.

Assume that a bit stream is generated from the information source, and each bit is independent of the previous bits. The prior probability is given by $P(x)$ $(x = 0, 1)$, and the probability that x falls in the range of x and $x + dx$ is calculated by $P(x)\, dx$. The bit is detected at the receiver with the conditional probability density function $P(x|y)$ where y is the output parameter at the receiver. Due to the signal impairment caused by noise and distortion, there are two classes of error as shown in Fig. 4.3: the decoded signal is "1" when "0" is sent out, $P(1|0)$, and the decoded signal is "0" when "1" is sent out, $P(0|1)$. Since either "1" or "0" is randomly generated at the transmitter, the prior probability becomes even, that is $P(0) = P(1) = \frac{1}{2}$. Therefore, the bit error rate, BER, is written as

$$\mathrm{BER} = \frac{1}{2}P(1|0) + \frac{1}{2}P(0|1). \tag{4.1}$$

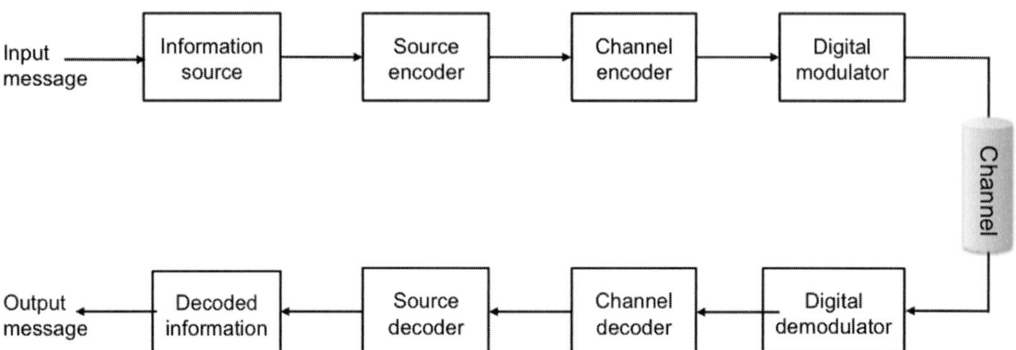

Figure 4.1 Block diagram of a digital communication system.

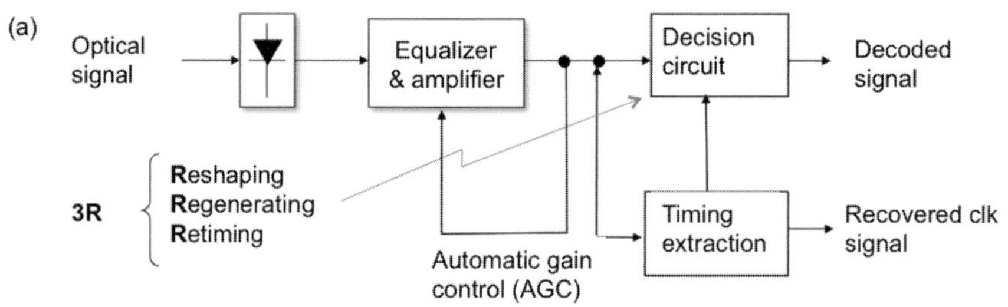

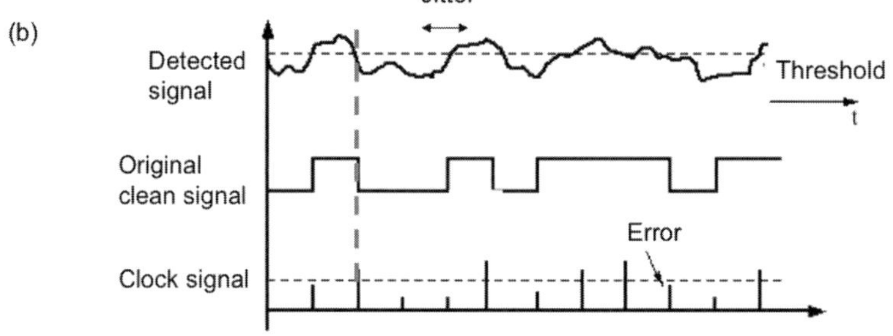

Figure 4.2 (a) Functional block of a 3R repeater and (b) signal waveforms.

The output parameter at the receiver y fluctuates due to additive noise such as the shot noise. As the average number of photons arriving at the receiver increases, the arrival rate is approximated as Gaussian. In this case the conditional probability density function $P(x|y)$ is of the form

$$P(y|x) = \frac{1}{\sqrt{2\pi}\sigma} \exp\left\{-\left(\frac{y-I}{\sqrt{2}\sigma}\right)^2\right\}, \tag{4.2}$$

where σ denotes the standard deviation, and m is the mean value. There are two Gaussian functions for data "1" and "0" having averages and standard deviations $(I_1,\ \sigma_1)$ and $(I_0,\ \sigma_0)$, respectively. Assume that the decision threshold is set at $y = D_{th}$, the bit error

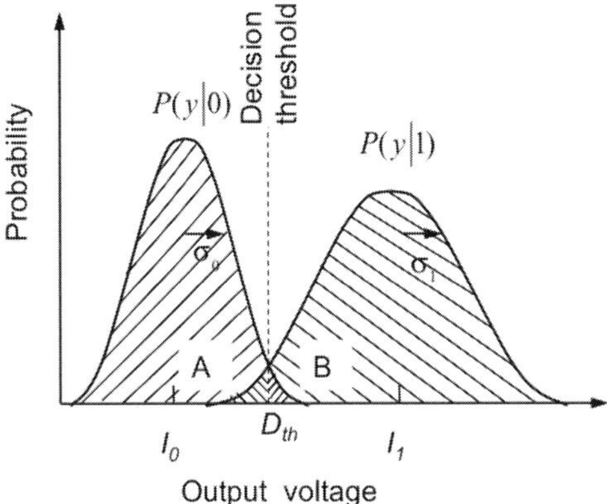

Figure 4.3 Probability density functions for the received binary signals.

rate is given by

$$
\begin{aligned}
\text{BER} &= \frac{1}{2} \int_{D_{th}}^{\infty} \frac{1}{\sqrt{2\pi}\,\sigma_0} \exp\left\{-\left(\frac{y-I_0}{\sqrt{2}\sigma_0}\right)^2\right\} dy + \frac{1}{2} \int_{-\infty}^{D_{th}} \frac{1}{\sqrt{2\pi}\,\sigma_1} \exp\left\{-\left(\frac{y-I_1}{\sqrt{2}\sigma_1}\right)^2\right\} dy \\
&= \frac{1}{2} Q\left(\frac{D_{th}-I_0}{\sigma_0}\right) + \frac{1}{2} Q\left(\frac{I_1-D_{th}}{\sigma_1}\right)
\end{aligned}
\tag{4.3}
$$

where the Q-function is defined as

$$
Q(\alpha) = \frac{1}{\sqrt{2\pi}} \int_{\alpha}^{\infty} \exp(-\xi^2/2)d\xi.
\tag{4.4}
$$

The threshold value D_{th} is determined so that the BER is minimized, and this requirement results in equating two areas A and B in Fig. 4.3:

$$
Q\left(\frac{D_{th}-I_0}{\sigma_0}\right) = Q\left(\frac{I_1-D_{th}}{\sigma_1}\right)
\tag{4.5}
$$

yields the threshold value

$$
D_{th} = \frac{I_0\sigma_1 + I_1\sigma_0}{\sigma_0 + \sigma_1}.
\tag{4.6}
$$

Finally, the BER is obtained as

$$
\begin{aligned}
\text{BER} &= Q(\gamma) \\
\gamma &= \frac{I_1 - I_0}{\sigma_0 + \sigma_1}
\end{aligned}
\tag{4.7}
$$

where γ is defined as the Q-factor. The Q-function is plotted as a function of γ in Fig. 4.4. The dotted curve shows the approximation. BER values of 10^{-9} and 10^{-12} are obtained with $\gamma = 6$ and 7, respectively, which are typically claimed to be "error free" in optical fiber transmission. When the error function is introduced, defined by

$$
\text{erfc}(x) = 1 - \int_{-x}^{x} \frac{1}{\sqrt{\pi}} \exp(-y^2)dy,
\tag{4.8}
$$

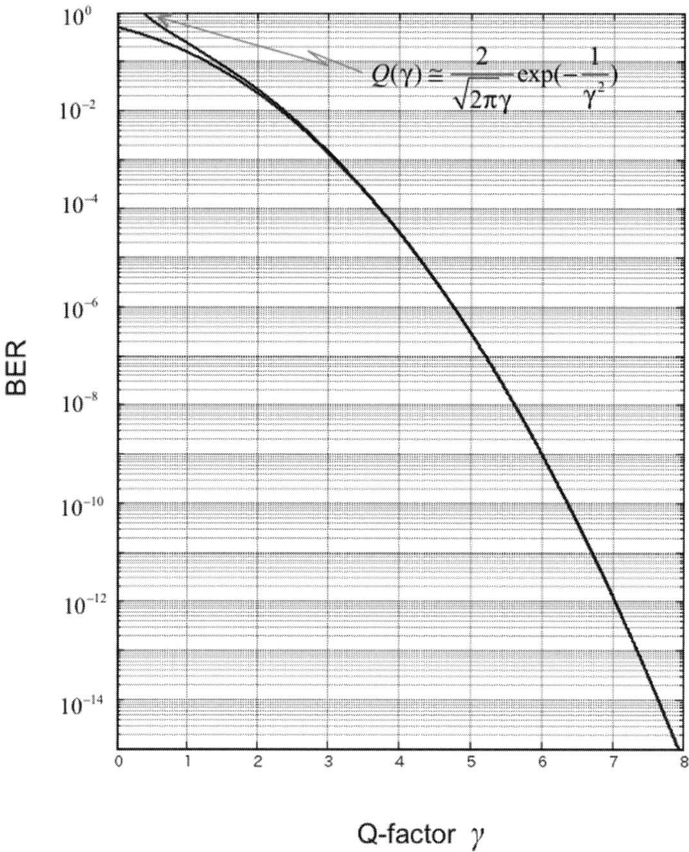

Figure 4.4 Q-function versus Q-factor γ.

the BER is rewritten as

$$\mathrm{BER} = \frac{1}{2}\mathrm{erfc}\left(\frac{\gamma}{\sqrt{2}}\right). \tag{4.9}$$

4.2   Modulation

4.2.1   Modulation formats

In optical fiber communications, there are two schemes of optical modulation, analog
and digital modulation. When either an analog or a digital electrical signal is supplied to
the amplitude modulator, the envelope of the optical carrier yields the replica waveform
of the supplied signal as shown in Figs. 4.5(a) and (b), respectively. The optical carrier
is modulated to map the information data onto its amplitude, phase, or frequency or
alternatively a combination of its amplitude and phase. In Fig. 4.6 the waveforms of
on-off-keying (OOK) or amplitude-shift-keying (ASK), phase-shift-keying (PSK), and

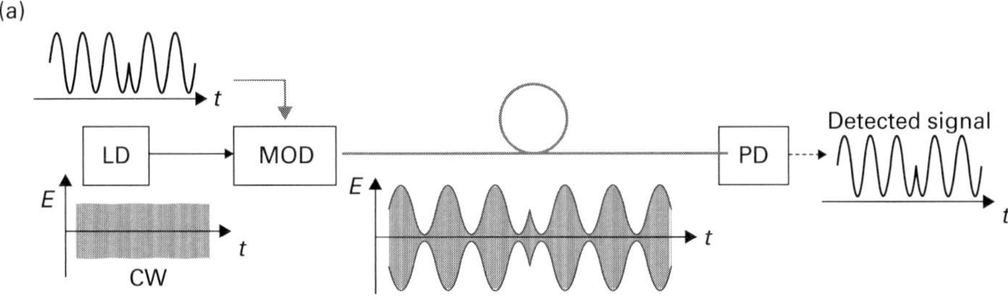

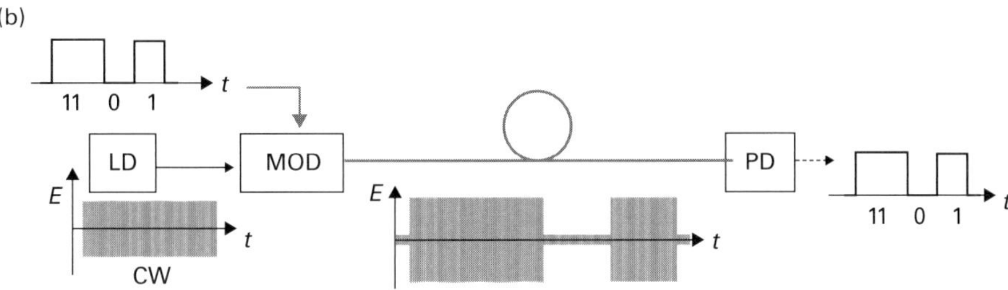

Figure 4.5 Modulation schemes for (a) analog and (b) digital information data.

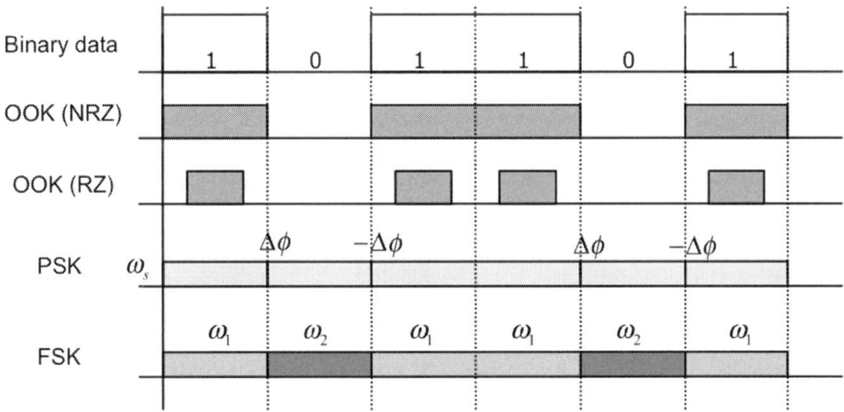

$\Delta\phi$: Phase change

Figure 4.6 Waveforms of on-off-keying (OOK) or amplitude-shift-keying (ASK), phase-shift-keying (PSK), and frequency-shift-keying (FSK) for binary data.

frequency-shift-keying (FSK) for binary digital data are illustrated. In OOK, a pulse represents data "1", and no pulse is transmitted for data "0". There are also non-return-to-zero OOK (NRZ-OOK) and return-to-zero OOK (RZ-OOK). In RZ-OOK the duty ratio represents the time duration of the data for one bit time duration. For the OOK optical signal in general, a simple photon counting device is used for detection. For the

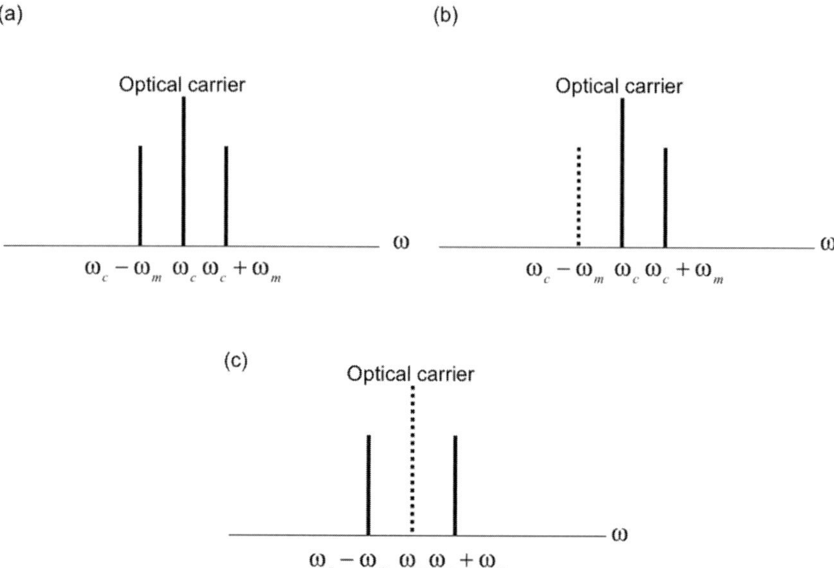

Figure 4.7 (a) Double sideband signal, (b) single sideband signal, and (c) carrier-suppressed signal.

FSK optical signal, an optical filter passing optical carriers of angular frequency either ω_1 or ω_2 discriminates between the digital data. For the PSK optical signal, a phase reference will be required at the receiver to detect whether the optical carrier suffers from the phase shift $\Delta\phi$ or not. The modulation format most commonly used in optical fiber communications has been OOK.

There has been growing interest in applying multi-level modulation formats to increase the information capacity per symbol transmitted in one bit time duration, leading to higher spectral efficiency. PSK or a combination of ASK and PSK such as QAM format and OFDM, which has been widely used in wireless communications, serves this purpose. As laser diodes have gained better stability in the oscillation frequency, and as a consequence the linewidth has become narrower, and as furthermore powerful digital signal processing has enabled compensating carrier frequency offset (CFO), thus the opportunity has arisen to apply PSK in commercial optical fiber transmission systems.

4.2.2   On-off-keying modulation

When the monochromatic carrier of angular frequency ω_c is amplitude-modulated with the sinusoidal signal of the angular frequency ω_m, the electric field is written as

$$E(t) = \sqrt{P}\{1 + m\cos\omega_m t\}\, e^{j\omega_c t}$$
$$= \sqrt{P}\{\exp(j\omega_c t) + m\exp[j(\omega_c + \omega_m)t] + m\exp[j(\omega_c - \omega_m)t]\} \quad (4.10)$$

where P denotes the peak intensity, and m is the modulation depth. There are two sidebands on both sides of the carrier as shown in Fig. 4.7(a), and this is called a double

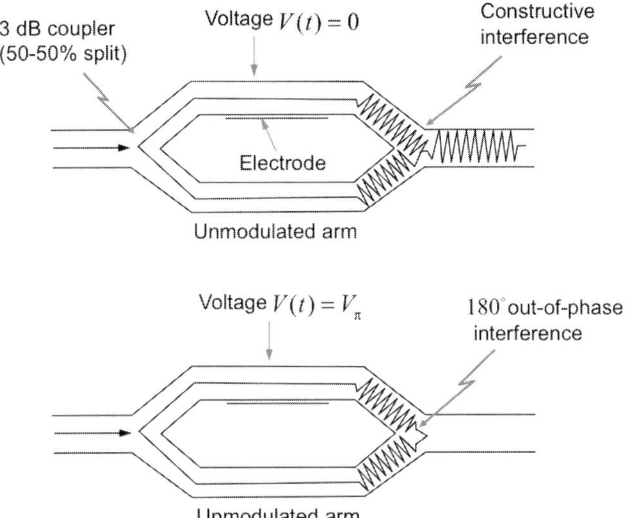

Figure 4.8 Configuration of a Mach–Zehnder interferometer (MZI) modulator.

sideband (DSB) signal. Figure 4.7(b) shows the single sideband (SSB) signal, expressed as

$$E(t) = \sqrt{P}\{\exp(j\omega_c t) + m\exp[j(\omega_c - \omega_m)t]\}. \tag{4.11}$$

Figure 4.7(c) shows the carrier-suppressed signal:

$$\begin{aligned} E(t) &= \sqrt{P}\{1 + m\cos\omega_m t\}\,e^{j\omega_c t} \\ &= m\sqrt{P}\{\exp[j(\omega_c + \omega_m)t] + \exp[j(\omega_c - \omega_m)t]\} \end{aligned} \tag{4.12}$$

By detecting the amplitude-modulated signal with a photodetector, the square-law detection yields the envelope of the optical carrier, and the output photocurrent is given by

$$I(t) = |E(t)|^2 = \left|\sqrt{P}(1 + m\cos\omega_m t)e^{j\omega_c t}\right|^2. \tag{4.13}$$

If the modulation depth is sufficiently small, that is $m \ll 1$, the photocurrent is reduced to the original signal waveform as

$$I(t) \cong P(1 + 2m\cos t). \tag{4.14}$$

As an example of an optical modulation technique, let us consider the normal Mach–Zehnder interferometer (MZI) type of optical modulator shown in Fig. 4.8, which is the most commonly used type of modulator, particularly for high bit rate optical transmission systems. It consists of two identical arms having an electrode on one arm, coupled with a 3-dB coupler on both ends. Note that a traditional space-optic MZI uses beam splitters instead of the 3-dB coupler. Suppose that the refractive index changes in proportion to the applied voltage $V(t)$, the optical carrier passing underneath the electrode suffers from the phase shift $\phi(t)$, while the optical carrier passes through the other arm without any phase shift. Without the applied voltage, the input optical power is split in half by the 3-dB coupler at the input and combined again by the input by the 3dB coupler, and consequently the input power appears at the output port as it is. When the voltage is applied so that a π phase shift happens to the optical carrier on one arm, destructive interference occurs at the 3-dB couple at the output, and no output appears. This is the operation principle of OOK modulation using the MZI modulator.

The applied voltage $V(t)$, and the optical carrier passing underneath the electrode with phase shift $\phi(t)$ are expressed as

$$\phi(V) = \pi \frac{V}{V_\pi}, \quad V(t) = V_b + V_m \sin \omega_m t, \tag{4.15}$$

where V_π and V_b are the half-wavelength and the dc bias voltage, respectively. The input lightwave is split into two arms by the first 3-dB coupler and combined together by the second 3-dB coupler. Without the phase shift on one arm, the input lightwave is reconstructed at the output port due to the constructive interference between the two lightwaves. For the phase shift $\phi(t) = \pi$ on one arm, the lightwaves will destructively interfere with each other, resulting in no output. It is convenient to use a scattering matrix to express the transfer function of the MZI modulator. The overall scattering matrix $[S]$ is given by

$$[S] = [S]_{3dB}[S]_{MZI}[S]_{3dB}. \tag{4.16}$$

Given the scattering matrix of the 3-dB coupler from Eq. (3.63),

$$[S]_{3dB} = \frac{1}{\sqrt{2}} \begin{bmatrix} 1 & j \\ j & 1 \end{bmatrix} \tag{4.17}$$

and the phase shift of the modulator section

$$[S]_{MZI} = \begin{bmatrix} \exp(j\phi) & 0 \\ 0 & 1 \end{bmatrix}. \tag{4.18}$$

By plugging Eqs. (4.17) and (4.18) into Eq. (4.16), the overall scattering matrix is expressed as

$$V = 0 \, (\phi = 0) \text{ and } V = V_\pi \, (\phi = \pi) \text{ or } [S] = \frac{1}{2} \begin{bmatrix} \exp(j\phi) - 1 & j\exp(j\phi) + j \\ j\exp(j\phi) + j & -\exp(j\phi) + 1 \end{bmatrix}. \tag{4.19}$$

For the input from Port 1 the output is obtained:

$$\begin{bmatrix} E_{out1} \\ E_{out2} \end{bmatrix} = [S] \begin{bmatrix} 1 \\ 0 \end{bmatrix} \exp(j\omega_c t)$$

$$= \frac{1}{2} \begin{bmatrix} \exp(j\phi) - 1 \\ j\exp(j\phi) + j \end{bmatrix} \exp(j\omega_c t) = \begin{bmatrix} j \sin\frac{\phi}{2} \exp(j\frac{\phi}{2}) \\ \cos\frac{\phi}{2} \exp(j\frac{\phi}{2}) \end{bmatrix} \exp(j\omega_c t). \tag{4.20}$$

For $V = 0 \, (\phi = 0)$ and $V = V_\pi \, (\phi = \pi)$ and $V = 2V_\pi \, (\phi = 2\pi)$ in Eq. (4.15), respectively, the electric field at the output Port 1 and Port 2 swings between zero and one. As shown in Fig. 4.9 this demonstrates OOK modulation for binary digital data. The output

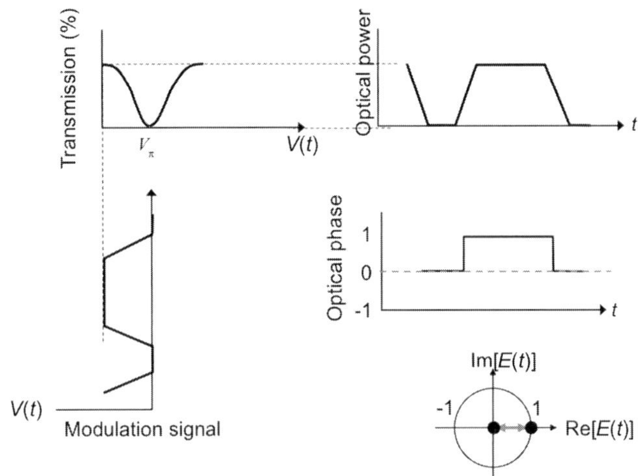

Figure 4.9 Intensity transfer function of the normal MZI modulator (Port 2) plotted as a function of applied voltage.

intensity is obtained from Eq. (4.20) as

$$\begin{bmatrix} |E_{out1}|^2 \\ |E_{out2}|^2 \end{bmatrix} = \begin{bmatrix} \sin^2(\phi/2) \\ \cos^2(\phi/2) \end{bmatrix}. \tag{4.21}$$

For the dc bias voltage $V_b = V_\pi/2$, the intensity-modulated signals are obtained as

$$
\begin{aligned}
I_{out1} &= \frac{1}{2}\left\{1 + 2\frac{\pi V_m}{V_\pi}J_1\left(\frac{\pi V_m}{V_\pi}\right)\sin\omega_m t\right\} \\
&\cong \frac{1}{2}\left(1 + \frac{\pi V_m}{V_\pi}\sin\omega_m t\right) \\
I_{out2} &= \frac{1}{2}\left\{1 - 2\frac{\pi V_m}{V_\pi}J_1\left(\frac{\pi V_m}{V_\pi}\right)\sin\omega_m t\right\} \\
&\cong \frac{1}{2}\left(1 - \frac{\pi V_m}{V_\pi}\sin\omega_m t\right).
\end{aligned}
\tag{4.22}
$$

In the case with $V_b = V_\pi$ the electric field-modulated signal is obtained as is described in Section 4.7.1.

4.2.3 Electro-optic effect

In general the vectorial electric flux density $\mathbf{D}$ is expressed using the electric field and the non-linear polarization

$$\mathbf{D} = \varepsilon_0\left(1 + \chi^{(1)}\mathbf{E}\right) + \mathbf{P}_{NL}$$

$$\mathbf{P}_{NL} = \varepsilon_0\left(\chi^{(2)} \vdots \mathbf{EE} + \chi^{(3)} \vdots \mathbf{EEE} + \cdots\right) \tag{4.23}$$

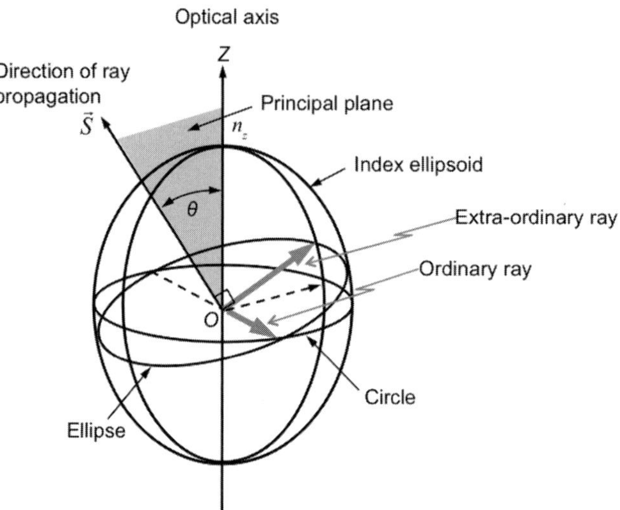

Figure 4.10 Index ellipsoid.

where $\chi^{(i)}$ ($i = 1, 2, \ldots$) is the ith order permittivity taking tensor form. The refractive index changes in proportion with the applied voltage V as a result of the second-order non-linearity, that is the electro-optic effect or Pockels effect of crystals. The index ellipsoid in principal coordinates, as in Fig. 4.10, is written as

$$\frac{x^2}{n_x^2} + \frac{y^2}{n_y^2} + \frac{z^2}{n_z^2} = 1, \tag{4.24}$$

where n_x, n_y and n_z are the principal refractive indices. In an isotropic medium all three values of refractive index are equal, $n_x = n_y = n_z$, while this is not the case with an anisotropic medium. There are always two refractive indices for plane waves propagating along an arbitrary direction $\vec{S}$ in Fig. 4.10, depending on their polarization direction: the refractive indices of the ordinary ray and for the extraordinary ray. Let us consider the propagation of an electromagnetic field in a crystal in the presence of a dc applied electric field. The electro-optic effect is traditionally defined in terms of the change in impermeability tensor. Under the dc applied electric field the index ellipsoid is deformed as

$$\left(\frac{1}{n_x^2} + r_{1k}E_k\right) x^2 + \left(\frac{1}{n_y^2} + r_{2k}E_k\right) y^2 + \left(\frac{1}{n_z^2} + r_{3k}E_k\right) z^2 = 1, \tag{4.25}$$

where r_{ij} is the element of a 6×3 electro-optic tensor. Although there are, in general, 18 independent coefficients of the tensor, some of these elements are zero or identical. The form of the tensor r_{ij} can be derived from symmetry considerations of the crystal such as the inversion symmetry, which dictate which of the 18 coefficients are zero as well as the relationship between the remaining non-zero coefficients. $E_k (k = 1, 2, 3)$ is a component of the applied electric field, and summation over repeated indices k is assumed. Here 1, 2, and 3 customarily correspond to the principal dielectric axes

x, y, and z. This new index ellipsoid reduces to the unperturbed ellipsoid when $E_k = 0$. A set of principal axes under the applied field is found by a coordinate rotation. The principal-axis transformation takes a quadratic form

$$\Delta\left(\frac{1}{n^2}\right)_i = \sum_{j=1}^{3} r_{ij} E_j, \tag{4.26}$$

and

$$\Delta n_i \cong -\frac{n_i^3}{2} \sum_{j=1}^{3} r_{ij} E_j \quad (i = x, y, z). \tag{4.27}$$

For a specific crystal the electro-optic tensor is available to calculate the change in refractive index. Consider the specific example of lithium niobate crystal, LiNbO$_3$, which has been widely used for the material of optical modulators. It has a crystal symmetry of $3m$, and the coefficients are in the form

$$[r_{ij}] = \begin{bmatrix} 0 & -r_{22} & r_{13} \\ 0 & r_{22} & r_{13} \\ 0 & 0 & r_{33} \\ 0 & r_{51} & 0 \\ r_{51} & 0 & 0 \\ -r_{22} & 0 & 0 \end{bmatrix}$$

$$n_x = n_y = n_o, \quad n_z = n_e. \tag{4.28}$$

We now consider the case with the electric field along the z-axis E_3. The index ellipsoid is written as

$$\left(\frac{1}{n_o^2} + r_{1k3}E_3\right) x^2 + \left(\frac{1}{n_o^2} + r_{13}E_3\right) y^2 + \left(\frac{1}{n_e^2} + r_{33}E_3\right) z^2 = 1. \tag{4.29}$$

The refractive indices along the new principal axes are

$$n_x = n_o - \frac{1}{2}n_o^3 r_{13} E_3,$$

$$n_y = n_o - \frac{1}{2}n_o^3 r_{13} E_3, \tag{4.30}$$

$$n_z = n_e - \frac{1}{2}n_e^3 r_{33} E_3.$$

We consider the propagation of a TE mode (see Fig. 3.4) along the y-axis under the applied electric field along the z-axis, shown in Fig. 4.11. The phase retardation which the lightwave suffers is

$$\delta\phi_z = \frac{2\pi}{\lambda} L \left(\frac{1}{2}n_e^3 r_{33} E_3\right) = \frac{\pi L}{\lambda d} n_e^3 r_{33} V_z, \tag{4.31}$$

where L and d denote the length of the modulator and the thickness of the waveguide. The half-wavelength voltage V_π can be calculated for given V_z and d by setting $\delta\phi_z = \pi$.

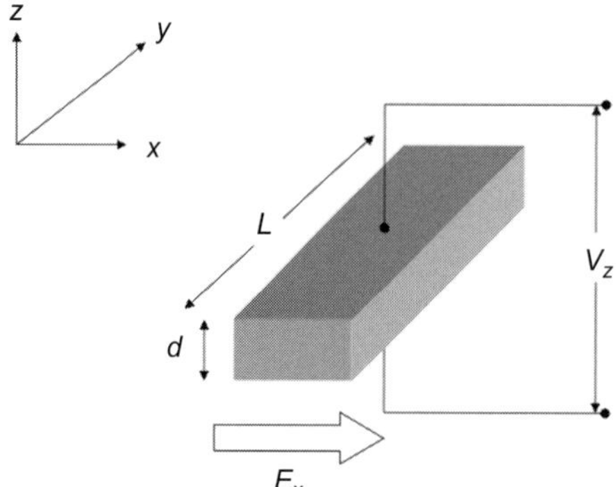

Figure 4.11 Lightwave propagation in LiNbO$_3$ crystal along the y-axis under an applied electric field along the z-axis.

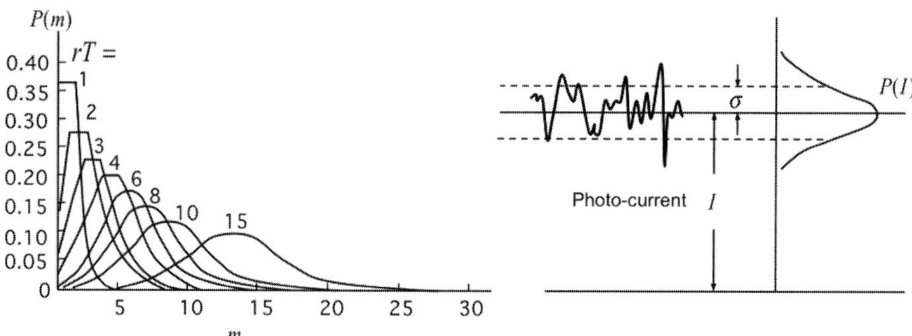

Figure 4.12 Poisson process representing the random arrival of photons. (More in Section 4.3.)

4.3   Additive noise

In the bit error rate regime of 10^{-15} to 10^{-9} of a practical transmission system, that is, a Q-factor in the range $6 \leq \gamma \leq 8$, additive noise such as thermal noise of the receiver circuit, shot noise, and optical amplifier noise can be modeled as a Gaussian process.

4.3.1   Shot noise

The photodetector generates an electric current, photo-induced current or photocurrent, which is proportional to the optical power. The origin of the noise is a sequence of randomly generated impulse currents. The arrivals are completely random, and the arrival of photons at the receiver for the time period T is modeled by a random Poisson process, shown on the left in Fig. 4.12, given by

$$P(m) = \frac{(rT)^m \exp(-rT)}{m!}, \tag{4.32}$$

where m is the number of photons, and the arrival rate r is expressed using the optical power P as

$$r = \frac{P}{hf}. \tag{4.33}$$

As a consequence the photo-induced current fluctuates instantaneously, as shown on the right in Fig. 4.12, as it is the summation of impulses of photons as depicted. The mean and its variance are equal and are given by

$$\overline{m} = \sum_{m=0}^{\infty} mP(m) = rT$$

$$\left\langle (m - \overline{m})^2 \right\rangle = rT. \tag{4.34}$$

The average photocurrent I during the one bit time period T_B and the variance σ_{shot} at the receiver due to shot noise are obtained as

$$I = \left(\eta \frac{q}{T_B} \right) \langle m \rangle = \mathbf{R}P$$

$$\sigma_{shot}^2 = \left(\eta \frac{q}{T_B} \right)^2 \langle m - \overline{m} \rangle^2 = 2qIB_e \tag{4.35}$$

where P is the power of the received signal, and q is the charge of electron. $\mathbf{R}$ is the responsivity of the photodetector by

$$\mathbf{R} \equiv \eta \frac{q}{hf} \tag{4.36}$$

where η $(0 \leq \eta \leq 1)$ is the quantum efficiency of the photodetector. The electrical bandwidth of the receiver B_e in baseband units actually ranges from $1/2T_B$ to $1/T_B$, and the minimum value $B_e = 1/2T_B$ is adopted. The optical bandwidth of the signal is determined by the filter in front of the receiver, and the passband is usually set as $B_{opt} = 1/T_B$, in order to prevent signal distortion. Since there is a relation $B_{opt} = 2B_e$, the minimum electrical bandwidth B_e is equal to $1/2T_B$.

4.3.2 Thermal noise

Electric components and circuits after the photodetector, such as the front-end amplifier, generate thermal noise or Johnson noise due to thermal agitation of electrons in a register. The power spectral density is nearly constant throughout the frequency spectrum of interest, and it becomes zero at a temperature of absolute zero. The variance of thermal noise at absolute temperature Υ (in Kelvin), within the bandwidth B_e is given by

$$\sigma_{th}^2 = \frac{4k_B \Upsilon}{R_L} B_e \ [\text{A}^2], \tag{4.37}$$

where k_B is Boltzmann's constant (1.38×10^{-23} J/K), and R_L is the resistance of the load.

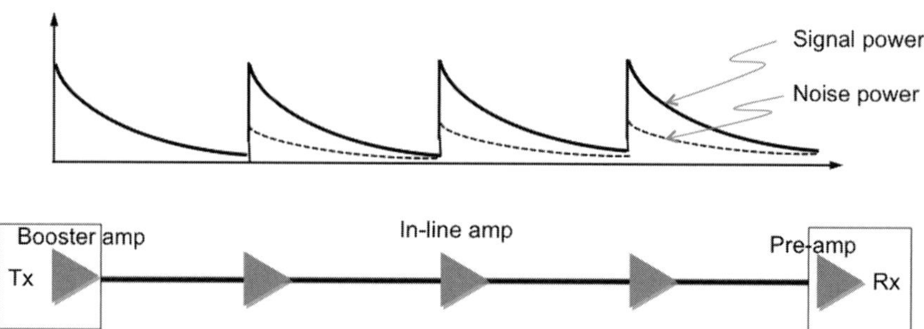

Figure 4.13 Optical amplifiers for a booster amplifier, in-line amplifier, and pre-amplifier.

4.3.3   Optical amplifier noise

Optical amplifiers have a large impact on optical fiber transmission systems. They can compensate for the loss of optical fiber transmission lines, thus reducing the number of electrical repeaters. A major economical advantage is the ability to amplify simultaneously a number of WDM signals. In a long-haul transmission line, shown in Fig. 4.13, the optical amplifiers, indicated by solid triangles, are used as booster amplifiers at the transmitter, in-line amplifiers, and pre-amplifiers at the receiver. The in-line amplifier is called a 1R repeater. Typically, a 1R repeater is inserted every 80–100 km in long-haul transmission systems. An erbium-doped fiber amplifier (EDFA) is most commonly used for the amplifier in the conventional band (C-band) of the spectral region 1530–1565 nm. This is because the transition from the metastable state to the ground state of an EDFA falls into the C-band. More details of the EDFA are given in Section 5.5. Each time the signal passes through the optical amplifier, the amplifier noise is accumulated, resulting in SNR degradation. Therefore, after a few passes of the 1R repeaters, the optical signal is regenerated by an electrical 3R repeater which has the three functions of reshaping, retiming, and regeneration.

It should be remembered that the optical amplifier can maintain the intensity level of the optical signal, but the output signal-to-noise ratio, SNR_{out}, after amplification degrades from SNR_{in} at the input by a factor of the noise figure NF by

$$\text{SNR}_{out} = \frac{\text{SNR}_{in}}{\text{NF}}. \tag{4.38}$$

An EDFA is described as a two-energy-level diagram in Fig. 4.14. More details of EDFA will be described in Section 5.5.1. Assume that the input light is under the coherent state or shot-noise limit, then the input SNR becomes

$$\text{SNR}_{in} = n. \tag{4.39}$$

From Appendix 4.1, the SNR at the output is obtained as

$$\text{SNR}_{out} = \frac{(Gn)^2}{2G(G-1)nn_{sp}} \cong \frac{n}{n_{sp}}, \tag{4.40}$$

where n_{sp} denotes the spontaneous emission coefficient. This derivation is based upon the master equation which models the state transition between the three possible states

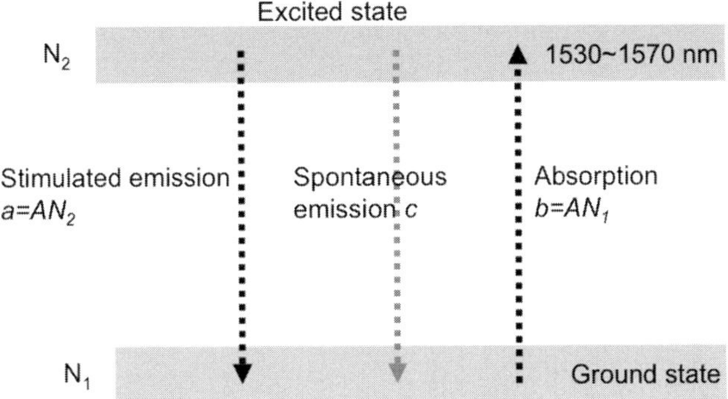

Figure 4.14 Energy level diagram of an optical amplifier.

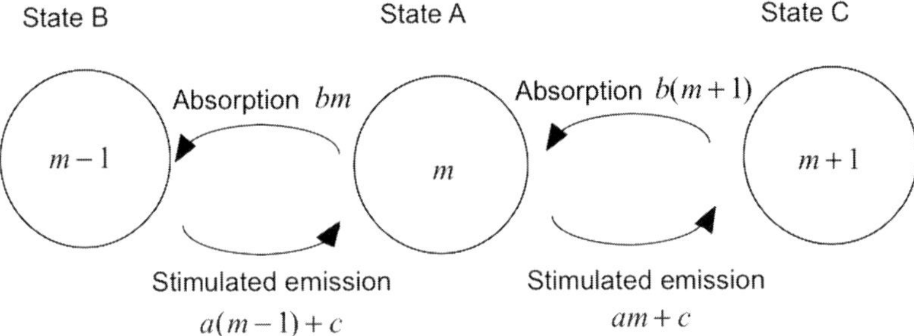

Figure 4.15 State transition diagram of stimulated emission and absorption in an optical amplifier [1]. © Reprinted by permission of Shokodo.

of photons, as shown in Fig. 4.15 [1]. The noise figure of the optical amplifier is obtained as

$$\mathrm{NF} = \frac{SNR_{in}}{SNR_{out}} = 2n_{sp}. \tag{4.41}$$

The spontaneous emission coefficient n_{sp} is unity and more. When the ideal case $n_{sp} = 1$ is considered, the noise figure becomes $\mathrm{NF} = 3\ dB$. The amplified photocurrent and its variance are

$$I = Gn\left(\eta \frac{q}{T_B}\right) = \mathbf{R}GP \tag{4.42}$$

$$\sigma_{amp}^2 = \left(\eta \frac{q}{T_B}\right)^2 2G(G-1)nn_{sp} = 4\mathbf{R}^2 GP(G-1)n_{sp}hfB_e. \tag{4.43}$$

A chain of optical amplifiers will be important in a long-haul transmission line. The overall noise figure of cascaded optical amplifiers is dealt with in Problem 4.5.

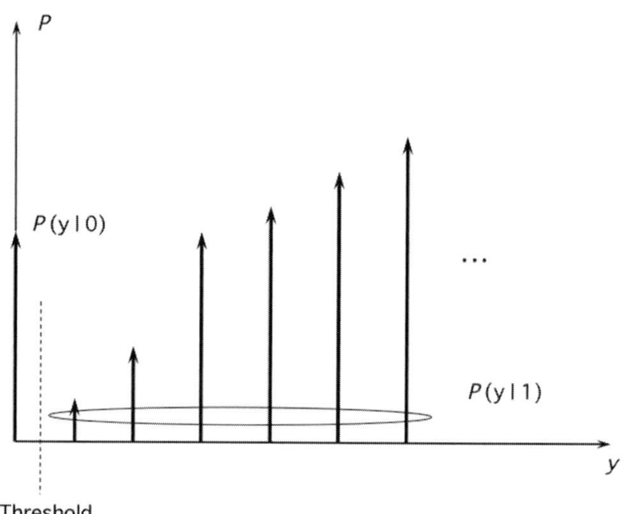

Figure 4.16 Photon-counting receiver [2]. © Reprinted by courtesy of P. E. Green, Jr.

4.4   Quantum limit of OOK direct detection

The receiver sensitivity M is defined by the average number of photons to achieve a given bit error rate as

$$M_{ph} = r T_B = \frac{P}{2hf B_e}. \tag{4.44}$$

The minimum value of M_{ph} in an idealized system is defined to be the quantum limit. We will consider the quantum limit of an OOK and a direct-detection (DD) receiver, so-called intensity modulation direct-detection (IM-DD). Assume that the receiver is a photon-counting receiver, as shown in Fig. 4.16 [2], and it can count up to one without introducing noise. Note that the probability distributions for logical data "0" and "1" are definitely non-Gaussian, so the Q-function does not apply. Assume that the threshold of the photon count is set below unity, the probability of error of this optimum receiver is

$$\text{BER} = \frac{1}{2} \text{Prob} \, [1 \, \text{photon arrived but 0 photon was sent}]$$

$$+ \frac{1}{2} \text{Prob} \, [0 \, \text{photon arrived but 1 photon was sent}] \tag{4.45}$$

In an ideal case with no additive noise in the transmission system, the first term should become zero. Note that in practice this would only be the case with no thermal noise at a temperature of absolute zero Kelvin. From the random Poisson process shown in Fig. 4.12, however, one can see that there is a chance that no photon will arrive even if the mean value of photons that is sent is not zero. Therefore, only the second term on the right-hand side remains. Suppose that the threshold for photon counting is set below unity as shown, the BER is reduced to

$$\text{BER} = 0 + P(0) = \frac{1}{2} \exp(-rT). \tag{4.46}$$

For example, for the targets of BER $= 10^{-12}$ and 10^{-9} the average required numbers of photons per bit "1" are calculated to be $M_{ph} = 27$ and 20. This number represents the so-called quantum limit bounded by the shot noise of the photodetector. When the average power is considered, the probabilities of "mark" and "space" are taken into account and these probabilities are assumed even, the photon count in the quantum limit is reduced to half the value, 14 and 10, respectively for BER $= 10^{-12}$ and 10^{-9}. This ideal quantum limit of OOK IM-DD is a useful marker against which to compare what is possible with real receivers.

4.5 Bit error rate of OOK transmission systems

In general the variances of photocurrents for data "1" and "0" sent from the IM-DD transmitter by taking into account the additive noise are

$$\sigma_1^2 = \sigma_{th}^2 + \sigma_{sh}^2 + \sigma_{amp}^2, \tag{4.47}$$

and

$$\sigma_0^2 = \sigma_{th}^2. \tag{4.48}$$

The variances are summarized as

$$\sigma_{shot}^2 = 2q I B_e, \tag{4.49}$$

$$\sigma_{th}^2 = \frac{4k_B Y}{R_L} B_e \ [A^2], \tag{4.50}$$

and

$$\sigma_{amp}^2 = 4\mathbf{R}^2 G(G-1) P n_{sp} h f B_e. \tag{4.51}$$

Provided that the photocurrent for the bit "0" can be neglected, and $\sigma_1^2 \gg \sigma_0^2$, the signal-to-noise ratio (SNR) becomes equal to the square of the Q-factor γ^2, that is SNR $\cong \gamma_2$. This is only the case when the signal power is large enough to guarantee $\sigma_{sh}^2 \gg \sigma_{th}^2$.

We will consider the sensitivity of an optically pre-amplified receiver, for which the optical amplifier noise is dominant over shot noise and thermal noise $\sigma_{amp}^2 \gg \sigma_{sh}^2, \sigma_{th}^2$. As the signal power is given by Eq. (4.42), the BER is obtained as

$$\mathrm{BER} = Q \left(\sqrt{\frac{(G\mathbf{R}P)^2}{4\mathbf{R}^2 G(G-1) P n_{sp} h f B_e}} \right)$$

$$\leq Q \left(\frac{1}{2} \sqrt{\frac{P}{h f B_e}} \right) = Q \left(\sqrt{\frac{M_{ph}}{2}} \right) \quad (n_{sp} = 1) \tag{4.52}$$

where the receiver sensitivity M_{ph} is defined in Eq. (4.44).

For example, to attain BER $= 10^{-12}$ or 10^{-9}, Q-factors of $\gamma \cong 7$ and $\gamma \cong 6$ are required, respectively, and the receiver sensitivities are $M_{ph} = 98$ and $M_{ph} = 72$.

Without the optical pre-amplifier, when the thermal noise becomes dominant, the Q-factor γ is calculated as

$$I_1 = \mathbf{R}P$$

$$\sigma_0^2 = \sigma_1^2 = \frac{4k_B\Upsilon}{R_L}B_e$$

$$\gamma = \frac{I_1}{\sigma_0 + \sigma_1} = \frac{\mathbf{R}P}{2}\sqrt{\frac{R_L}{4k_B\Upsilon B_e}}. \tag{4.53}$$

The "shot noise limit," at which only the shot noise is dominant over the thermal noise and the optical amplifier noise is

$$\mathrm{BER} = Q\left(\sqrt{\frac{I^2}{2qIB_e}}\right) = Q(\sqrt{M_{ph}}). \tag{4.54}$$

Compared to the receiver sensitivity of the pre-optical amplifier in Eq. (4.52) it gains by 3 dB.

4.6   Coherent detection of OOK transmission systems

The coherent receiver provides raw data on the electric field phase, amplitude and polarization, which the digital processor analyzes using special algorithms. Since the quantum limit will not be achieved owing to the presence of thermal noise and shot noise in IM-DD systems, a technique to improve receiver sensitivity is needed. Since the mid-2000s the challenge has been to increase the transmission capacity further, but there is a strict requirement that the signal bandwidth be squeezed into the 50 GHz frequency interval of the existing WDM system. Therefore, multi-level phase-shift-keyed and/or amplitude-shift-keyed modulation formats can improve the spectrum efficiency. In addition, it has become common practice to utilize polarization division multiplexing (PDM) to double the transmission capacity. Given the powerful digital signal processing (DSP) now available, coherent detection techniques, so-called digital coherent techniques, are being revisited because coherent receivers are capable of measuring phase and polarization as well as amplitude. Actually, application-specific integrated circuits (ASICs) are fabricated to do the processing, and they will vastly simplify the design and operation if they become available for 100 Gb/s and work on a real-time basis. Instead of needing to determine dispersion along a transmission line, then figuring out how to offset the dispersion, installers can plug in a coherent receiver with digital signal processing to clean up the signal. However, the purpose of digital coherent transmission differs from its predecessors of the mid-1980s in that it is intended to equalize not statically but dynamically the impairment of multi-level modulated optical signals, produced by chromatic and polarization dispersions, with the aid of DSP. Coherent detection was known as super-heterodyne detection and was used in radio receivers back in the early

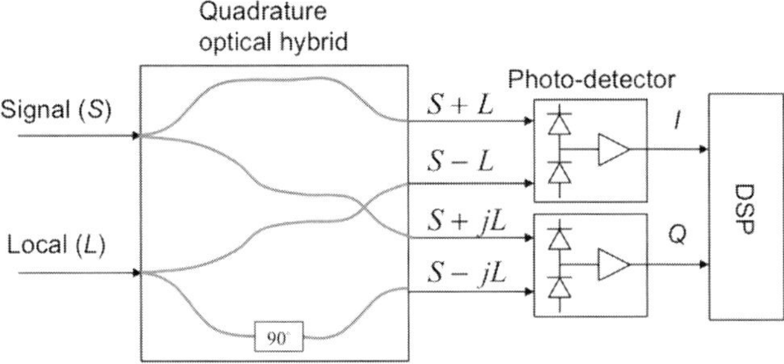

Figure 4.17 Schematic diagram of coherent detection.

1920s.[1] As this detection scheme exploits the nature of monochromatic waves, the linewidth of the optical carrier has to be sufficiently narrow, compared with the spectrum of the signal bandwidth.

The key behind coherent detection is to improve the receiver sensitivity by mixing the incoming optical signal with another local light signal from a local oscillator laser. Figure 4.17 illustrates the scheme of coherent heterodyne detection to generate in-phase (I) and quadrature-phase (Q) intermediate frequency (IF) signals using a balanced photodetector. In the quadrature optical hybrid, a $\pi/2$-phase shifter is inserted into the second arm for the Q-channel. Assume that the polarizations of the two lightwaves are aligned perfectly, the electric fields of the incoming OOK signal and the local oscillator light are written as

$$E(t) = \sqrt{Pd(t)}\exp\{j(\omega_c t + \phi)\},$$
$$E_{LO}(t) = \sqrt{P_{LO}}\exp(j\omega_{LO}t). \tag{4.55}$$

where $\omega_c \geq \omega_{LO}$ is assumed, and

$$d(t) = \begin{cases} 1 & \text{for data ``1''} \\ 0 & \text{for data ``0''.} \end{cases} \tag{4.56}$$

For the I-channel two outputs are detected with the photodetectors after the 3-dB coupler, and the photocurrent is obtained as

$$\begin{aligned}
I_{ph1} &= \Re\left(\frac{1}{\sqrt{2}}\right)^2 |E(t) + E_{LO}(t)|^2 \\
&= \frac{\Re}{2}\left[Pd(t) + P_{LO} + 2\sqrt{PP_{LO}}d(t)\cos\{(\omega_c - \omega_{LO})t + \phi\}\right], \\
I_{ph2} &= \Re\,|E(t) - E_{LO}(t)|^2 \\
&= \frac{\Re}{2}\left[Pd(t) + P_{LO} - 2\sqrt{PP_{LO}}d(t)\cos\{(\omega_c - \omega_{LO})t + \phi\}\right]
\end{aligned} \tag{4.57}$$

The balanced detection can eliminate the dc terms by subtracting I_{ph2} from I_{ph1} and detects only the beat term between the two photocurrents, and thus the average difference

[1] The word "heterodyne" is derived from the Greek roots "hetero" meaning different and "dyne" meaning power. The original heterodyne technique was pioneered by the Canadian inventor Reginald Fessenden in 1900. The super-heterodyne principle was revisited in 1918 by US Army Major Edwin Armstrong in France during World War I.

of the photocurrents of the I-channel is

$$I_I = 2\Re\sqrt{PP_{LO}}d(t)\cos\{(\omega_c - \omega_{LO})t + \phi\}]. \tag{4.58}$$

For the Q-channel the quadrature component is obtained in the same manner as for the I-channel as

$$I_Q = 2\Re\sqrt{PP_{LO}}d(t)\sin\{(\omega_c - \omega_{LO})t + \phi\}]. \tag{4.59}$$

In a homodyne receiver, $\omega_c - \omega_{LO} = 0$, and the baseband I and Q signals are obtained. For an OOK signal and homodyne detection, the average photocurrents for logical data "1" and "0" are obtained as

$$I_1 - I_0 = \begin{cases} 2\Re\sqrt{PP_{LO}} & \text{for data "1"} \\ 0 & \text{for data "0"} \end{cases} \tag{4.60}$$

By making the power P_{LO} of the local oscillator much larger than that of the optical signal, $P_{LO} \gg P$, typically $0\,\text{dB m}$ and $-20\,\text{dB m}$ or less, respectively, P can be neglected compared to P_{LO} and further P and $\sqrt{PP_{LO}}$ in the computation of σ_1. As a consequence, the shot noise becomes dominant for both data "1" and "0," and the amplifier noise and thermal noise can be neglected. Hence, the overall noises are reduced to:

$$\sigma_1^2 = \sigma_0^2 = 2q\Re P_{LO}B_e. \tag{4.61}$$

The Q-factor for the IM-DD transmission system is obtained as

$$\gamma = \frac{I_1 - I_0}{\sigma_1 + \sigma_0} = \frac{2\Re\sqrt{PP_{LO}}}{2\sqrt{2q\Re P_{LO}B_e}} = \sqrt{M_{ph}}. \tag{4.62}$$

This shows that the homodyne detection can realize the "shot noise limit" in Eq. (4.54). Compared with the BER in IM-DD, the receiver sensitivity is increased by 3 dB. To attain $\text{BER} = 10^{-12}$ and 10^{-9}, the required receiver sensitivities are $M_{ph} = 49$ and $M_{ph} = 36$, respectively. It is noteworthy that if $\omega_c - \omega_{LO} \neq 0$ heterodyne detection is applied, and the Q-factor decreases by a factor of $\sqrt{2}$ to $\sqrt{M_{ph}/2}$. This is because the signal photocurrent I decreases by 3 dB.

4.7   Phase-shift-keying modulation transmission systems

4.7.1   Binary phase-shift-keying modulation

We consider again the MZI modulator. For the dc bias voltage $V_b = V_\pi$, shown in Fig. 4.18, the phase shift $\phi(V)$ becomes

$$\phi(V) = \pi + \frac{\pi V_m}{V_\pi}\sin\omega_m t \tag{4.63}$$

$$\begin{bmatrix} E_{out1} \\ E_{out2} \end{bmatrix} = \frac{1}{2}\begin{bmatrix} \exp(j\phi) - 1 \\ j\exp(j\phi) + j \end{bmatrix}\exp(j\omega_c t) \cong \begin{bmatrix} -1 - j\dfrac{\pi V_m}{2V_\pi}\sin\omega_m t \\ \dfrac{\pi V_m}{2V_\pi}\sin\omega_m t \end{bmatrix}\exp(j\omega_c t). \tag{4.64}$$

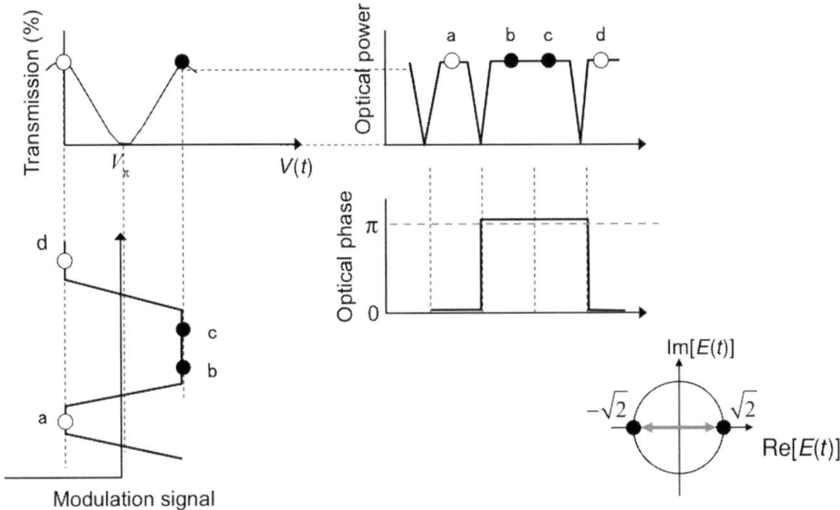

Figure 4.18 Binary PSK (BPSK) modulation where the dc bias voltage is set at $V_b = V_\pi$.

The intensities of the outputs are given by

$$I_{out1} = |E_{out1}|^2 \cong 1 + \left(\frac{\pi V_m}{V_\pi}\right)^2 \sin^2 \omega_m t$$

$$= 1 + \frac{1}{2}\left(\frac{\pi V_m}{V_\pi}\right)^2 (1 - \cos 2\omega_m t) \tag{4.65}$$

$$I_{out2} = |E_{out2}| \cong J_1 \left(\frac{\pi V_m}{V_\pi}\right)^2 \sin^2 \omega_m t$$

$$= \frac{1}{2}\left(\frac{\pi V_m}{V_\pi}\right)^2 (1 - \cos 2\omega_m t). \tag{4.66}$$

The output electric fields swing between the positive points (a, d) and the minus points (b, c) in amplitude, exhibiting a replica of the input modulation signal, and hence this yields the binary PSK modulation. It is interesting that the intensity $|E_{out}|^2$ does not have the ω_m frequency component but only has the $2\omega_m$ frequency component.

4.7.2 M-ary phase-shift-keying modulation

Multi-level modulation formats such as M-ary quadrature amplitude modulation (M-QAM) and M-ASK-PSK could be a solution to increasing the symbol rate without increasing the signal bandwidth. M-QAM and M-ASK-PSK have been widely used in wireless communications. With recent progress in low phase and intensity noise laser diodes, multi-level modulation formats are attracting much attention for optical fiber communications. One of the main motivations is to increase the transmission capacity within the bandwidth constraint. When increasing the bit rate or symbol rate, more spectrum is required to maintain spectral efficiency, so a possible solution would be to adopt

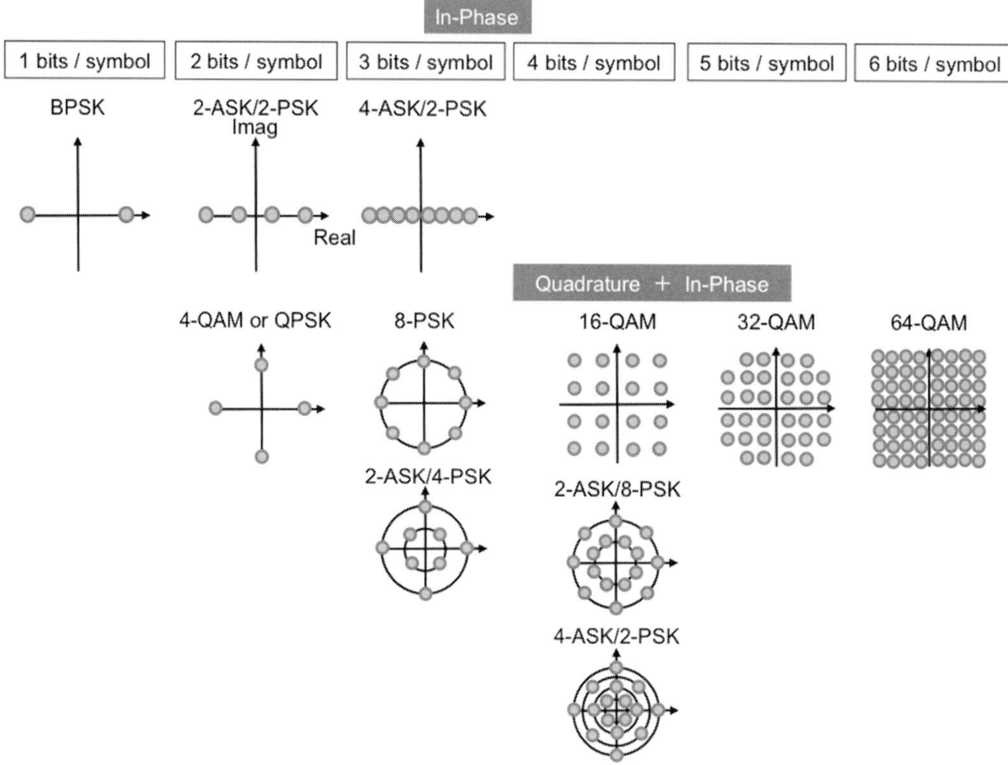

Figure 4.19 Examples of constellation maps where the number of symbols represents the first digit of each format label [5]. © Reprinted by permission of IEEE.

a multi-level modulation format without increasing the symbol rate, thus squeezing the signal bandwidth in a given frequency bandwidth.

Figure 4.19 shows examples of constellations of multi-level signals that utilize both the real part and the imaginary part of the field. Comprehensive tutorials on M-ary modulation formats can be found in the literature [3–5]. The number of symbols represents the first digit of each format label in the cases with PSK and QAM. These constellations generally carry a different number of bits per symbol, depending on the number of symbols M. A constellation can carry a maximum of $\log_2 M$ information bits per symbol. In general, a multi-level signal can be generated using an in-phase "sin"/quadrature (I)-phase "cos" (Q) vector modulator. A dual-drive MZI (DD-MZI) I-Q modulator having two separate electrodes on both arms, as shown in Fig. 4.20(a), serves the purpose. The I-Q modulator is driven by an arbitrary waveform generator comprising a symbol mapper and a lookup table (LUT) for polar coordinate transformation, followed by digital-to-analog converters (DACs). Alternatively, a nested parallel type of DD-MZI modulator without DACs can be used, shown in Fig. 4.20(b). For example, a quad-parallel DD-MZM has been proposed as a 16-QAM modulator, consisting of two parallel DD-MZI modulators with 3 dB amplitude difference, as shown in Fig. 4.20(b).

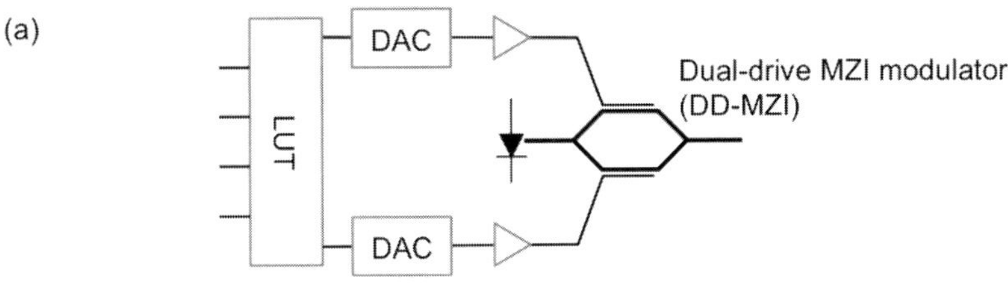

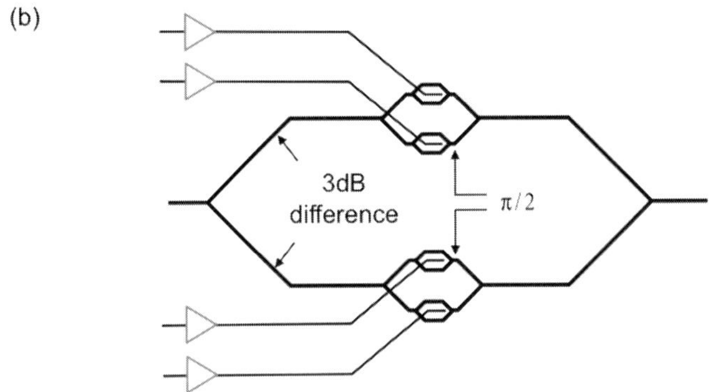

Figure 4.20 (a) Dual-drive MZI (DD-MZI) modulator having two separate electrodes on both arms and (b) quad-parallel DD-MZM proposed as a 16-QAM modulator.

This configuration does not need DACs, but the modulator configuration is more complicated. In the DD-MZI modulator in Fig. 4.20(a), the two optical fields acquire phase shifts ϕ_1 and ϕ_2, controlled by the applied phase modulation voltages $V_1(t)$ and $V_2(t)$ applied to the electrodes on both arms. Finally, the two fields interfere at the outputs, resulting in the desired constellation.

In Fig. 4.20(a) the scattering matrix of the DD-MZI in Eq. (4.18) is modified to

$$[P] = \begin{bmatrix} \exp(j\phi_1) & 0 \\ 0 & \exp(j\phi_2) \end{bmatrix} \tag{4.67}$$

where

$$\phi(V_i) = \pi \frac{V_i(t)}{V_\pi} \quad (i = 1, 2). \tag{4.68}$$

The output from Port 1 is obtained as

$$\begin{bmatrix} E_{out1} \\ E_{out2} \end{bmatrix} = [S] \begin{bmatrix} 1 \\ 0 \end{bmatrix} \exp(j\omega_c t)$$

$$= \frac{1}{2} \begin{bmatrix} j \sin \frac{\Delta\phi}{2} \exp\left(-j\frac{\phi_1 - 3\phi_2}{2}\right) \\ \cos \frac{\Delta\phi}{2} \exp(j\phi) \end{bmatrix} \exp(j\omega_c t) \tag{4.69}$$

$$\Delta\phi = \phi_1 - \phi_2, \quad \phi = \frac{\phi_1 + \phi_2}{2}.$$

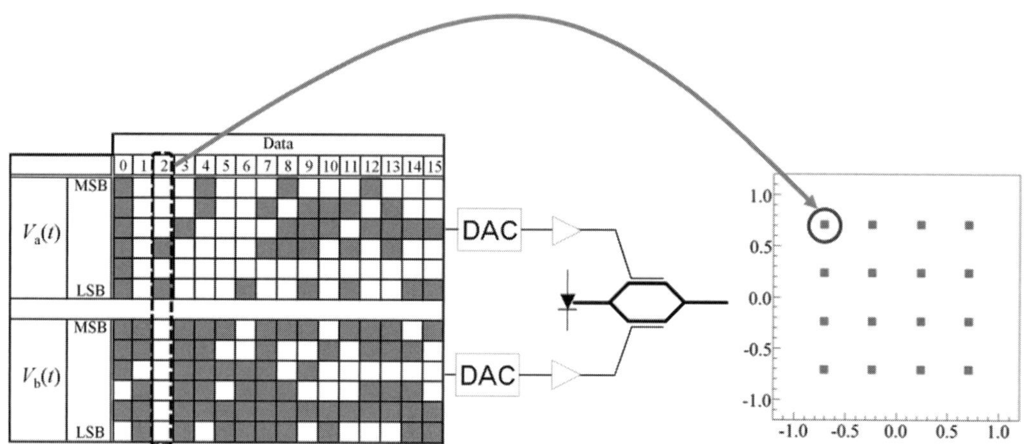

6bit-DAC LUT for each constellation point

Figure 4.21 Example of the LUT bit map for a 16-QAM signal [7]. © Reprinted by permission of the Optical Society of America.

The output (from Port 2) is expressed as

$$|E_{out2}(t)|\exp\{j\phi(t)\} = \frac{1}{2}\cos\frac{\Delta\phi}{2}\exp\left(j\frac{\phi_1 + \phi_2}{2}\right) \tag{4.70}$$

and

$$\frac{V_1(t) - V_2(t)}{2V_\pi}\pi = \cos^{-1}\{|E(t)|\}$$

$$\frac{V_1(t) + V_2(t)}{2V_\pi}\pi = \phi(t). \tag{4.71}$$

Using polar coordinate transformation, the following expressions for the voltages $V_1(t)$ and $V_2(t)$ are determined as [7]:

$$V_1(t) = \frac{V_\pi}{\pi}\left\{\phi(t) + (-1)^n\cos^{-1}|E_{out2}(t)| + 2\pi n\right\}$$

$$V_2(t) = \frac{V_\pi}{\pi}\left\{\phi(t) - (-1)^n\cos^{-1}|E_{out2}(t)|\right\} \quad (n = 0, 1). \tag{4.72}$$

Figure 4.21 shows an example of the LUT bit map for a 16-QAM signal, its constellation map is shown in Fig. 4.22.

4.7.3   Bit error rate of binary phase-shift-keying

The main advantage of using binary PSK (BPSK) instead of OOK comes from an improvement in receiver sensitivity of 3 dB. This can be understood intuitively from the constellation maps in Figs. 4.9 and 4.18, showing that the symbol spacing for BPSK is increased by $\sqrt{2}$. The electric field of a M-ary PSK signal is expressed as

$$E(t) = \sqrt{P}\exp\left\{j\omega_c t + \frac{2\pi}{M}(i - 1)\right\}, \quad i = 1, \ldots, M. \tag{4.73}$$

Table 4.1 Q-factors of intensity and PSK modulations in heterodyne and homodyne detection

Detection	Q-factor γ	Quantum limit for $BER = 10^{-12}$
Direct detection		
IM with photon counter	$0.5\exp(-M_{ph})$	27
IM with optical pre-amplifier	$\sqrt{M_{ph}/2}$	98
Coherent detection		
IM heterodyne	$\sqrt{M_{ph}/2}$	98
IM homodyne	$\sqrt{M_{ph}}$	49
PSK heterodyne	$\sqrt{2M_{ph}}$	25
PSK homodyne	$\sqrt{4M_{ph}}$	13

16-QAM

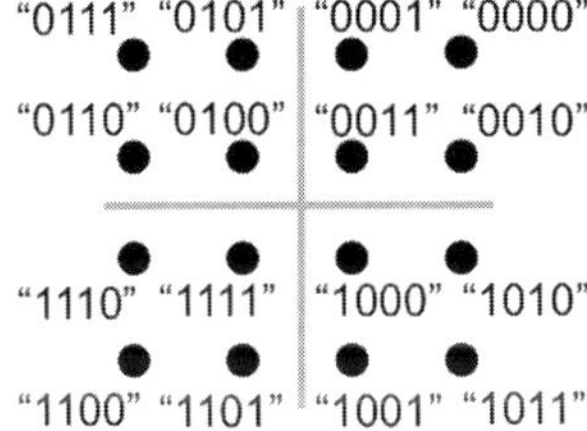

Figure 4.22 Constellation map of a 16-QAM signal.

In order to detect information carried by the phase of the electric field with the square-law detection of the photodetector, a phase reference has to be provided. There are two alternatives: one is coherent detection using a local oscillator, described in Section 4.6, and the other is a differential detection scheme without using a local oscillator, described in this section. When coherent detection is used, for a BPSK signal, $M = 2$, the average photocurrent in homodyne detection is obtained as

$$I_1 - I_{-1} = \begin{cases} 2\Re\sqrt{PP_{LO}} & \text{for data "1"} \\ -2\Re\sqrt{PP_{LO}} & \text{for data "}-1\text{".} \end{cases} \tag{4.74}$$

Therefore, the Q-factor becomes

$$\gamma = \frac{4\Re\sqrt{PP_{LO}}}{2\sqrt{2q\,\Re P_{LO}B_e}} \leq \sqrt{4M_{ph}}. \tag{4.75}$$

As a consequence, receiver sensitivity for the homodyne BPSK signal gains 6 dB, compared to the homodyne OOK signal. The Q-factor of the heterodyne BPSK signal is $\sqrt{2M_{ph}}$. It is worth mentioning that the threshold level for the decision is always fixed at zero for BPSK, while in the OOK receiver the optimum threshold level has to be determined according to the noise probability density distribution function of the logical mark and space. Table 4.1 summarizes the Q-factors of intensity and PSK modulations in heterodyne and homodyne detection.

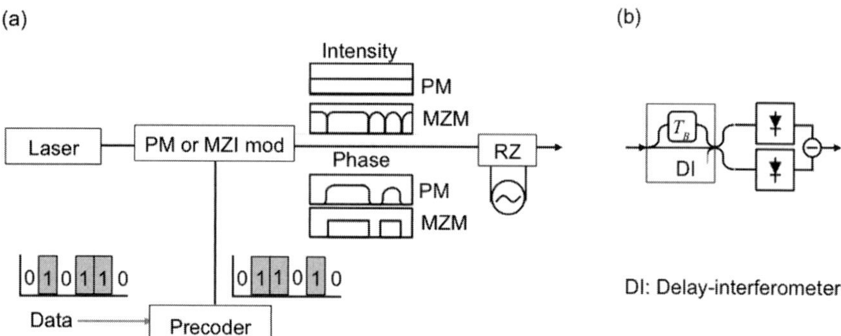

Figure 4.23 Configurations of RZ-DBPSK (a) transmitter and (b) balanced detection at the receiver using a one-bit delay interferometer [5]. © Reprinted by permission of IEEE.

In the case with either the absence of an optical phase reference at the receiver or phase ambiguity in the generation of a coherent local reference, the phase reference has to be provided by the signal itself. In the differential detection scheme each bit acts as a phase reference for another bit, which is at the heart of all differential BPSK (DBPSK) formats. The data sequence D_n is first encoded differentially with C_n at the transmitter, and D'_n is decoded at the receiver by

$$C_n = D_n \oplus C_{n-1}$$
$$D'_n = C_n \oplus C_{n-1}. \tag{4.76}$$

where the initial bit $C_{-1} = 1$, and $\oplus$ denotes the exclusive-OR operation (XOR).

$\{D_n\}$: 10111001010
$\{C_n\}$: 100101110011
$\{D'_n\}$: 10111001010

In Fig. 4.23 the DBPSK transmitter and balanced detection at the receiver using a one-bit delay interferometer are shown. The electric fields forming each output of the balanced photodetector at the receiver are expressed by

$$E(t) = \sqrt{P} \exp[\, j(\omega_c t - \phi(t))]$$
$$E(t - T_B) = \sqrt{P} \exp[\, j\{\omega_c(t - T_B) - \phi(t - T_B)\}]. \tag{4.77}$$

The photocurrents after square-law detection are obtained as

$$I_{ph1} = \frac{\Re}{2} |E(t) + E(t - T_B)|^2$$
$$= \frac{\Re}{2}[2P + 2P \cos\{\omega_c \tau + \phi(t) - \phi(t - T_B)\}]$$
$$I_{ph2} = \frac{\Re}{2} |E(t) - E(t - T_B)|^2$$
$$= \frac{\Re}{2}[2P - 2P \cos\{\omega_c \tau + \phi(t) - \phi(t - T_B)\}], \tag{4.78}$$

Table 4.2 Differential output photocurrent D'_n

$\phi(t)$	$\phi(t - T_B)$	D'_n
0	0	1
0	π	-1
π	0	-1
π	π	1

Finally, subtraction of the photocurrents obtained above yields

$$2\Re P \cos\{\omega_c T_B + \phi(t) - \phi(t - T_B)\}. \tag{4.79}$$

Therefore, the output photocurrent represents the differential output photocurrent D'_n shown in Table 4.2. The phase reference is obtained by inserting the delay-interferometer (DI) by the signal itself before the photodetector in Fig. 4.23. The balanced detection acquires the differential current between the outputs from each photodetector.

4.8 Stimulated Brillouin scattering

The third part of the triple broadband service offered by a PON system, RF video, occupies the transmission window 1550–1560 nm, as shown in Fig. 2.4. The quality of the analog RF signal is gauged by the carrier-to-noise ratio (CNR). The CNR is the difference, in decibels, between the amplitude of the RF signal and the amplitude of noise. In the presence of a number of signals of analog TV channels, non-thermal noise such as composite and inter-modulation noise, will be dominant. The CNR improves as the optical power of the video signal increases. However, the stimulated Brillouin scattering (SBS) caused by the fiber non-linearity imposes a constraint on the signal power. The description of optical fiber transmission systems under the assumption of linearity will be valid as long as the systems are operated at a relatively low bit rate (<2.5 Gb/s) at moderate input power (a few milliwatts) launched into the fibers for a short distance (<20 km). Therefore, fiber non-linearity has been thought to be irrelevant to PONs, but the SBS limit restricts the maximum launch power to maintain the CNR to around 50 dB. The CNR of video signal degrades drastically as the length and the launch power increase. For example, the CNR decreases by nearly 10 dB after 50 km propagation if the launch power exceeds the limit by only 2–3 dBm.

There are two categories of non-linear effects in optical fibers. The origin of the first effect is the power-dependent refractive index of silica, the optical Kerr effect. Self-phase modulation (SPM), cross-phase modulation (XPM), and four-wave mixing (FWM) are in this category. The other category includes stimulated Raman scattering (SRS) and SBS. SRS and SBS are induced by the interaction of lightwaves with molecular vibrations. In quantum mechanics the interaction is described as scattering of a photon. When the scattering is induced in a coherent manner, we use the terminology "stimulated." SRS arises from an optical photon, while an acoustic photon is involved in inducing SBS. There is a fundamental difference between the optical Kerr effect and stimulated scattering. In stimulated scattering, the energy of the lightwave is transferred to the molecular vibrations, and the frequency of the incident lightwave is down-shifted to a longer wavelength. In contrast, no energy transfer occurs in the optical Kerr effect.

In SBS the field of the optical signal acts as the pump and generates an acoustic wave due to electrostriction. The acoustic wave induces a refractive index grating and scatters the lightwave through Bragg diffraction. The scattered light is so-called Stokes light, and its frequency is down-shifted. This process obeys the laws of conservation of both energy and momentum, and hence the frequencies (energy) and the wave vectors (momentum) of the three waves are related as

$$\delta f_B = f_p - f_s$$
$$\mathbf{k}_B = \mathbf{k}_p - \mathbf{k}_s, \tag{4.80}$$

where f_p and f_s are the frequencies, and $\mathbf{k}_p$ and $\mathbf{k}_s$ are the wave vectors of the pump and Stokes waves, respectively. The frequency shift δf_B becomes maximum when the Stokes wave directs backward, and it is given by [7]

$$\mathbf{k}_s = -\mathbf{k}_p, \quad |\mathbf{k}_B| = 2\left|\mathbf{k}_p\right|$$
$$\frac{2\pi\,\delta f_B}{v_a} = 2\frac{2\pi n_0}{\lambda_p}$$
$$\delta f_B = \frac{2n_0 v_a}{\lambda_p} \tag{4.81}$$

where v_a denotes the velocity of the acoustic wave. With the value $v_a = 5940$ [m/s], the frequency shift at the pump wavelength of 1550 nm, the Brillouin shift is calculated to be $\delta f_B = 11.1$ [GHz]. The gain of SBS is dependent on the linewidth Δf_p of the pump according to

$$g_B = \frac{\Delta f_B}{\Delta f_B + \Delta f_p} g_{B0}, \tag{4.82}$$

where Δf_p is the bandwidth of the Brillouin gain. Typically, $g_{B0} = 4.1 \times 10^{-11}$ [m/W], which is a few hundred times larger than the gain of SRS. For a pump of very narrow linewidth, the gain is reduced to $g_B \cong g_{B0}$ while, on the contrary, the gain diminishes for $\Delta f_p \gg \Delta f_B$. Phase modulation on the incident lightwave to reduce the SBS gain is based on Eq. (4.82).

In Fig. 4.24 the output powers from both ends of a single-mode fiber are shown when the pump laser is operated either in a single-frequency or a dual-frequency mode with a mode interval of 20 MHz [9]. With a single-frequency input to the fiber, the non-linear reflection and transmission characteristic of SBS was observed for input powers greater than about 6 mW. However, with the laser operating in the dual-frequency configuration, no deviation from optical linearity in either the forward or backward direction was observed up to 90 mW, the maximum available in this experiment. From Fig. 4.24 the SBS limit restricts the maximum power to about 8 dBm to maintain the CNR around 50 dB. The SBS suppression techniques, including the phase modulation technique, have

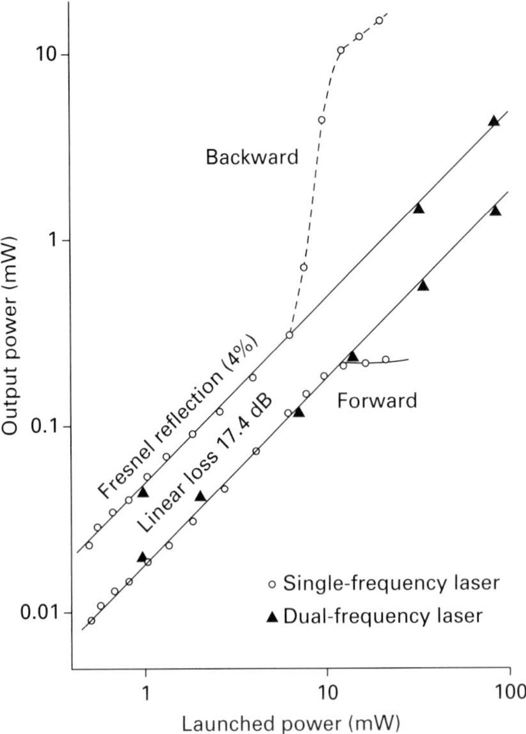

Figure 4.24 Output power from forward and backward ends of a 31.6 km long single-mode fiber pumped by a 1320 nm Nd:YAG laser [9]. © Reprinted by permission of *Journal of Optical Communication.*

increased the launch power to 20 dBm and 16 dBm for transmission distances of 20 km and 50~60 km, respectively, in the EPON system of 32-split ratio.

Appendix 4.1

Assume that the energy of a photon is equal to the energy gap between the ground state and the excited state. The transition probability per unit time a of the emission of photon a is

$$a = AN_2 \tag{A4.1.1}$$

where A is the Einstein coefficient, and N_2 is the number of atoms in the upper state. The transition probability per unit time of the absorption b of a photon is

$$b = AN_1 \tag{A4.1.2}$$

where N_1 is the number of atoms in the ground state. The probability of spontaneous emission c is independent of the number of atoms. Let us consider the transition between three states of photons $m - 1$, m, and $m + 1$ in the optical amplifier. Assuming the probability $P(m)$ of m photons, we consider the simplified transition model in Fig. 4.15, and the master equation is derived as

$$\frac{dP_m}{dt} = -[(a+b)m+c]P_m + [a(m-1)+c]P_{m-1} + b(m+1)P_{m+1}. \quad \text{(A4.1.3)}$$

In the case that the optical bandwidth is set as $B_{opt} = 1/T_B$ as described in Section 4.3.1, the probability of spontaneous emission becomes $c = a$. The master equation is rewritten as

$$\frac{dP_m}{dt} = -[(a+b)m+a]P_m + [a(m-1)+a]P_{m-1} + b(m+1)P_{m+1}. \quad \text{(A4.1.4)}$$

The average number of photons is given by

$$\overline{m} = \sum_{m=o}^{\infty} m P_m, \quad \text{(A4.1.5)}$$

by applying

$$\frac{d\overline{m}}{dt} = \sum_{m=o}^{\infty} m \frac{dP_m}{dt}. \quad \text{(A4.1.6)}$$

The above differential equation is written

$$\frac{d\overline{m}}{dt} = -(a+b)\sum_{m=o}^{\infty} m^2 P_m - a \sum_{m=o}^{\infty} m P_m + a \sum_{m=1}^{\infty} m(m-1)P_{m-1}$$

$$+ a \sum_{m=1}^{\infty} m P_{m-1} + b \sum_{m=o}^{\infty} m(m+1)P_{m+1}$$

$$= (a-b)\overline{m} + a. \quad \text{(A4.1.7)}$$

For the initial condition $\overline{m} = n$ at $t = 0$, the solution obtained is

$$\overline{m} = Gn + (G-1)n_{sp} \quad \text{(A4.1.8)}$$

where the gain G and the spontaneous emission coefficient n_{sp} are expressed as

$$G = \exp[(a-b)t]$$
$$n_{sp} = \frac{a}{a-b}. \quad \text{(A4.1.9)}$$

Note that n_{sp} becomes unity when the ground state becomes empty, while when a large portion of atoms are unexcited and remain in the ground state, n_{sp} goes to infinity. The first term shows the input n photons are amplified by a factor of G. The second term is the so-called amplified spontaneous emission (ASE) noise in which the spontaneous emission is stimulated and amplified by a factor of $(G-1)$.

To evaluate the noise the variance of m, $\langle (m-\overline{m})^2 \rangle$ has to be derived. From the differential equation for $\overline{m^2}$ is derived from Eq. (A3.4.3) as

$$\frac{d\overline{m^2}}{dt} = 2(a-b)\overline{m^2} + (a+b+2c)\overline{m} + c \quad \text{(A4.1.10)}$$

The variance of the case with $c = a$ is obtained as

$$\langle (m - \overline{m})^2 \rangle = \langle m^2 \rangle - \overline{m}^2$$
$$= Gn + (G-1)n_{sp} + 2G(G-1)nn_{sp} + (G-1)^2 n_{sp}^2 \quad \text{(A4.1.11)}$$

The optical power is proportional to the square of the electric field. Thus the noise field beats against the signal and against itself, giving rise to the third and fourth terms referred to as the amplified signal-to-spontaneous beat noise and amplified spontaneous-to-spontaneous beat noise, respectively. The first and second terms are regarded, respectively, as the amplified signal-to-vacuum field beat noise and amplified spontaneous-to-vacuum field beat noise. Since the gain is sufficiently large, $G \gg 1$, and the signal power is much larger than the spontaneous emission power, $n \gg n_{sp}$, only the third term in the variance is significant:

$$\langle (m - \overline{m})^2 \rangle \cong 2G(G-1)nn_{sp}. \quad \text{(A4.1.12)}$$

Problems

4.1 Poisson random process. Given the Poisson distribution in Eq. (4.32), calculate the mean and the variance.

4.2 Decision threshold. Derive the threshold value D_{th} in Eq. (4.6). For the case of high SNR, it can be assumed that

$$(I_1 - I_0)^2 \gg \frac{2(\sigma_1^2 - \sigma_0^2)\ln(\sigma_1/\sigma_0)}{\sigma_0^2 \sigma_1^2}.$$

4.3 Power penalty. Let P_1 denote the optical power received during a logical data 1 bit, and P_0 the power received during a logical data 0 bit without any system impairment. The BER is given by

$$\text{BER} = Q\left(\frac{\mathbf{R}(P_1 - P_0)}{\sigma_1 + \sigma_0}\right).$$

Assume that

$$P_0 = 0, \ \sigma_0 = 0, \ \text{BER} = Q\left(\frac{\mathbf{R}P_1}{\sigma_1}\right).$$

After the transmission the noise σ_1 increases by a factor of 2: $\text{BER} = Q(\mathbf{R}P_1/\sigma_1) = Q(\mathbf{R}P_1'/2\sigma_1)$.
Calculate the power penalty defined by $P_1' - P_1$ for $\text{BER} = 10^{-9}$.

4.4 LiNbO$_3$ MZI modulator. Calculate the half-wavelength voltage V_π along the z-axis for z-polarized light propagating along the y-axis, in order to obtain the phase-shift of π at $\lambda = 1550$ nm. Use the values $n_e = 2.20$, $r_{33} = 30.8 \times 10^{-12}$ [m/V].

4.5 Cascaded optical amplifiers.

(1) Derive the noise figure when the loss element is placed in front of the amplifier with gain G and noise figure F. The loss is given by ε $(0 < \varepsilon \leq 1)$ where $\varepsilon = 0$ implies no loss and $\varepsilon = 1$ implies 100% loss.
(2) Derive the noise figure when the loss element is placed after the amplifier.
(3) Derive the noise figure for the optical amplifier chain with two amplifiers, with gains G_1 and G_2 and noise figures F_1 and F_2.
(4) Derive the noise figure for the optical amplifier chain with two amplifiers, with gains G_1 and G_2 and noise figures F_1 and F_2 , where a loss element with loss ε is introduced between the first and second amplifiers.

4.6 Quantum limit using ideal receiver (photon counter) for IM-DD.
(1) How many photons are required per one bit at the bit rate of 10 Gb/s to achieve the bit error rate (BER) of 10^{-12}? Assume that the threshold of the receiver is set below unity photon count.
(2) Calculate the receiver sensitivity [dBm] at the bit rate of 10 Gb/s at the wavelength of 1550 nm,

$$k_B = 1.38 \times 10^{-23}\,[\text{J/K}], \ T = 300\,\text{K}, \ R_L = 50\,[\Omega], \ B_e = 5\,[\text{GHz}]$$

5   Enabling techniques

5.1   Star coupler

An optical splitter is a key optical device in PON systems, which splits the optical signal power evenly into all the output ports. In the PON field plant shown in Fig. 1.30, a $1 \times 8 \sim 1 \times 32$ optical splitter is placed on an electric pole, connecting the distribution optical cable in the air and the drop wire to the customer premises. A $1 \times N$ optical splitter can be part of an $N \times N$ optical star coupler. For example, a 16×16 star coupler with four-stage topology is illustrated in Fig. 5.1 [1], and the dotted line denotes a 1×16 optical splitter. The star coupler can be constructed by cascading 3-dB couplers in the perfect shuffle topology. The 3-dB coupler has two input and two output ports, and it splits the input power 50:50 to the output ports. The transfer matrix of a 3-dB coupler is given by Eq. (3.63) in Section 3.5.1. The number of 3-dB couplers required for the case with k-stage $1 \times N$ coupler is given by

$$N_{3dB\ coupler} = 2^k - 1, \quad k = \log_2 N \tag{5.1}$$

and the splitting loss per output port is given by

$$\text{Splitting loss} = 3k \ [\text{dB}]. \tag{5.2}$$

For a 1×16 optical splitter, $k = 4$, the number of 3-dB couplers required is 32, and the splitting loss per output port is 12 dB.

There are two types of these devices: fiber and silica planar lightwave circuit (PLC). The fused biconic taper fiber 3-dB coupler shown in Fig. 5.2(a) is fabricated from two separate fibers by fusing the coupling region together. The tapered section on both sides of the coupling region is long enough that incident power from either of the left-hand ports couples to the fibers on the right-hand ports with light reflection to the other left-hand port. Star couplers with up to 32 ports have been possible using fused tapered fiber 3-dB couplers [2]. Advantages are the low-loss easy coupling with the optical fiber transmission line and no polarization-dependent loss. In Fig. 5.2(b) the silica-based PLC star coupler is shown [3]. This optical splitter integrated with the coupler has been developed for testing the optical distribution network (ODN). For testing, the OTDR is used at the wavelength of 1650 nm as shown in Fig. 2.4. The splitter has reflection-type couplers at the splitter outputs, made using multilayered dielectric filters. It is compact but fibers have to be attached to both ends of the input and output ports. There is not much difference in the loss characteristics between the two types of coupler.

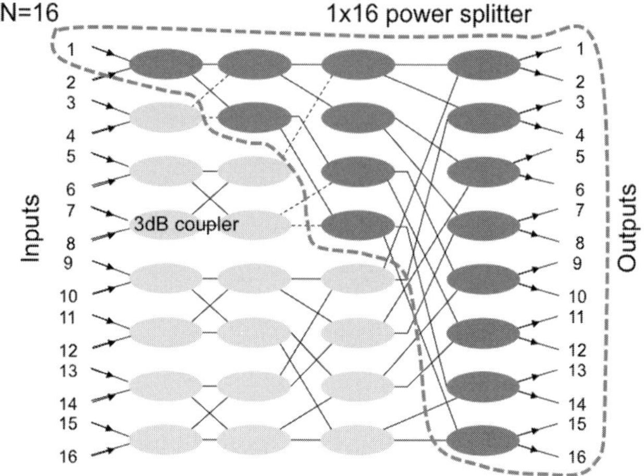

Figure 5.1 16×16 star coupler. The dotted line indicates a 1×16 optical splitter [1]. © Reprinted by courtesy of P. E. Green, Jr. and modified.

The insertion loss of the commercially available 1×16 star coupler, for example, is about 13~14 dB, including excess loss of 1~2 dB in both fiber type and PLC coupler. The polarization dependent loss is as small as 0.3 dB. Consider a passive double star configuration configured with 1×4 optical splitter in the central office and 1×8 optical splitter in the outside plant. As described in Section 10.3.2, for example, in order to test in-service fiber cables in the ODN, optical couplers have to be inserted between the 1×8 splitter and the ONUs, and the output at the wavelength of 1650 nm from the optical time-domain reflectometer (OTDR) placed at the OLT is launched in the coupler. Note that this OTDR signal has to be cut off in front of the ONU and the OLT to pass only signals at 1310 nm and 1550 nm. Compared with the separate device configuration, the integrated device will provide ease of handling because the input ends of the couplers and the splitter are on opposite sides.

5.2   Broadband light source

5.2.1   Mode-locked laser diode

A short-pulse laser is one of the key devices used in the transmitter of OCDMA systems, regardless of whether the system uses time-spread or spectral en/decoding. In time-spread OCDMA, an optical code sequence consists of a number of chips, and each chip itself has to be a short pulse for high bit rate systems. Spectral encoding needs an ultra-wide spectrum of short optical pulse. The mode-locked laser diode is often used as the light source due to its handling ease, compactness, high-speed operation as well as its availability in the wavelength region of 1550 nm. Let us consider the resonance condition of the Fabry–Pérot cavity shown in Fig. 5.3. When the light is incident from both sides of the half mirror, the relation between reflectance and transmittance is

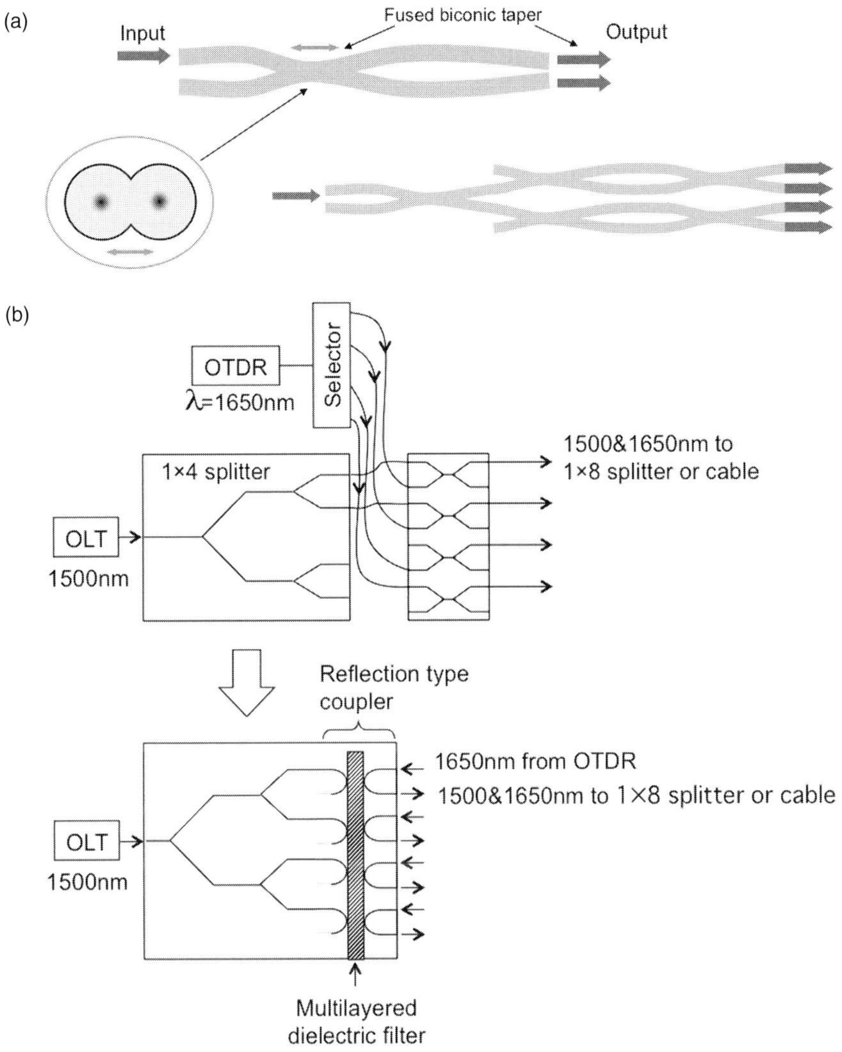

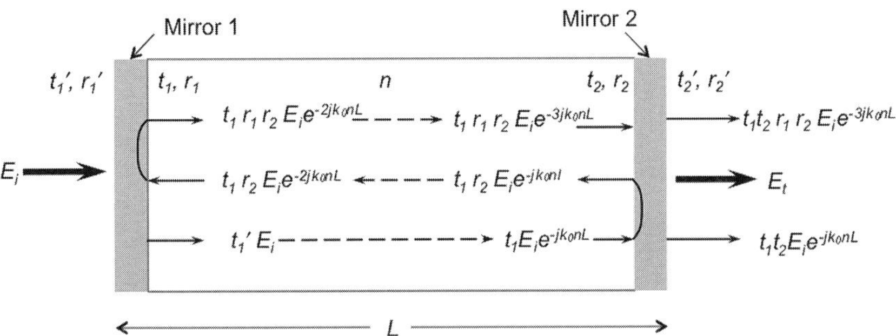

Figure 5.2 (a) Fused biconic tapered fiber 3-dB coupler and (b) planar lightwave circuit (PLC) based star coupler with in-service OTDR testset. Modified after [3].

Figure 5.3 Fabry–Pérot cavity.

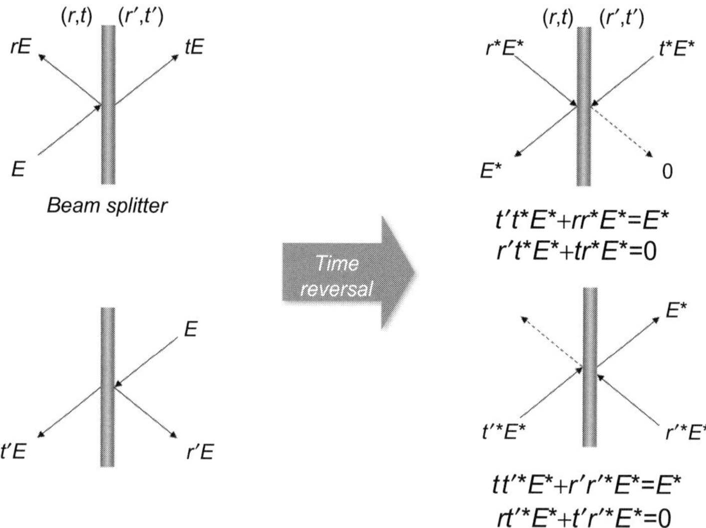

Figure 5.4 Stokes' relationship.

derived from Stokes' theorem. Imagine that in Fig. 5.4 the time is reversed so that r^*E^* and t^*E^* retrace their paths and recombine to the outgoing field E^* at the same half mirror. Provided that the properties of the dielectric half mirror are invariant under the time reversal operation, the following relations are obtained,

$$t^*t' + r^*r = 1$$
$$t^*r' + r^*t = 0 \tag{5.3}$$

where r and t are the reflectance and transmittance in amplitude. In general, they take complex values, and for the lossless case the reflectance and transmittance become real. Then, Eq. (5.3) is reduced to

$$tt' + rr = 1$$
$$tr' + rt = 0. \tag{5.4}$$

Therefore,

$$r' = -r. \tag{5.5}$$

From the energy conservation law,

$$|E|^2 = |rE|^2 + |tE|^2$$
$$t^*t + r^*r = 1. \tag{5.6}$$

For the lossless case,

$$t^2 + r^2 = 1. \tag{5.7}$$

This leads to

$$tt' = t^2 \quad \text{and then } t = t'. \tag{5.8}$$

With these relations $r' = -r$, $t = t'$, the electric field of the incident lightwave is

$$E_i(z, t) = \sqrt{P_0} \exp\left[j\left(\omega_0 t - k'z\right)\right] \tag{5.9}$$

where k' is the complex wave number of the medium having refractive index n, gain g, and loss α,

$$k' = nk_0 + j\frac{1}{2}(g - \alpha). \tag{5.10}$$

The transmitted electric field is

$$\begin{aligned}
E_t(z, t) &= t_1 t_2 e^{-jnkL}\left(1 + r_1 r_2 e^{-j2nkL} + r_1 r_2 e^{-j4nkL} + \cdots\right) E_i(z, t)\\
&= \frac{t_1 t_2 e^{-jnkL}}{1 - r_1 r_2 e^{-j2nkL}} E_i(z, t).
\end{aligned} \tag{5.11}$$

Assume that $t_1 = t_2 = t$, $r_1 = r_2 = r$,

$$\frac{E_t}{E_i} = \frac{t^2 e^{-jnkL}}{1 - r^2 e^{-j2nkL}}. \tag{5.12}$$

The power transmission coefficient is obtained as

$$\begin{aligned}
T &= \frac{1}{1 + \frac{4R}{(1-R)^2}\sin^2\left(k_0 n L\right)}\\
&= \frac{1}{1 + 4(F/\pi)^2 \sin^2\left(k_0 n L\right)}
\end{aligned} \tag{5.13}$$

where the finesse F is defined by

$$F \equiv \frac{\pi\sqrt{R}}{1 - R}. \tag{5.14}$$

Here, it is assumed that the half mirrors on both sides of the cavity are identical. Lasing will occur if the following condition is satisfied:

$$1 - r^2 e^{-j2nkL} = 0. \tag{5.15}$$

From the lasing condition, the real and imaginary parts must satisfy

$$\begin{aligned}
r^2 e^{(g-\alpha)L} &= 1\\
e^{-j2nk_0 L} &= 1.
\end{aligned} \tag{5.16}$$

From the first equation, the relation between the gain and the loss has to be $g > \alpha$, and from the second equation the phase-matching condition imposes

$$\begin{aligned}
2nk_0 L &= 2\pi m \; (m = 1,\, 2,\, 3,\, \ldots),\\
\Delta\omega \equiv \omega_{m+1} - \omega_m &= \frac{\pi c}{nL}.
\end{aligned} \tag{5.17}$$

The total electric field of the longitudinal modes[1] within the gain bandwidth which can oscillate at some arbitrary position within the cavity is written

$$E_m(t) = \sum_m \sqrt{P_{0m}}\,\exp[\,j\{(\omega_0 + m\,\Delta\omega)t + \phi_m\}]. \qquad (5.18)$$

There are two approaches to obtain a stable output without fluctuations in the amplitude and phase [4].

- One approach is single-mode oscillation in that only one longitudinal mode is selectively amplified. For example, a short cavity makes the frequency interval $\Delta f(=\Delta\omega/2\pi)$ large so that all the other modes are outside the gain bandwidth.
- The other approach is mode-locking in that the phases of all the longitudinal modes are forced to lock each other.

When $\sqrt{P_{0m}} = 1$, $\phi_m = 0$ for all the modes, for simplicity, the total electric field is expressed by the superposition of all the forward and backward propagating modes

$$\begin{aligned}
E_m(t) &= \sum_{m=-(N-1)/2}^{(N-1)/2} \exp[\,j(\omega_0 + m\,\Delta\omega)t] \\
&= e^{j\omega_0 t}\,\frac{\sin(N\Delta\omega t/2)}{\sin(\Delta\omega t/2)}
\end{aligned} \qquad (5.19)$$

where N denotes the number of modes locked.

The average output laser power $|E_m(t)|^2$ is given by

$$P_{out}(t) \propto \frac{\sin^2\left(\frac{N\Delta\omega}{2}t\right)}{\sin^2\left(\frac{\Delta\omega}{2}t\right)}. \qquad (5.20)$$

The individual pulse width, defined as the time from the peak to the first zero point, is given by the inverse of the gain bandwidth ω_{gain}

$$\tau = \frac{2\pi}{N\Delta\omega} \cong \frac{2\pi}{\omega_{gain}}. \qquad (5.21)$$

The above equation suggests that the larger the gain bandwidth becomes, the shorter the pulse width becomes. The output power for the case with $N=5$ is illustrated in Fig. 5.5.

There are two techniques of mode-locking. Active mode-locking refers to the case where the modulator is externally driven while, on the other hand, passive mode-locking refers to the case where the pulse forms its own modulation through non-linearities. Mode-locking is achieved using an external driven intra-cavity loss modulator. The modulator is driven at a frequency Δf corresponding to the longitudinal mode spacing. In the time domain the modulator acts as a gate which opens once per round trip transit time. The initial laser radiation becomes shorter each time it experiences the round trip in the cavity. Eventually a short pulse is formed with the frequency components synchronized with the timing of the gate opening.

---

[1] The longitudinal mode is the oscillation mode having the designated frequency of the laser. It contrasts with the transverse mode which has a designated electric field distribution in the cross-section vertical to the cavity direction.

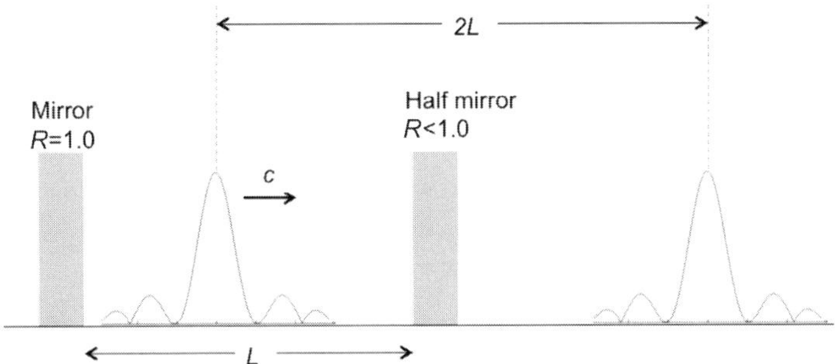

Figure 5.5 Output power for the case with $N = 5$ [4]. © Reprinted by courtesy of A. Yariv.

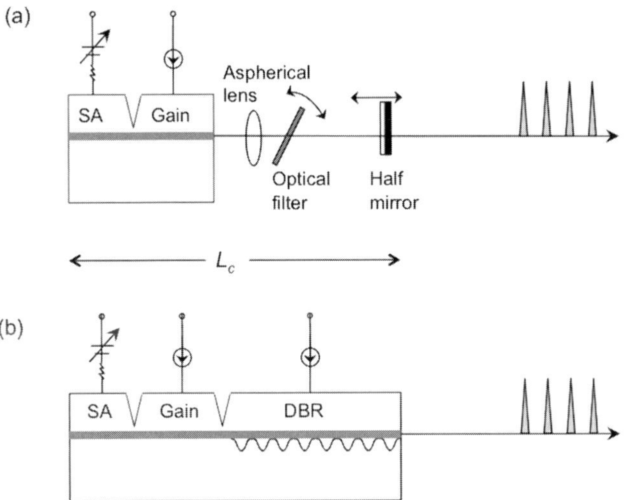

Figure 5.6 Schematics of (a) an external-cavity MLLD (EC-MLLD) and (b) a monolithically integrated MLLD.

In passive mode-locking the externally driven modulator is replaced by a non-linear optical element such as a saturable absorber whose loss decreases with increasing laser pulse intensity. Figure 5.6 shows schematics of (a) an external-cavity MLLD (EC-MLLD) and (b) a monolithically integrated MLLD. In the EC-MLLD, the device chip is a two-section laser diode having an active layer consisting of a forward-biasing gain section and a reverse-biasing saturable absorber section. One of the facets on the saturable absorber is cleaved, and one of the facets on the gain section is anti-reflection coated. A partial transmission mirror with a reflectivity of 0.8 is placed as an external cavity reflector. The repetition frequency is determined by the round trip time between the cleaved facets of the saturable absorber and the external half-mirror. An ultra-short pulse circulating in the laser will modulate the intracavity loss, which in turn will modulate the pulse. The optical filter used as the wavelength selector allows tuning of the wavelength. The monolithically integrated MLLD consists of three sections of the saturable absorber, the gain, and the distributed Bragg feedback reflector (DBR), as

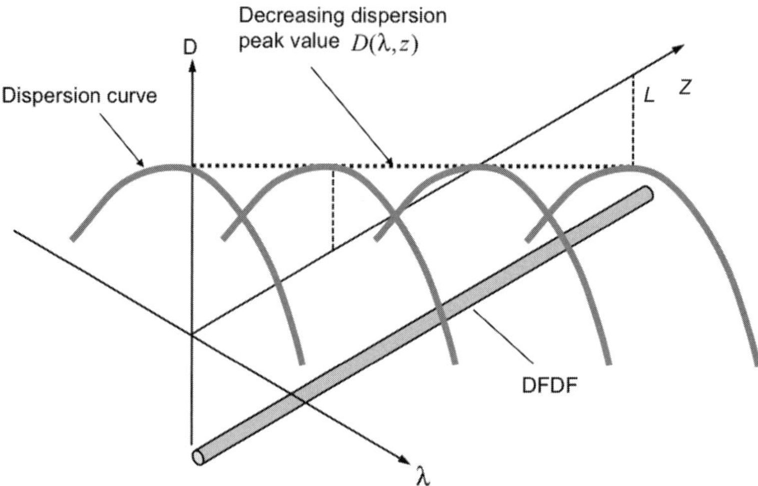

Figure 5.7 Dispersion curve along the z-axis of dispersion-flattened and decreasing fiber (DFDF). Modified after [6].

shown in Fig. 5.6(b). The cavity is formed between the cleaved facet of the saturable absorber and the DBR. An optical modulator such as an electro-absorption modulator (EAM) can be integrated in between the gain section and DBR.

5.2.2 Supercontinuum light source

A supercontinuum (SC) emission having an ultra-broadband spectrum is generated with a high-intensity pump light via non-linearity in a special class of optical fiber [5]. When a mode-locked laser is used as the pump light source, replicas of a short pulse are generated over a few tens of terahertz. Such an SC can be a versatile light source because by wavelength-demultiplexing the emission spectrum, each longitudinal mode serves as thousands of monochromatic continuous-wave light sources for DWDM. 1000 channel DWDM transmission with an interval of 12.5 GHz was demonstrated using the SC pumped by the MLLD with repetition frequency of 12.5 GHz.

For application to a broadband light source, a flat top profile of the emission spectrum is desirable. An exotic phenomenon is the interplay between the dispersion and the optical Kerr effect of "dispersion-flattened and decreasing" fiber (DFDF) with chromatic dispersion D that decreases from a positive to a negative value along the z-axis as shown in Fig. 5.7 [6]. The dispersion of DFDF is expressed as

$$D(\lambda, z) = D_0 \left(1 - \frac{z}{L_0}\right) + \frac{D_2}{2}(\lambda - \lambda_{peak})^2 \quad (D_0 > 0). \tag{5.22}$$

Figure 5.8 shows the evolution of the intensity spectrum of an SC pulse during propagation through the DFDF where the parameters are $D_0 = 6$ ps/nm/km, $D_2 = -0.0002$ ps/nm^3/km, and $L_0 = 600$ m [7]. The pump pulse has a $\sec h^2$ shape in the intensity waveform and is a Fourier transform limited pulse, which is described in Section 3.7.

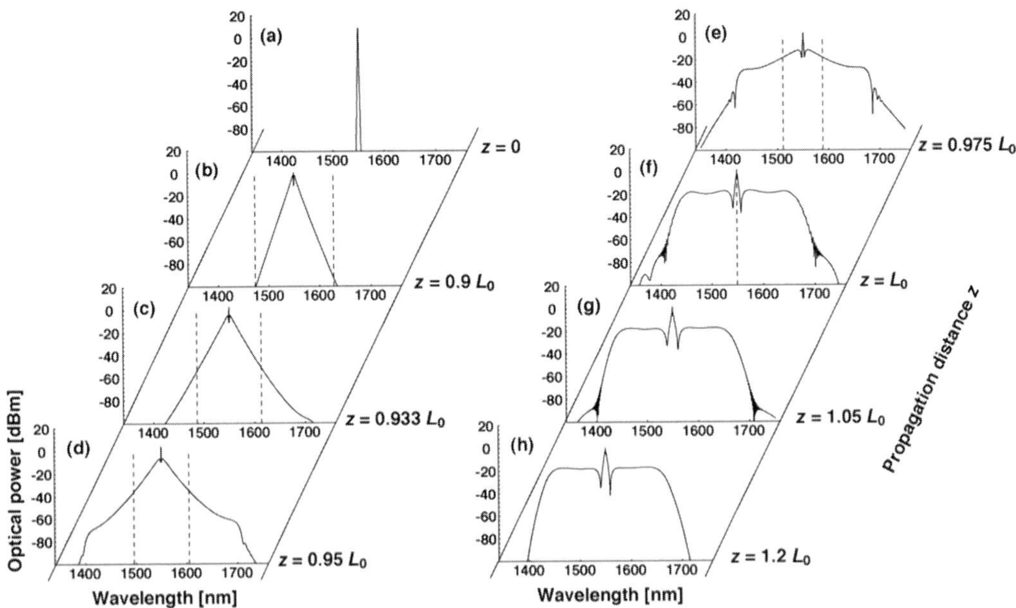

Figure 5.8 Evolution of the intensity spectrum of an SC pulse during propagation through a DFDF [7]. © Reprinted by permission of the Optical Society of America.

The effective peak power of the pump pulse, which is defined as the product of peak power P_0 and non-linear coefficient γ, is $\gamma \cdot P_0 = 5.84$ km^{-1}. The full width at half-maximum (FWHM) of the pump pulse is $T_{FWHM} = 4$ ps. The pump wavelength λ_0 is set equal to the peak wavelength $\lambda_{peak} = 1550$ nm. The spectrum of the pump pulse in Fig. 5.8(a) develops as it propagates. Initially, the spectrum broadens with z owing to a pulse compression process that is similar to adiabatic soliton compression. The pulse is compressed to approximately 50 fs at a propagation distance $z = 0.975L_0$. After $z = 0.975L_0$, the spectrum begins to become rectangular and flat. The spectrum continues to change even after $z = L_0$, where the chromatic dispersion becomes normal for all wavelengths, because the waveform of the propagating field is still intense enough to produce non-linearity. At $z = 1.05L_0$, the spectrum eventually achieves good top flatness and thereafter ceases to change its shape.

5.3   Optical time-gating

Optical time-gating can be constructed by a non-linear optical loop mirror (NOLM), as shown in Fig. 5.9. OCDM signals after decoding with target decoder include auto-correlation of the desired optical code and cross-correlations of other interfering codes. The time gate extracts only a desired auto-correlation main lobe, thereby rejecting the sidelobes of the auto-correlation and the other interfering codes that fall outside the gate interval. This operation is equivalent to bandpass filtering in the frequency domain

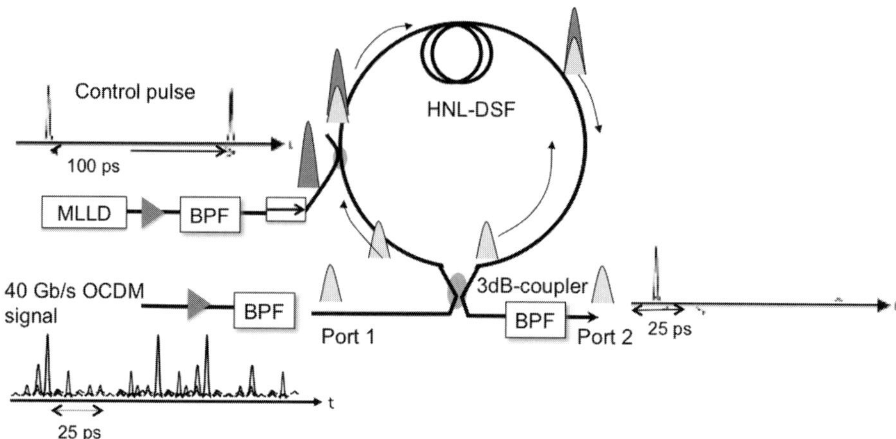

Figure 5.9 Configuration of optical time-gating using a non-linear loop mirror (NOLM) [8]. © Reprinted by permission of IEEE.

of wireless CDMA. The analogy between wireless CDMA with bandpass filtering and OCDMA employing time-gating is illustrated in Fig. 6.3. The NOLM uses a hundred to a few hundreds meter long highly non-linear dispersion-shifted fiber (HNL-DSF). It has two input ports for the OCDM signal and control pulse. The input OCDM signal is launched and split into clockwise and counterclockwise signals by passing the 3-dB coupler, and both the signals pass through the 3-dB coupler once again. Only when a high-intensity control pulse is launched from the arm on Port 1, does the output appear from Port 2. Otherwise, the input signal is reflected back to Port 1. This is due to the cross-phase modulation (XPM) induced by the control pulse. Without the control pulse, no non-linear effect occurs, and the configuration is reduced to a fiber loop mirror. It is referred to as a "mirror" because the input signal goes back to the same input port as if it were reflected back to the original port. The operation principle will be described later in this section.

Let us look at the experimental results of 40 Gb/s 40-channel WDM, 4-code OCDM signals [8]. Four-channel 40 Gb/s signals are encoded with four different 3-chip QPSK pulse code sequences with chip interval of 5 ps and multiplexed. The code extends 15 ps and fits in the 25 ps time slot at 40 Gb/s. The control pulse was a 1.5 ps width pulse train from the MLLD at a repetition rate of 10 GHz at 1562.5 nm, and the average power of the control pulse was around 10 dB m. The 40 Gb/s signal is time-gated every four times from Port 2 in order to demultiplex to 10 Gb/s data. Figures 5.10(a) and (b) show the measured eye diagrams after optical decoding of WDM channel 1 at 1533.0 nm and channel 39 at 1563.1 nm. (The multiple interference noise of the other three unmatched codes severely distorts signal-to-noise ratio as shown in the measured eye diagrams without time-gating. By optical time-gating the main lobe of the matched auto-correlation waveform, interference noise outside the time-gate window is rejected.) The wavelengths of the signal and the control pulses have to be carefully chosen so that the walkoff between the signal and the control pulses is minimized. As the zero dispersion wavelength is 1549.8 nm, the wavelengths of the control pulses are set at 1562.5 nm for

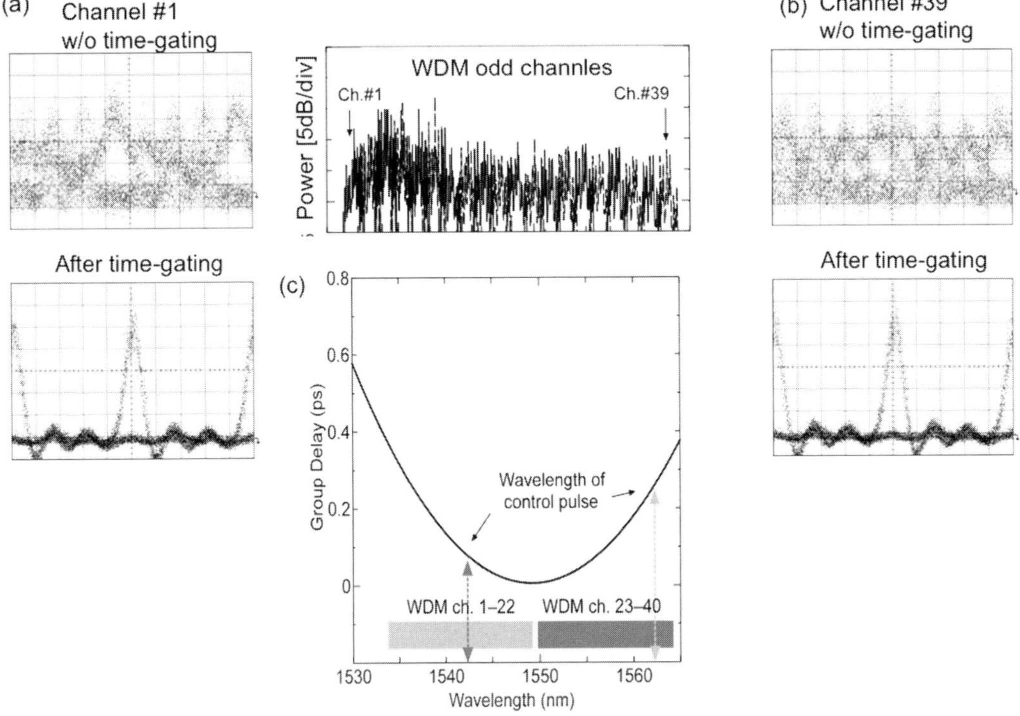

Figure 5.10 Measured eye diagrams after optical decoding of (a) WDM channel 1 at 1533 nm and (b) channel 39 at 1563.1 nm [8]. © Reprinted by permission of IEEE.

gating of WDM channel 1 to channel 22 (1533–1549 nm) and at 1542.2 nm for gating of WDM channel 23 to channel 40 (1550–1564 nm), respectively. The walkoff between signals at 1533.0 nm and 1563.1 nm and the control pulses is less than 300 fs, and this is much smaller than the pulse widths of control and signal pulses as shown in the inset (c).

We will derive the transfer matrix of the NOLM. From Section 4.2.2, the overall transfer matrix $[S]_{NOLM}$ is given by

$$[S]_{NOLM} = [S_2]_{3dB}[S]_{NL}[S_1]_{3dB}. \tag{5.23}$$

Note that the OCDM signal passes the 3-dB coupler twice, which is expressed as the cascaded 3-dB coupler in cross-coupled fashion. Therefore, $[S_2]_{3dB}$ becomes the transpose of $[S_1]_{3dB}$, that is $[S_2]_{3dB} = [S_1]_{3dB}^T$. The transfer matrix of the 3-dB coupler is given by Eq. (3.63) as

$$[S_1]_{3dB} = [S_2]_{3dB} = \frac{1}{\sqrt{2}} \begin{bmatrix} 1 & j \\ j & 1 \end{bmatrix} \tag{5.24}$$

and the phase shift of the modulator section is

$$[S]_{MZI} = \begin{bmatrix} \exp(j\phi) & 0 \\ 0 & 1 \end{bmatrix}. \tag{5.25}$$

Here the phase shift ϕ is induced by the cross-phase modulation (XPM) due to the optical Kerr effect of the fiber, driven by the high-power control pulse which is launched from the arm, given by

$$\phi = \gamma P_{peak} L_{eff} \tag{5.26}$$

where γ denotes the non-linear coefficient, P_{peak} the peak power of the control, and L_{eff} the effective length of the HNL-DSF. We ignore the phase shift incurred by the counterclockwise wave because it is pulsed and thus the non-linear effect is negligible, compared to that of the clockwise wave. The analysis takes into account the non-linear phase shift incurred by the counterclockwise wave as well as the counter-measure to mitigate the effect [9]. By plugging Eqs. (5.24) and (5.25) into Eq. (5.23), the overall scattering matrix is obtained as

$$
\begin{aligned}
[S]_{NOLM} &= \frac{1}{2} \begin{bmatrix} j & 1 \\ 1 & j \end{bmatrix} \begin{bmatrix} \exp(j\phi) & 0 \\ 0 & 1 \end{bmatrix} \begin{bmatrix} 1 & j \\ j & 1 \end{bmatrix} \\
&= \frac{1}{2} \begin{bmatrix} j\exp(j\phi)+j & -\exp(j\phi)+1 \\ \exp(j\phi)-j & j\exp(j\phi)+j \end{bmatrix}.
\end{aligned}
\tag{5.27}
$$

Note that for only the input from Port 1, the outputs from Ports 1 and 2 are obtained

$$
\begin{aligned}
\begin{bmatrix} E_{out1} \\ E_{out2} \end{bmatrix} &= [S]_{NOLM} \begin{bmatrix} 1 \\ 0 \end{bmatrix} = \frac{1}{2} \begin{bmatrix} j\exp(j\phi)+j \\ \exp(j\phi)-j \end{bmatrix} \exp(j\omega_c t) \\
&= \begin{bmatrix} \cos\frac{\phi}{2}\exp(j\frac{\phi}{2}) \\ j\sin\frac{\phi}{2}\exp(j\frac{\phi}{2}) \end{bmatrix} \exp(j\omega_c t).
\end{aligned}
\tag{5.28}
$$

The intensity of the electric fields is written as

$$
\begin{bmatrix} |E_{out1}|^2 \\ |E_{out2}|^2 \end{bmatrix} = \begin{bmatrix} \cos^2(\phi/2) \\ \sin^2(\phi/2) \end{bmatrix} = \frac{1}{2} \begin{bmatrix} 1+\cos\phi \\ 1-\cos\phi \end{bmatrix}.
\tag{5.29}
$$

Without the control pulse, no non-linear phase shift due to the XPM occurs, $\phi = 0$, Eq. (5.29) is rewritten as

$$
\begin{bmatrix} |E_{out1}|^2 \\ |E_{out2}|^2 \end{bmatrix} = \begin{bmatrix} 1 \\ 0 \end{bmatrix}.
\tag{5.30}
$$

For the case with the non-linear phase shift $\phi = \pi$,

$$
\begin{bmatrix} |E_{out1}|^2 \\ |E_{out2}|^2 \end{bmatrix} = \begin{bmatrix} 0 \\ 1 \end{bmatrix}.
\tag{5.31}
$$

The output does not appear from Port 1 but appears only from Port 2. In Fig. 5.11 the measured and theoretical transmitted power $|E_{out2}|^2$ from Port 2 are plotted as a function of the power of the control signal P_{peak} [9]. On increasing the control power the multi-period transfer function is obtained. The deviation from the ideal sinusoidal curve is mainly due to fluctuation of the polarization.

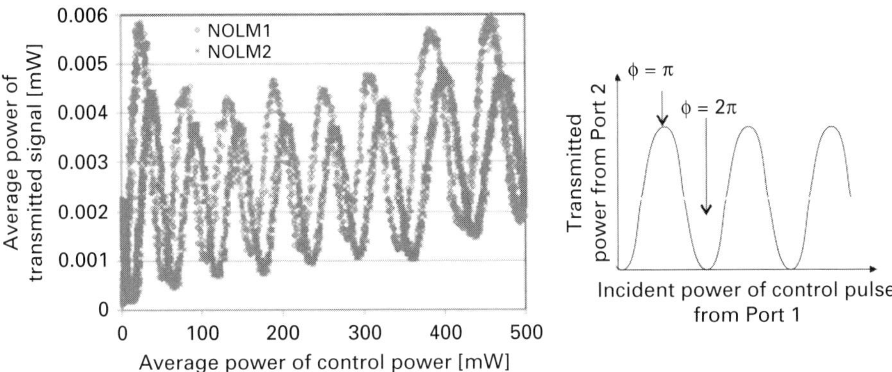

Figure 5.11 Transfer function as a function of optical power of the control signal [9]. © Reprinted by permission of IEEE.

5.4 Optical thresholding

Optical thresholding is crucial to enable data-rate detection to achieve a practical OCDMA system to mitigate multiple interference noise (MAI) because data-rate detection does not use optical time-gating. There are several candidate techniques. The second-harmonic generation (SHG) in periodically polled lithium niobate (PPLN) has exhibited a low operation power, but its operation is polarization dependent [10]. A NOLM, discussed in Section 5.3, is able to suppress the pedestal of a decoded pulse with low operation power, but it might not be suitable for optical thresholding since the power transfer function of the NOLM does not have a steep thresholding characteristic but has a sinusoidal transfer function. The technique used for optical thresholding is super-continuum (SC) generation in normal dispersion-flattened fiber (DFF) [11]. The generation mechanism of SC is described in Section 5.2.2. The optical signal after the optical decoder is amplified to an appropriate intensity level so that the auto-correlation peak of the decoded signal can generate the SC but the cross-correlation of interfering codes does not generate SC due to the low peak power. A fractional frequency component off the peak of the SC spectrum, which is generated solely by the desired signal, is filtered out by the optical bandpass filter (OBPF) and detected as shown in Fig. 5.12. Note that the center wavelength of OBPF locates slightly off the wavelength of the signal. In this way the interference noise is thresholded out from the detected signal. In Fig. 5.13(a) the measured spectra of the original pulse and of the generated SC are shown. As the input power increases, the SC becomes broader. The transmitted power from the OBPF as a function of the input power and the dispersion of the DFF, respectively, are shown in Figs. 5.13(b) and (c). The dispersion curve exhibits good flatness within ± 1 ps/nm over the spectral region of 70 nm. By comparing the eye diagrams of the signal with and without thresholding in Fig. 5.14, a clear eye opening is seen with thresholding, due to a significant noise reduction.

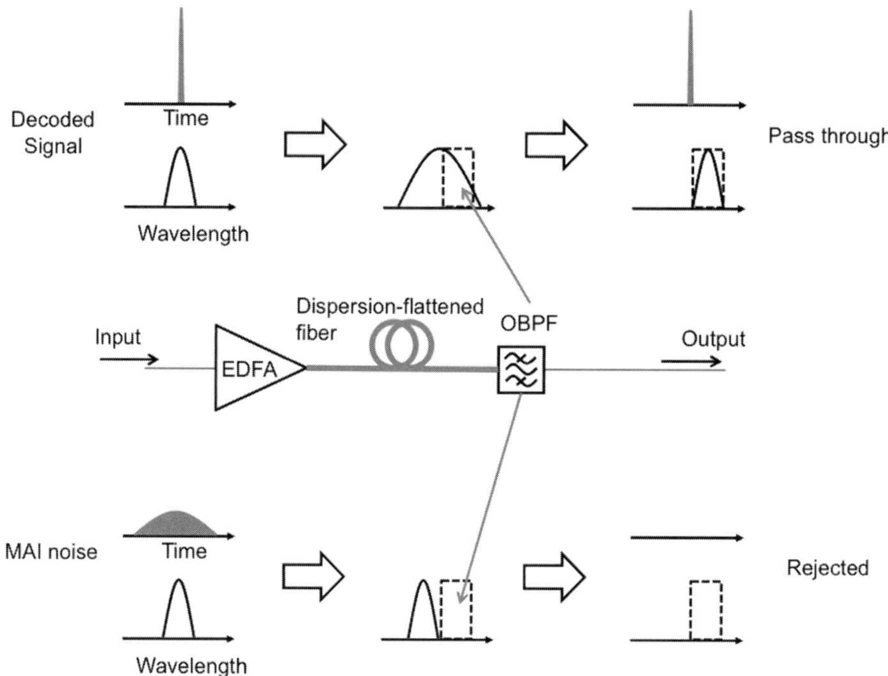

Figure 5.12 Operation principle of the SC-based optical thresholder [11]. © Reprinted by permission of the Optical Society of America.

5.5   Erbium-doped fiber amplifier

Optical fiber amplifiers can compensate for signal attenuation due to fiber loss during propagation in an optical fiber without optical-to-electrical and electrical-to-optical (O-E-O) conversion. In a WDM transmission system, optical fiber amplifiers play versatile roles such as booster amplifier at the transmitter, in-line amplifier as the 1R repeater, and pre-amplifier at the receiver. The capability of simultaneous amplification of a number of optical signals at different wavelengths can drastically drive down the cost of a WDM transmission system with repeaters. An advantage of the fiber amplifier over the semiconductor optical amplifier (SOA) is the low coupling loss from/to the transmission fiber link because of good matching of its circular near field pattern to one of the transmission fiber links and the ease of handling.

An erbium-doped fiber amplifier (EDFA), a special class of fused-silica fiber doped with Er^{3+} ions, has been widely utilized in commercial systems because the gain bandwidth covers the conventional band (C-band) in the spectral region 1530–1565 nm, 4.2 THz bandwidth which is the transmission window of conventional fused-silica optical fiber. It is just a lucky coincidence that the most transparent window is located in the spectral region around 1550 nm where the energy band of the erbium ion Er^{3+} in fused silica fiber creates the desired energy gap.

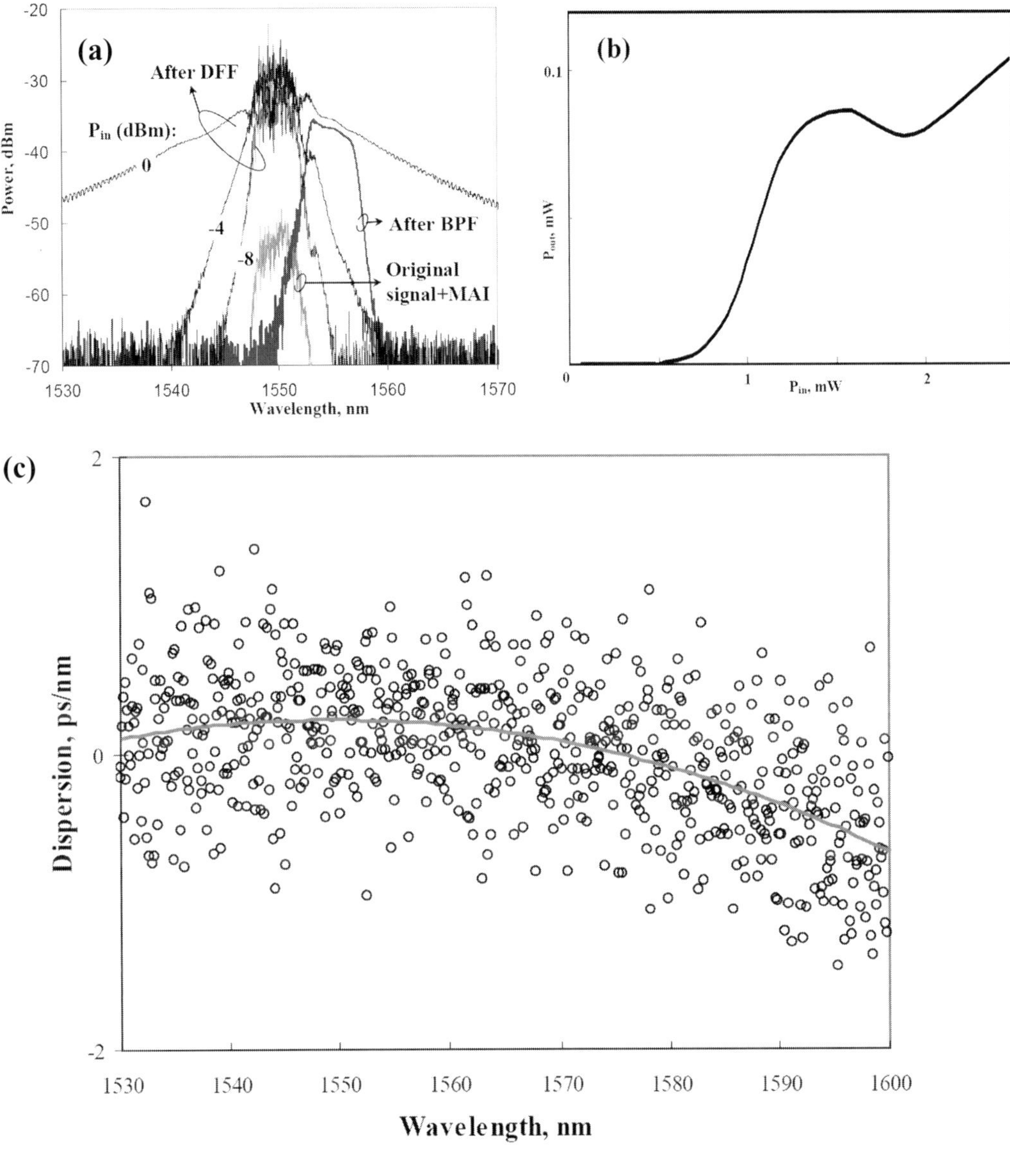

Figure 5.13 (a) Measured spectra of the original pulse and the generated SC with different input power and signal after BPF. (b) Power transfer function of the SC-based optical thresholder. (c) Dispersion characteristics of the DFF [11]. © Reprinted by permission of the Optical Society of America.

5.5.1 Operation principle

Figure 5.15 shows a model of the energy levels of erbium ion Er^{3+} in fused silica fiber [12]. This is a generic three energy level model. The wavelengths equivalent to the energy difference between the ground states E_1 and E_2 is 1530 nm, and that between

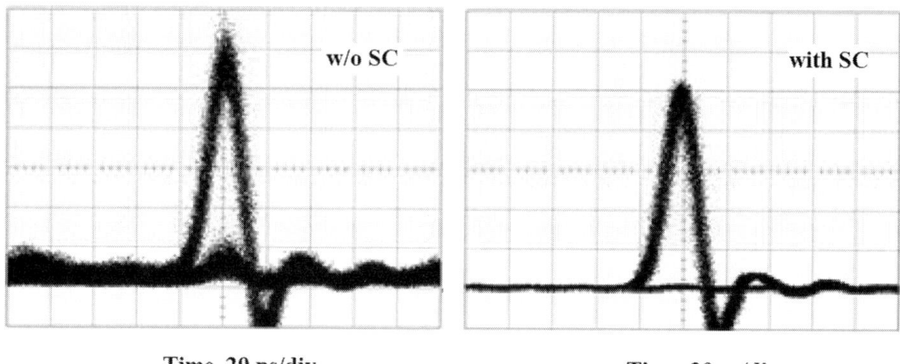

Figure 5.14 Eye diagrams with and without thresholding [11]. © Reprinted by permission of the Optical Society of America.

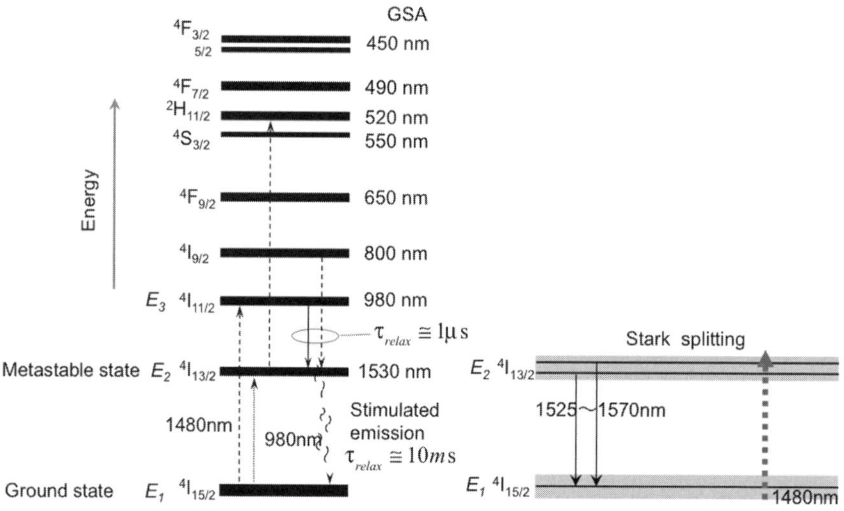

Figure 5.15 Model of the energy levels of erbium ion Er^{3+} of fused silica fiber [12]. © Reprinted by permission of IEEE.

E_1 and E_3 is 980 nm. Each energy band is smeared out in energy, called Stark splitting. This is due to the presence of an external static electric field. Note that each level does not consist of a finite number of well-defined lines but is a continuum, exhibiting the nature of an amorphous glass. Consequently, the E_2 band covers continuously the C-band of the most practical importance, the C-band ranging from 1530 nm to 1565 nm. E_2 is metastable, meaning that the transition from E_2 to the ground state has a long lifetime around 10 ms to decay to the ground state, compared with the shorter lifetime of 1 µs of level E_3. Therefore, Er^{3+} ions that are excited to the level E_3 will quickly transit to level E_2 by spontaneous emission. Thus, population inversion occurs

Table 5.1 Pumping schemes at 980 nm and 1480 nm

	980 nm	1480 nm
Pump efficiency	High 10.2 dB/mW	Low 5.9 dB/mW
Noise	Low	High
Fiber loss	High ~1 dB/km	Low ~0.2 dB/km
Pump laser diode	Commercially available 550 mW	Commercially available 500 mW

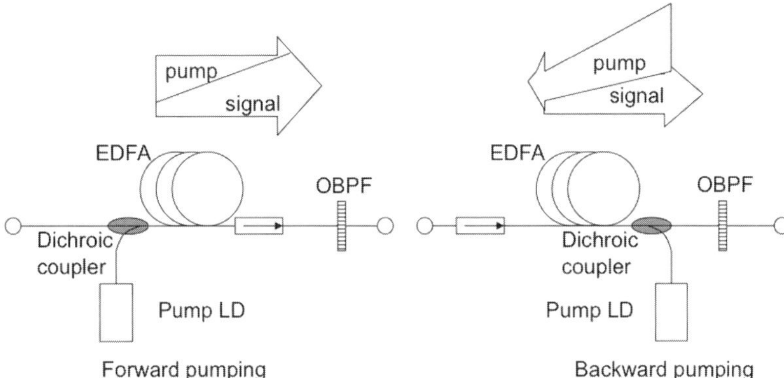

Figure 5.16 Two pump schemes, forward and backward pumps.

between levels E_2 and E_1, and light of frequency f_c in the band 1530–1565 nm can be amplified:

$$hf_c = E_2 - E_1. \tag{5.32}$$

5.5.2 Pump scheme

In Fig. 5.16 the two pump schemes, forward and backward pumps, are illustrated. The pumping light is coupled into the EDFA through a wavelength-selective coupler, a dichroic coupler, and rejected at the output by the optical bandpass filter which discriminates it from the signal light. An optical isolator is inserted at the output of the EDFA in order to suppress laser oscillation. The length of the EDFA is typically a few tens of meters. Laser diodes for pumping are available with wavelengths of either 980 nm or 1480 nm. Table 5.1 summarizes the advantages and disadvantages of two pump schemes at both wavelengths.

The pump at 980 nm is more favorable because of its higher pump efficiency, lower noise, and higher power of the pump laser. However, the pump at 1480 nm enables transmission of the pump light over a longer distance because of the lower fiber loss compared with the pump at 980 nm. A commercial EDFA has typically a gain of 30 dB and saturation output power of up to 30 dBm. The theoretical limit of the noise figure (NF) is 3 dB as described in Section 4.3.3, but the typical NF of actual EDFA is in the range 5–7 dBm. The typical gain spectrum of the EDFA is neither flat nor smooth, as

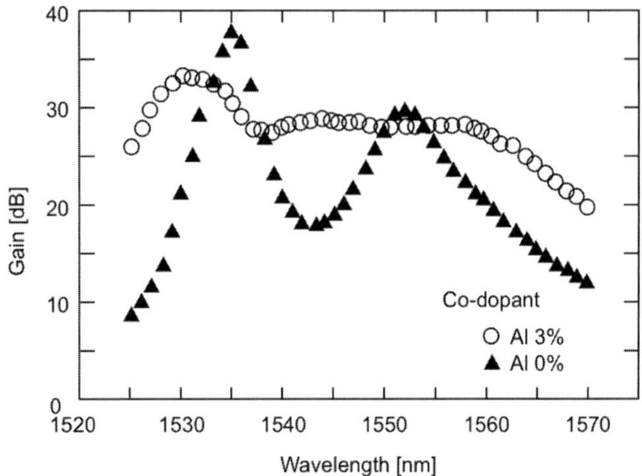

Figure 5.17 Gain profile of EDFA. Modified after [13].

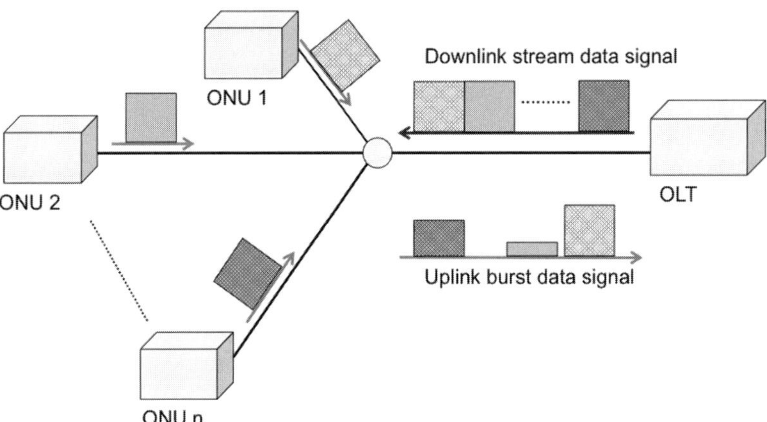

Figure 5.18 Uplink and downlink data signals in a PON system [15]. © Reprinted by permission of NTT.

shown in Fig. 5.17 [13]. The resonance peak of gain is located at 1532 nm. The gain flatness, a desirable characteristic for WDM signals, can be improved by co-doping with ionized aluminum [14]. A notch filter is used for smoothing the gain curve. The gain ripples of commercially available EDFA are suppressed within ±0.5 dB.

5.6   Burst-mode 3R receiver

The downlink signal from the OLT to an ONU is a stream data signal as shown in Fig. 5.18, and it is detected with a photodetector in continuous mode at each ONU. In the uplink from an ONU to the OLT in the PON, the data signal becomes bursty, and hence a burst-mode receiver with 3R capability is used at the OLT for uplink signal reception. It is recalled from Section 4.1 that the 3R function comprises reshaping

Table 5.2 Burst-mode formats of 10G-EPON and EPON

Required time	10G-EPON	EPON
On/off	512 ns	512 ns
Response time of TIA and LA	800 ns	400 ns
CDR	400 ns	400 ns
Minimum receiver sensitivity (Overload)	−28 dBm (−6 dBm)	−27 dBm (−6 dBm)

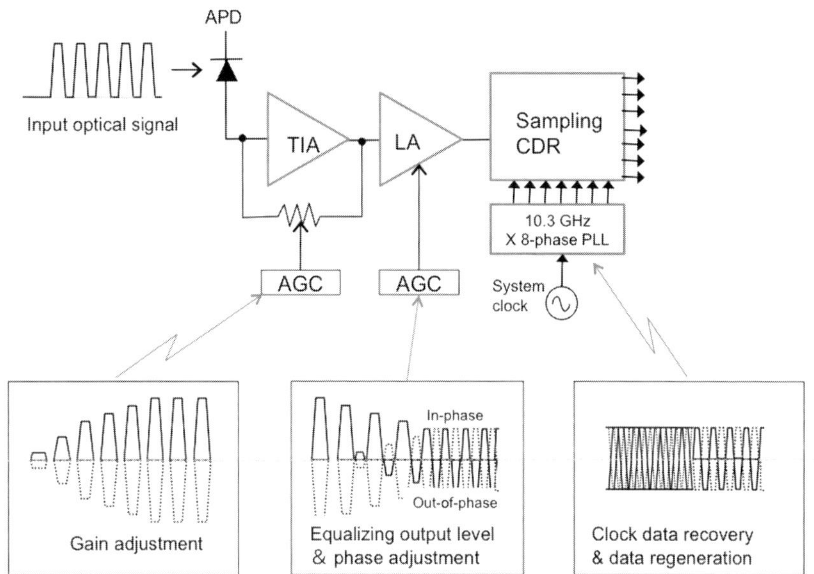

Figure 5.19 Functional diagram of a burst-mode receiver [15]. © Reprinted by permission of NTT and after [16].

including equalization and amplification, retiming, and regenerating. The burst-mode recover consists of a trans-impedance amplifier (TIA), a limiting amplifier (LA), and the clock and data recovery (CDR) circuit with either an avalanche photodiode (APD) or p-i-n PD as shown in Fig. 5.19 [15, 16]. In Table 5.2 the specifications of burst-mode formats of 10G-EPON and EPON, standardized respectively in IEE802.3ah and IEEE802.3av, are summarized. Typical events of data burst occur in the following ways. There is a chance that an ONU will send a sequence of data bits "1" after a long silence and repeat a similar event over again. Or the ONU will send a sequence of data bits "0." For 10G-EPON, 66 bits of logical data, either "1" or "0," are allowed. In such an event, the burst-mode receiver has to adjust the gain of the TIA and LA as well as the timing of the CDR circuit. The intensity of the burst signal varies, depending on where the ONU is located, and the intensity from an ONU far from the OLT is weaker than the intensity from an ONU near the OLT. Therefore, the receiver must be sensitive over a large dynamic range from minimum values to overload.

The TIA, which is placed behind the APD, operates in one of two gain modes according to the input signal level. It operates with high gain for small burst inputs to lower

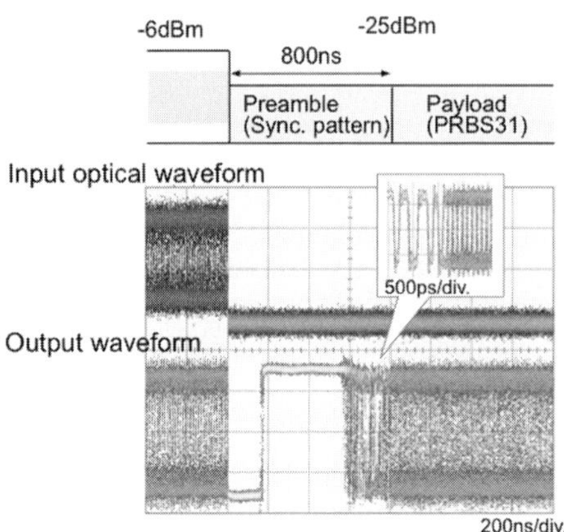

Figure 5.20 Measured waveforms for the input and output signals [17]. © IEEE by permission.

the thermal noise caused by the feedback resistor, controlled by the automatic gain controller (AGC). As the input photocurrent to the TIA increases, the TIA output voltage approaches the non-linear region because of saturation, and the TIA operates in low gain mode to amplify the signal without waveform distortion. Next, amplified signals with different voltages are guided to the LA. The LA acts as an equalizing amplifier and amplifies the input to a uniform voltage regardless of the input level. In particular, the LA must be able to compensate instantaneously for the offset level between a pair of differential signals. There are positive and negative gain amplifiers, and the negative output is fed back to the positive input and vice versa. Finally, amplified signals with constant amplitude from the LA are launched to the CDR. The clock recovery circuit (CRC) has to complete instantaneously phase synchronization of input signals which may be out of phase. The over-sampling CDR circuit produces eight 10.3 GHz multi-phase clocks, and these eight clocks are shifted sequentially by 45° in phase. The incoming data burst can be retimed by phase at 82.5 GS/s equivalent rate phase-shifted clocks. Another important requirement is a fast response time. Obviously, as the overhead time decreases, the efficiency of link bandwidth utilization improves. This is because the overhead time includes the guard time and preamble time. The guard time is the time gap between upstream packets from different ONUs, and the preamble is the time required for the receiver to settle and completely synchronize for each burst input. The preamble reflects the response speed of the receiver to burst input. Experimental results of input (received) burst optical waveform and output waveform from the 10.3G burst-mode optical receiver are shown in Fig. 5.20 [17]. The input burst signal consists of first and second packets with an average power of −6 and −25 dBm, respectively. Each packet has a preamble period (800 ns) followed by the payload data. There is no guard time between two packets for the worst condition. The preamble period depends on the difference in the averaged power of the first and second packets, and when this difference becomes large, a longer preamble period is required. Here, the preamble period of 800 ns is used to cover a wide burst-mode dynamic range of more than 24 dB. A clear eye opening is maintained over the 30 to 6 dB m dynamic range even with the transient response of the received signal (not shown in Fig. 5.20).

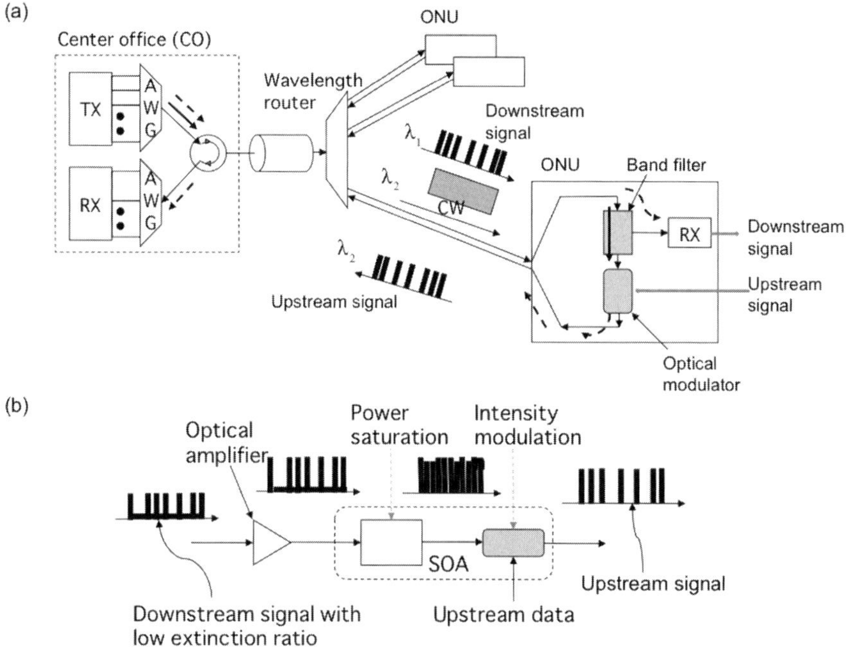

Figure 5.21 Two approaches to colorless ONU: (a) seeding unmodulated light from the OLT and (b) a remodulation scheme using the downlink optical signal [18]. © Reprinted by permission of IEEE.

5.7   Colorless technique

WDM-PON is one of the promising candidate technologies for the next-generation PON. Reduction of the system cost, particularly the cost of ONUs, is a vital concern for successful deployment. As described in Section 2.2.2, each ONU in WDM-PON is assigned a different wavelength. If a light source with a different wavelength is placed at each ONU, the ONU itself, as well as maintaining the temperature of the light source, becomes expensive, and the system might lose its competitiveness with respect to cost. A solution to this problem is the "colorless" ONU approach, which refers to the ONU being wavelength independent. The colorless approach requires a broadband optical gain medium such as an inexpensive Fabry–Pérot laser diode (FPLD) and a semiconductor optical amplifier (SOA). In the case of the FP-LD, the polarization states of downlink signals need to be adjusted to maximize the gain of the FPLD. A colorless light source scheme using FPLD is depicted in Figs. 2.26 and 2.20. To realize colorless ONU, there are two options: one is seeding an unmodulated light from the OLT, modulating it with the uplink signal, and looping back the modulated uplink signal to the OLT, shown in Fig. 5.21(a), and the other is a re-modulation scheme, which reuses the downlink optical signal for uplink transmission, shown in Fig. 5.21(b) [18]. The data of the downlink signal are erased to nearly continuous light and modulated with the uplink signal.

A special class of SOA, a reflective semiconductor optical amplifier (RSOA) which has a half-reflective facet on one side of the SOA, can be used for the re-modulation

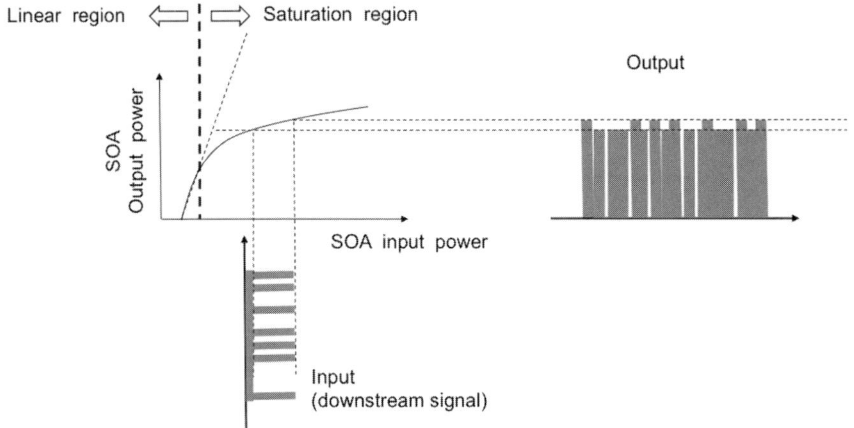

Figure 5.22 Input–output characteristic of a reflective SOA (RSOA) used in data rewrite [18]. © Reprinted by permission of IEEE.

scheme using the downlink optical signal. The optical power received at the ONU is divided into the RSOA and the receiver by an optical coupler in front of the RSOA and receiver. The RSOA serves as both a power limiter and modulator. The downlink signal is amplified linearly so that the power falls in the saturation region of the RSOA. Figure 5.22 shows the input–output characteristic of the RSOA, which indicates that the SOA output power saturates as the input power (i.e., the downlink signal power) increases [18]. As a result, the difference between the mark and space levels of the NRZ OOK-modulated downlink signal is considerably reduced in the RSOA output. Thus, the data pattern of the downlink signal can be almost erased, and simultaneously a quasi-continuous wave is modulated with the uplink data by injecting current according to the uplink data. A distinct advantage of the re-modulation scheme over the seeding scheme using FPLD is that it can be done with a single wavelength for downlink and uplink. In addition, the gain bandwidth of the SOA is around 50 nm, which fully covers the entire C-band. The re-modulation scheme is rather simple and potentially low cost, however, the performance of the uplink signal could be seriously deteriorated if the RSOA operates in the unsaturated regime due to the uncompressed thick level of data "1." Thus, the optical power of the downlink signal incident on the RSOA should be large enough to ensure its operation in the saturated regime which, in turn, would limit the power budget and scalability of WDM-PON. To circumvent this problem, the use of Manchester coding, instead of the conventional NRZ-OOK modulation format, in the downlink signal can improve the performance of the uplink signal [19]. Manchester coding uses a line code in which the encoding of each data bit has at least one transition and occupies the same time as shown in Fig. 5.23. Unlike the NRZ format, the Manchester-encoded signal has a negligible amount of low-frequency components within the bandwidth of the upstream receiver. Consider G-PON in which the bit rate of the uplink is 1.24 Gb/s, half of 2.5 Gb/s of the downlink. As a result, the influence of the downlink signal on the uplink signal is minimized, and the performance of the uplink signal can be improved even when the optical power level of the downlink signal incident on the RSOA is well below the saturation regime of the RSOA.

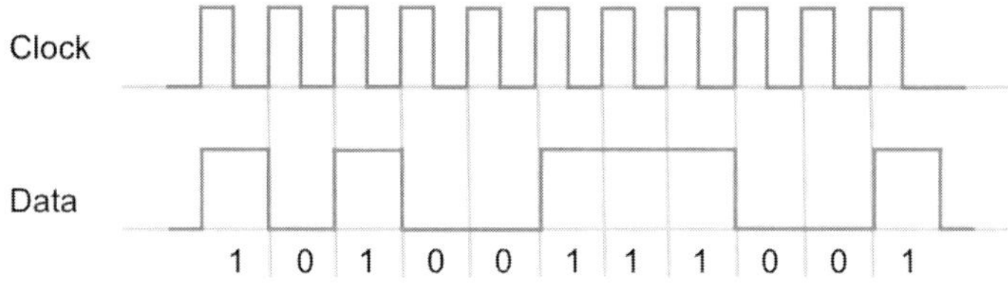

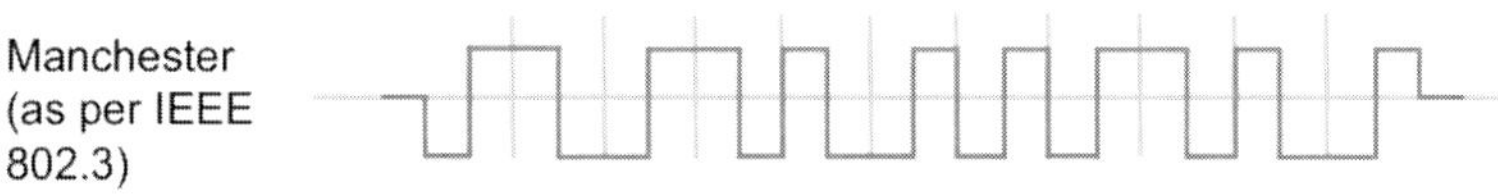

Original data XOR	clock	= Manchester value
0	0	0
0	1	1
1	0	1
1	1	0

Figure 5.23 Manchester coding.

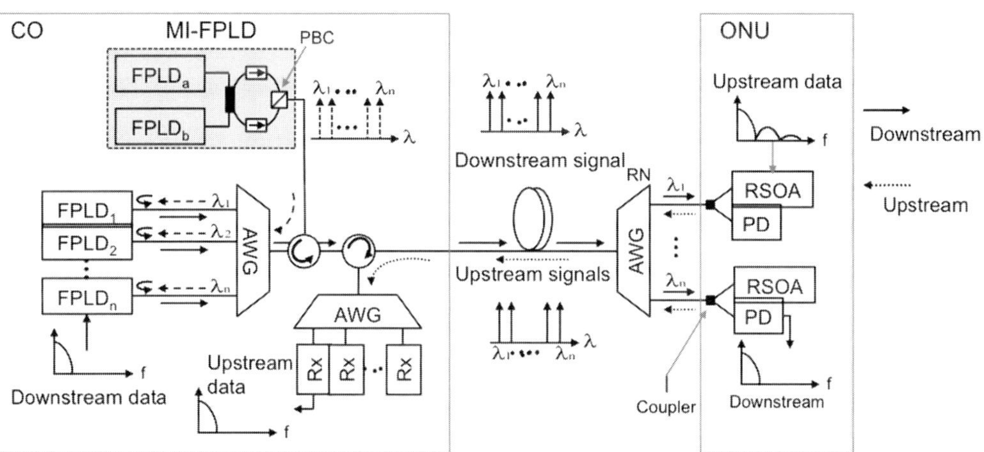

Figure 5.24 Experimental setup of uplink transmission using a MI-FPLD at OLT and reflective SOA (RSOA) at ONU [20]. © Reprinted by permission of IEEE. Note: the broadcast signal on the downlink is omitted for simplicity.

Figure 5.24 shows the experimental setup for uplink transmission using an MI-FPLD at the OLT and reflective SOA (RSOA) at the ONU. The MI-FPLD shown in Fig. 2.18 is used [20]. The data rate of the uplink is set at 122 Mb/s so that the null frequency of the upstream data spectrum coincides with the RIN peak located at around 120 MHz as shown in Fig. 2.20. This would minimize the effect of noise peaks on the performance of the upstream baseband signal while revealing the effect of the low-frequency noise near dc. The received power at the photodetector was fixed at 25 dB m. The measured BERs are plotted as a function of the input power to the RSOA in Fig. 5.25. The result shows that the

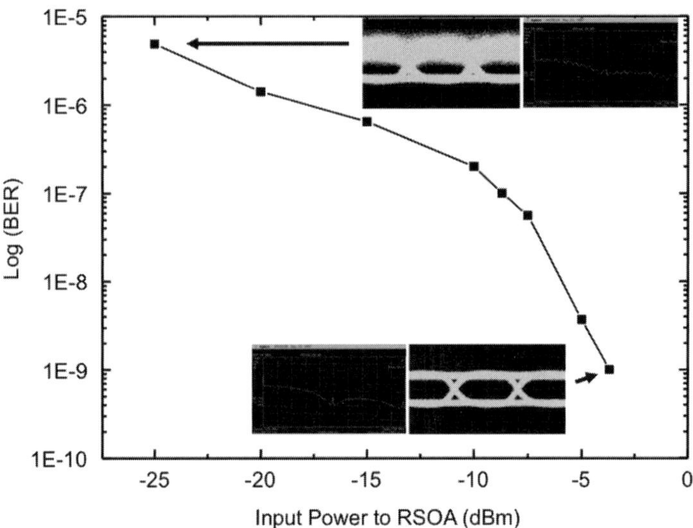

Figure 5.25 Measured BERs as a function of the input power to the RSOA [20]. © Reprinted by permission.

BER of the upstream data is greatly enhanced as the optical power to the RSOA increases.

Problems

5.1 Star coupler. Calculate the number of 3-dB couplers required and the theoretical insertion loss of a 1×16 star coupler.

5.2 Fabry–Pérot etalon. Calculate the finesse for $R = 0.9,\ 0.99$.

5.3 Optical spectrum analyzer using Fabry–Pérot etalon. Design the Fabry–Pérot etalon which can measure the emission spectrum of the laser.

(1) Calculate the frequency interval δf of the longitudinal modes of a Fabry–Pérot laser with cavity length $L_{cavity} = 100$ cm and refractive index of the cavity $n = 1.0$.

(2) Assume that the gain bandwidth $\Delta f_{gain} = 1.5$ GHz. Determine the Δf_{FWHM} of the etalon having the resolution of $1/10$ of the laser mode. Calculate the finesse F of the etalon for $2nL_{etalon} = 20$ cm and the power reflection coefficient R to obtain the required finesse.

5.4 Non-linear loop mirror (NOLM).

(1) Consider the case for a control pulse with a high intensity, and assume that the phase difference induced by the control pulses due to the XPM between the CW and CCW probe pulses is $\Delta\theta$. Derive the intensity of the transmission P_{out} at input Port 2.

(2) Show that there is only transmission and no reflection for the case with $\Delta\theta = \pi/2$.

Part II

6   OCDMA principles

Spread spectrum (SS) communication techniques date back to the early 1950s. SS techniques have been applied to communications, navigation, and test systems that are not possible with standard signal formats. Wireless CDMA is one of the successful applications of SS techniques. It improves the spectral efficiency by incorporating a number of unique features made possible by virtue of benign noise-like characteristics of the signal waveform. Notable among these is universal frequency reuse in that all subscribers in the network occupy a common frequency spectrum. Despite such success in using CDMA as a communications technique in radio, microwave, and millimeter wave bands, there are currently no commercial optical communication systems that use CDMA technology. Unfortunately, OCDMA has remained outside the mainstream of optical communications R&D since its proposal in the mid-1970s followed by experimental demonstrations in the 1980s. This has been due mainly to the immaturity of optical devices which are proprietary to OCDMA, such as the optical en/decoder and optical thresholding device. Recently, remarkable progress has been made in device technology, and demand has been growing for flexibility and scalability in multiple access. OCDMA now deserves a revisit as a powerful alternative to access technologies of TDM and WDM.

6.1   Spread spectrum communication

SS communication is the basis of CDMA. Before going on to OCDMA, let us briefly review wireless SS communication from a book written by Dixon [1]. CDMA has been widely deployed in second-generation (2G) cellular telephony, global positioning system (GPS), and LAN, IEEE802.11 (FH-SS) and 802.11b (DS-SS). There are two SS techniques: frequency hopping (FH)-SS and direct sequence (DS)-SS. In FH-SS, shown in Fig. 6.1, the frequency of the carrier is distributed in a bit time duration evenly in every available frequency channel, $\{f_1, f_2, \ldots, f_n\}$ with the same amount of power in each channel. The pattern of frequency hopping corresponds to the code sequence. The received frequency hopping signal is mixed with a locally generated replica, which is offset by a fixed intermediate frequency f_{IF}. When the code sequences of transmitter and receiver match each other, the multiplication $\{f_1, f_2, \ldots, f_n\} \times \{f_1 + f_{IF}, f_2 + f_{IF}, \ldots, f_n + f_{IF}\}$ produces the f_{IF} component. Any undesired received signal which is not a replica of the local reference remains spread in the frequency, and it can be rejected by the IF bandpass filter (IF BPF).

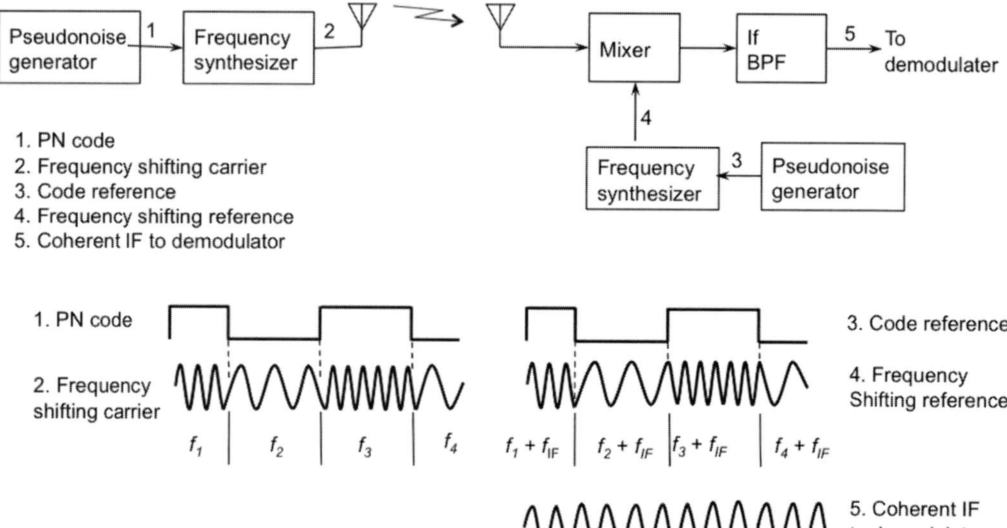

Figure 6.1 Frequency hopping spread spectrum (FH-SS) communication system [1]. © Reprinted by permission of John Wiley & Sons.

In DS-SS, shown in Fig. 6.2, the phase of the carrier shifts discretely by $\pm 90°$ in a bit time duration at the transmitter, according to the order predetermined by a known pseudo-random code sequence. Since the phase shift is typically much faster compared to the data bit time, the bandwidth of the encoded signal becomes much broader than the signal before the encoding. The spectrum bandwidth of the SS is typically 10^3 times broader than the data bandwidth. At the receiver, the received signal is multiplied with the same reference code, and the narrowband restored carrier passes through a bandpass filter. Undesired signals remain spread to their own bandwidth after the multiplication, and the bandpass filter can reject almost all the power of undesired signals.

The operation principle of SS communication is described by Shannon's channel capacity theorem [2] given by

$$C = W \log_2 \left(1 + \frac{S}{N} \right) = 1.44 W \log_e \left(1 + \frac{S}{N} \right), \tag{6.1}$$

where C and W, respectively, denote the channel capacity measured in units of bit per second and the bandwidth of the channel in units of Hz, S and N are the powers of the signal and noise, respectively. Given the signal-to-noise ratio of the signal, the error-free transmission capacity is determined by the bandwidth. Assume that the signal-to-noise ratio is sufficiently small, then the channel capacity is reduced to

$$C = 1.44 W \frac{S}{N} \tag{6.2}$$

using a Taylor series of the natural logarithm, also known as the Mercator series.

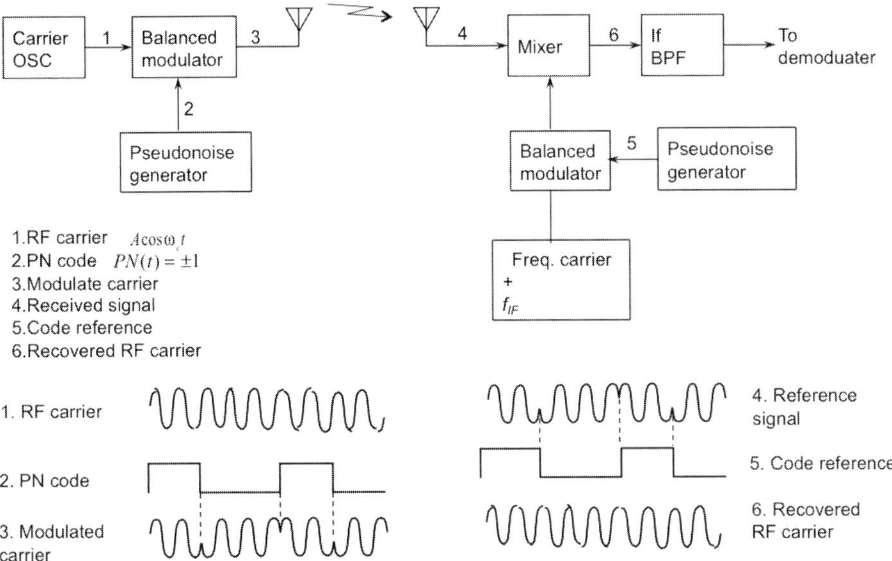

Figure 6.2 Direct sequence spread spectrum (DS-SS) communication system [1]. © Reprinted by permission of John Wiley & Sons.

For $x = 1.0$,

$$\log_e(1 + x) = x - \frac{1}{2}x^2 + \mathsf{ggg} + (-1)^{n-1}\frac{x^n}{n}. \tag{6.3}$$

Regardless of the value of signal-to-noise ratio, the desired channel capacity C can be achieved by increasing the bandwidth of the channel W. For example, for the case with $N/S = 10^2$, the required bandwidth of the channel is calculated as

$$W = \frac{CN}{1.44S} = \frac{100}{1.44}C = 69.4C. \tag{6.4}$$

This means that there is an encoding and decoding scheme to ensure the target channel capacity if a "sufficiently large bandwidth" is available. One can trade the bandwidth for an improvement in the bit error rate. This is a rationale that SS communication systems rely on. Abundant bandwidth was thought to be available in optical fiber communication, the rationale of SS communication initially motivated OCDMA.

6.2　Wireless CDMA versus optical CDMA

CDMA is a multiple access technique based upon SS communication. The multi-user interference can be treated as noise. It is called multiple access interference (MAI) noise. Assume that there are K subscribers in the system, the desired signal has power S, and the power of each $(K-1)$ interfering signal is also S. The signal-to-interference power in the presence of MAI noise and additive noise is

$$\frac{S}{N} = \frac{S}{(K-1)\,S + \sigma^2} = \frac{1}{(K-1) + \sigma^2/S} \tag{6.5}$$

where σ^2 [W] is the variance of additive noise power. We introduce the bit energy-to-noise density ratio, E_b/N_o in Eq. (6.6), which is commonly used in CDMA. The bit energy E_b [J] is defined as the desired signal power S divided by the data bit rate B [1/s], and the noise density N_0 [W/Hz] is defined as the noise power divided by the total bandwidth W:

$$\frac{E_b}{N_0} = \frac{S/B}{(K-1)\,S/W + \sigma^2/W} = \frac{W/B}{(K-1) + \sigma^2/S} = \left(\frac{W}{B}\right)\frac{S}{N} \tag{6.6}$$

where the ratio of data bit rate to total bandwidth W/B is referred to as the "process gain"

$$\text{Process gain} \equiv \frac{W}{B}. \tag{6.7}$$

Equation (6.6) tells that the signal-to-noise ratio is enhanced by a factor of processing gain. The number of subscribers supported is expressed as

$$K = 1 + \frac{W/B}{E_b/N_0} - \frac{\sigma^2}{S}. \tag{6.8}$$

As a consequence, the larger the process gain W/B becomes, and the smaller the bit energy-to-noise density ratio E_b/N_0, the more subscribers can be accommodated.

The operation principle of OCDMA employing time-spread code sequences is illustrated in Fig. 6.3(b) [3]. It is interesting to compare OCDMA with the forementioned wireless CDMA system employing SS technique in Fig. 6.3(a). An optical short pulse at the transmitter, having a much broader frequency spectrum than the data signal bandwidth, is spread over one bit duration T_B by encoding. At the receiver the desired time-spread signal is despread to an auto-correlation waveform with a sharp peak at the center by decoding if the code sequences between the encoder and decoder match. On the other hand, the multiple interference signals remain time-spread after decoding because there are mismatches between their code sequences and the desired sequence. There are two options for optical detection techniques after decoding, in order to improve the bit energy-to-noise density ratio. In Fig. 6.3(b) the optical time-gating is introduced which acts in the time domain as the counterpart of the bandpass filter in wireless CDMA. Another option is time-gating in the electrical domain after OE conversion. The optical time gate allows only the central peak of the auto-correlation waveform to pass and shuts out almost all the energy of interference signals outside the time window of the gate. The gating has to be operated within the chip time duration T_{chip}, for example, in the region of a few picoseconds in the case of the bit rate of 10 Gb/s. Even after the time-gating and the thresholding one will find that the MAI noise cannot be completely eliminated, and residual MAI noise within the time window of the gate is inevitably detected together with the desired auto-correlation peak. This causes the beat noise, which will be detailed in Section 6.5.2. When time-gating is adopted in OCDMA,

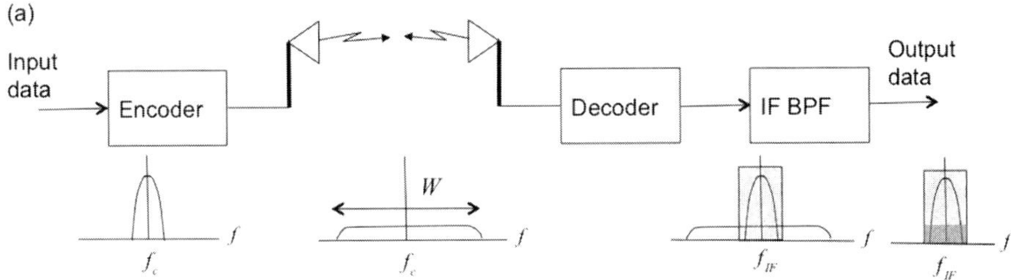

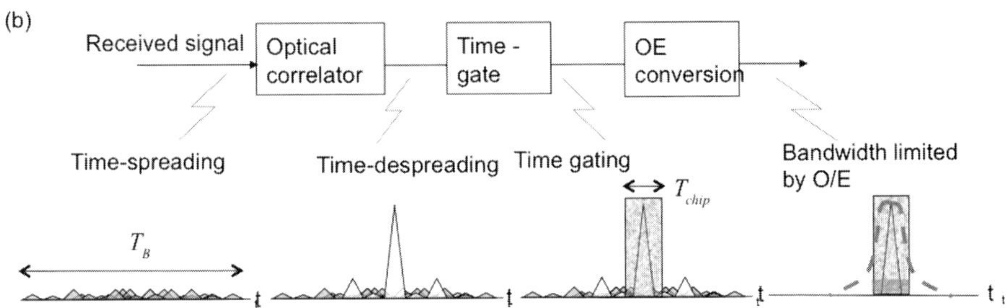

Figure 6.3 Analogy between wireless CDMA and OCDMA. (a) Wireless CDMA and (b) OCDMA system employing time-spread code sequence.

it is assumed that almost all the signal power is squeezed in the auto-correlation peak. Then, the bit energy-to-noise density ratio in Eq. (6.6) is rewritten by replacing W with $1/T_{chip}$ as

$$\frac{E_b}{N_0} = \frac{S T_B}{(K-1)\, S T_{chip} + \sigma^2 T_{chip}} = \frac{T_B/T_{chip}}{(K-1) + \sigma^2/S} \tag{6.9}$$

where T_{chip} denotes the one chip time duration, which is equal to the response time to the time-gate. The processing gain of OCDM is expressed as the ratio of one bit time duration T_B to the time window T_{chip} as

$$\text{Process gain} = \frac{T_B}{T_{chip}}. \tag{6.10}$$

The number of subscribers supported is obtained as

$$K = 1 + \frac{T_B/T_{chip}}{E_b/N_0} - \frac{\sigma^2}{S}. \tag{6.11}$$

It is interesting to observe that there is one-to-one correspondence in the bit energy-to-noise density ratio and the processing gain between wireless CDMA and OCDMA. As a consequence, the larger the process gain T_B/T_{chip} becomes and the smaller the bit energy-to-noise density ratio E_b/N becomes, the more subscribers can be accommodated. As T_B/T_{chip} normally corresponds to the number of chips, i.e., the code length, therefore a longer code sequence is preferable to accommodate more subscribers. As we will see in Section 6.5, a practical optical time-gate device is not yet commercially available, although an ultrafast optical time gate has been demonstrated using an NOLM.

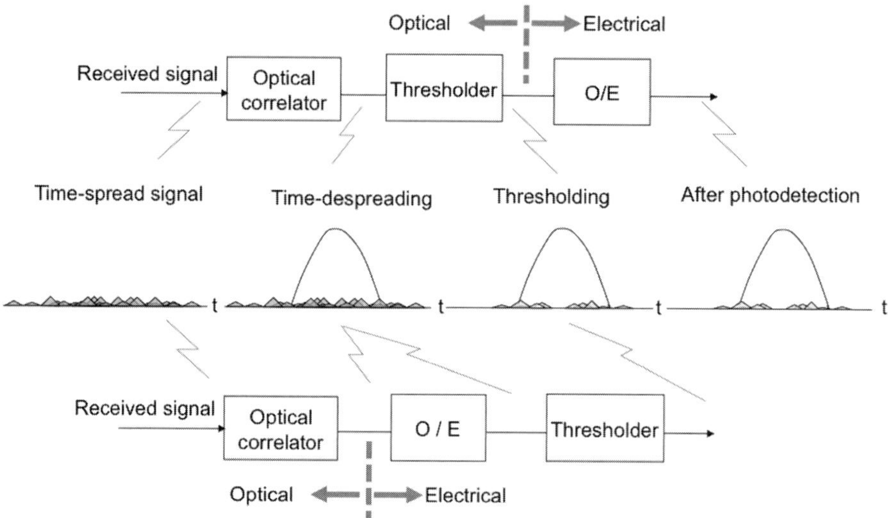

Figure 6.4 Receiver configuration for OCDMA without time-gating.

Another detection technique, shown in Fig. 6.4, does not use time-gating but a threshold is used. The choice of detection scheme with or without time-gating will depend on the auto-correlation waveform. The thresholder will be adopted for an auto-correlation waveform extending in the one bit time duration T_B, which is in contrast to that with a sharp central peak shown in Fig. 6.3(b). The thresholding can be performed either optically or electrically, respectively, before or after optical-to-electrical (OE) conversion. An advantage of the scheme without time-gating is that it does not require a high-speed optical device operating at the chip rate but it allows the use of a conventional optical receiver. Since the optical implementation of an OCDMA system without time-gating is more practically feasible than that using time-gating, it has been attracting much attention. Figure 6.5 depicts the transfer function of the threshold in intensity. The thresholding can significantly reduce the level of MAI noise even if it is not perfect. After being thresholded, the photocurrent is integrated over a bit time duration and guided to the decision circuit. The bit energy-to-noise density ratio of Eq. (6.6) is rewritten as

$$\frac{E_b}{N_0} = \frac{S/B}{(K-1)\,S_{int}/W_{PD} + \sigma^2/W_{PD}} \tag{6.12}$$

where S_{int} is the cross-correlation power of the interfering signal and W_{PD} is the bandwidth of the photodetector. Since the bandwidth is nearly equal to the bit rate, $W_{PD}/B \cong 1$, this means that the process gain is unity. Then, Eq. (6.12) is reduced to

$$\frac{E_b}{N_0} = \frac{1}{(K-1)\,S_{int}/S + \sigma^2/S} = \frac{S/S_{int}}{(K-1) + \sigma^2/S_{int}} \tag{6.13}$$

The number of subscribers supported is rewritten as

$$K = 1 + \frac{S/S_{int}}{E_b/N_0} - \frac{\sigma^2}{S_{int}}. \tag{6.14}$$

Table 6.1 Wireless CDMA and optical CDMA

	Wireless CDMA	Optical CDMA
Frequency carrier	Micrometer and millimeter wave	Infrared light
Encoding and decoding	RF frequency domain	Time domain and/or frequency domain
Code sequence	Direct sequence (DS), frequency hopping (FH)	Temporal or spectral waveform, wavelength-hopping
Transmission medium	Free space, linear	Optical fiber
Origin of signal impairments	Multi-path effect	Non-linear effect, chromatic, polarization and mode dispersions
Interference suppression	RAKE demodulation, etc.	Thresholding, time-gating
System capacity	Interference limited, soft capacity on demand	

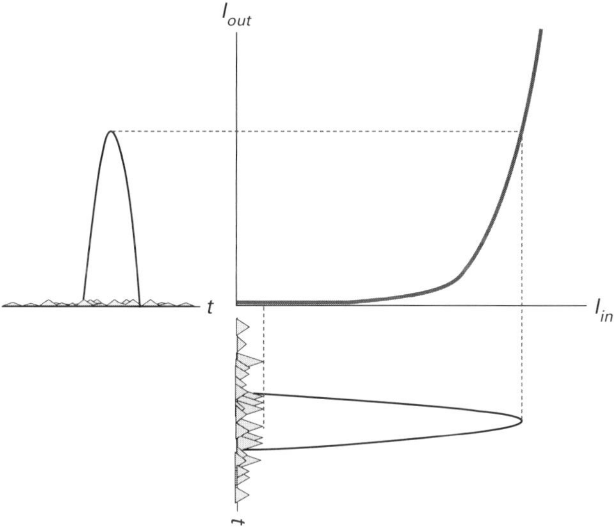

Figure 6.5 Transfer function of threshold in the intensity.

Since the power of the desired signal S becomes much larger than those of the $(K-1)$ interfering signals, i.e., $S \gg S_{int}$ due to thresholding, the optical thresholding provides the effective processing gain S/S_{int} by suppressing the MAI noise. The ratio S/S_{int} can be 100 or larger, and this value becomes comparable to the process gain T_B/T_{chip} using time-gating. Therefore, system performance similar to that of OCDMA with optical time-gating can be expected for OCDMA without time-gating.

In Table 6.1 various characteristics of OCDMA are compared with wireless CDMA, including the frequency carrier, techniques of encoding and decoding, code sequence, the transmission medium, interference, and system capacity. The most notable difference is in the transmission medium. Radio wave propagation in free space is free from dispersion and non-linearity, while optical CDMA is free from the multi-path effect as long as single-mode propagation is concerned, but it suffers from signal impairment due to chromatic and polarization dispersions. Both system capacities are interference limited by undesired codes.

6.3   Optical decoding based upon correlation

Correlation between the optical codes is used as a basis to distinguish the desired signal from interference signals in OCDMA. The correlation is based upon matched filtering in the time domain. In general, matched filtering is a detection technique in which the signal-to-noise ratio of the received signal is maximized. The impulse response of the correlator $h_d(t)$ is expressed in terms of the impulse response of the input signal waveform $h_e(t)$ as

$$h_d(t) = h_e(t_0 - t) \tag{6.15}$$

and its Fourier spectrum $H_d(\omega)$ is given by

$$H_d(\omega) = H_e(\omega)^* e^{-j\omega t_0} \tag{6.16}$$

where $H_c(\omega)$ is the Fourier spectrum of the input signal waveform $h_e(t)$. Then, the output of the correlator $\psi(t - t_0)$ becomes the auto-correlation function, and $\psi(t - t_0)$ is expressed by the convolution of the impulse responses of the encoder and the correlator:

$$
\begin{aligned}
\psi(t - t_0) &= \frac{1}{2\pi} \int_{-\infty}^{\infty} H_e(\omega) H_d(\omega) e^{j\omega t} d\omega \\
&= \frac{1}{2\pi} \int_{-\infty}^{\infty} |H_e(\omega)|^2 e^{j\omega(t - t_0)} d\omega \\
&= \int_{-\infty}^{\infty} h_e(t') h_e(t' - t + t_0) dt'.
\end{aligned}
\tag{6.17}
$$

Since the optical code sequence consists of N-chip pulses, it is convenient to express the correlation in discrete form

$$\psi(j) = \sum_{n=0}^{N-1} h_n h_{n-j}, \quad (-N + 1 \leq j \leq N - 1). \tag{6.18}$$

For better understanding, the auto-correlation waveform of a three-chip ($N = 3$) binary PSK optical code sequence $(0, 0, \pi)$ is illustrated in Fig. 6.6. It is noted that the correlation function of the N-chip code sequence extends over the time domain $(-N + 1, N - 1)$. In OCDMA this correlation can be performed in the optical domain using an optical decoder. A remarkable feature of optical decoding is that the optical decoder is the same device as the encoder, and the optical en/decoder is a passive optical device such as a super-structured fiber Bragg grating (SSFBG) decoder or planar lightwave circuit (PLC)-based multiport decoder, described in Chapter 7. The correlation operation is ultrafast because it takes the time of flight of the input encoded signal propagating in the decoder. For example, it takes 80 ps for the double optical path length of an 8 mm long SSFBG decoder.

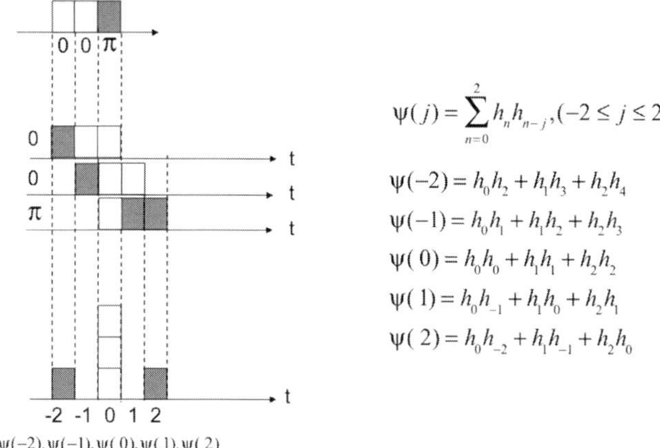

$$\psi(j) = \sum_{n=0}^{2} h_n h_{n-j}, (-2 \leq j \leq 2)$$

$$\psi(-2) = h_0 h_2 + h_1 h_3 + h_2 h_4$$

$$\psi(-1) = h_0 h_1 + h_1 h_2 + h_2 h_3$$

$$\psi(0) = h_0 h_0 + h_1 h_1 + h_2 h_2$$

$$\psi(1) = h_0 h_{-1} + h_1 h_0 + h_2 h_1$$

$$\psi(2) = h_0 h_{-2} + h_1 h_{-1} + h_2 h_0$$

Figure 6.6 Example of correlation and cross-correlation.

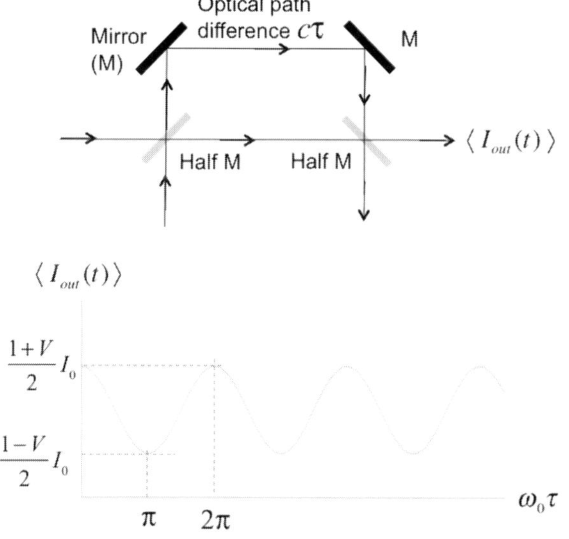

Figure 6.7 Mach–Zehnder interferometer and the coherence function.

6.4 Early stage of OCDMA

6.4.1 Coherence of the light source

Coherence of the light source is relevant to OCDMA. Here, the coherent nature of lightwaves is reviewed briefly. Let us assume a Mach–Zehnder interferometer (MZI) with optical path difference of $\tau (\tau = n\Delta L/c)$, with 50% coupling ratio of half mirror and perfect polarization alignment as shown in Fig. 6.7. When the MZI is excited by a lightwave of the form

$$E_{in}(t) = \sqrt{P_0} \exp\left[j\left(\omega_0 t + \phi(t)\right)\right], \tag{6.19}$$

the output optical intensity from such an interferometer is given by

$$I_{out}(t) = \frac{P_0}{2}[1 + \cos\{\omega_0\tau + \phi(t) - \phi(t - \tau)\}].\qquad(6.20)$$

In a common interferometer, the output light is detected by a relatively slow photodetector which averages $I_{out}(t)$. Thus, the average output reduces to

$$\langle I_{out}(t)\rangle = \frac{P_0}{2}[1 + V\cos(\omega_0\tau)].\qquad(6.21)$$

Here, the visibility V is defined by

$$V = \frac{I_{max} - I_{min}}{I_{max} + I_{min}} = \exp(-|\tau|/\tau_c)\qquad(6.22)$$

where τ_c denotes the coherence time of the light source. As shown in Fig. 6.7 $\langle I_{out}(t)\rangle$ in Eq. (6.21) swings between $1 + VI_0/2$ and $1 - VI_0/2$ as the path difference changes. The phase noise is given by the variance of $I_{out}(t)$, defined by [4]

$$\text{Variance} = \langle(I_{out}(t) - \langle I_{out}\rangle)(I_{out}(t - \tau) - \langle I_{out}\rangle)\rangle.\qquad(6.23)$$

This can be calculated as

$$\text{Variance} = \frac{P_0^2}{4}\exp\left(-2|T|/\tau_c\right)\left\{\cos^2(\omega_0\tau)\left[\cosh(2|\tau|/\tau_c) - 1\right]\right.$$
$$\left. + \sin^2(\omega_0\tau)\left[\sinh(2|\tau|/\tau_c) - 1\right]\right\}.\qquad(6.24)$$

This is quite a complicated function. Before discussing a few special cases, it should be emphasized that the detection system sees this variance only if its bandwidth far exceeds the linewidth of the source. This condition is met in most practical OCDMA systems: the linewidth of the source is much narrower than the bandwidth of the photodetector. Otherwise, the variance can be neglected. Next, we consider two extreme cases.

Case I: Incoherent limit ($|\tau| \gg \tau_c$)

$$V = \frac{I_{max} - I_{min}}{I_{max} + I_{min}} = 0\qquad(6.25)$$

$$\text{Variance}_{Incoherent} = \frac{P_0^2}{8}.\qquad(6.26)$$

Obviously, there is no interference in this case ($V = 0$), but phase noise is significant.

Case II: Coherent limit ($|\tau| \ll \tau_c$)

$$V = \frac{I_{max} - I_{min}}{I_{max} + I_{min}} = 1.\qquad(6.27)$$

Here, the result depends on the specific value of $\cos(\omega_0\tau)$.

(a) **Out-of-quadrature**

Here, the two arms of the interferometer are either in phase or 180° out of phase, and $\cos(\omega_0 \tau) = \pm 1$.

$$\text{Variance}_{\text{Coherent, Out-of-quadrature}} = \frac{P_0^2}{2}(|\tau|/\tau_c)^2. \tag{6.28}$$

(b) **In-quadrature**

Here, the two arms are either in phase or 90 degree out of phase, and $\cos(\omega_0 \tau) = 0$

$$\text{Variance}_{\text{Coherent, In-quadrature}} = \frac{P_0^2}{2}(|\tau|/\tau_c). \tag{6.29}$$

The noise in the in-quadrature case is stronger than the noise in the out-of-quadrature case because the interferometer has its highest sensitivity under in-quadrature conditions.

6.4.2 Coherence multiplexing

Coherence multiplexing exploits the coherent nature of the light source. It can be regarded as a root of OCDMA because it takes advantage of the coherent nature. One can use the optical path difference as a signature to distinguish a desired signal from interference in a multiple access system. In coherence multiplexing, the optical path difference of the encoder at the transmitter is set so that significant interference occurs only when the optical path differences of the encoder at the transmitter and of the decoder at the receiver match.

According to a comprehensive review article on early OCDMA by Sampson *et al.* [5], Cielo and Delisle proposed and demonstrated the use of coherence for multiplexing of communication signals in 1976 [6]. The architecture of coherence multiplexing is shown in Fig. 6.8. The en/decoder of the *i*th subscriber consists of interferometers with imbalance of optical paths. The other $(N-1)$ data signals are coupled through the $N \times N$ star coupler. Here, it is assumed that the optical path differences $\Delta\tau_{ie}$ at the transmitter and $\Delta\tau_{id}$ at the receiver are much longer than the source coherence time τ_c of the light source, i.e., $\Delta\tau_{ie} \gg \tau_c$, $\Delta\tau_{id} \gg \tau_c$ ($i = 1, 2, \ldots, N$) but it is also assumed that the differential delay is significantly less than the coherence time, i.e., $|\Delta\tau_{ie} - \Delta\tau_{id}| \ll \tau_c$. This condition will diminish crosstalk between the channels. Digital data are modulated to binary PSK signal by the phase modulator on one arm of the encoder. This does not produce an intensity modulation at the output of the encoder since the fields from the two paths combine incoherently because $\Delta\tau_{ie} \gg \tau_c$. Decoding is achieved by matching the delay of the decoder $\Delta\tau_{id}$ with $\Delta\tau_{ie}$ of the encoder.

There are four available optical paths from the transmitter to the receiver, including relative delays of 0, $\Delta\tau_{id}$, $\Delta\tau_{ie}$, and $\Delta\tau_{ie} + \Delta\tau_{id}$. The output photocurrent $I_{out}(t)$ is given by

$$I_{out}(t) \propto \left| e^{j\phi(t)} + e^{j\phi(t)}e^{j\Delta\tau_{id}} + e^{j\Delta\tau_{ie}} + e^{j(\Delta\tau_{ie}+\Delta\tau_{id})} \right|^2 \tag{6.30}$$

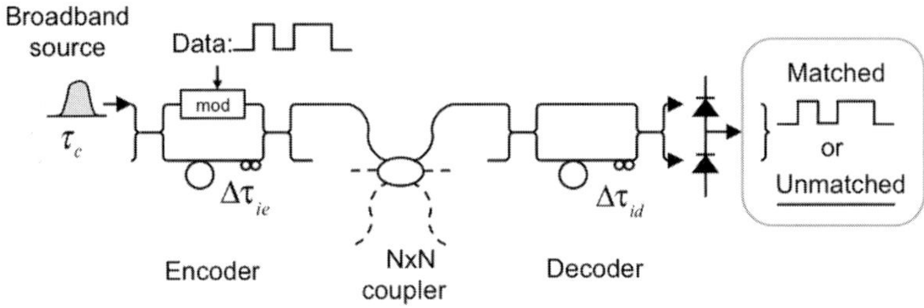

Figure 6.8 Architecture of coherence multiplexing [5]. © Reprinted by permission of Taylor & Francis.

where digital data $\phi(t)$ takes the value of either 0 or π. Since we assume $\Delta\tau_{ie} \cong \Delta\tau_{id}$, only the two fields within the coherence time interfere with each other coherently at the receiver: one field is phase modulated with the data, and the other is unmodulated, and the field with no path difference $e^{j\phi(t)}$ and the field experiencing the optical delay $\Delta\tau_{ie} + \Delta\tau_{id}$ have no chance to interfere with the other fields. The balanced detection will cancel out the dc components of the photocurrent, and consequently, Eq. (6.30) is reduced to

$$I_{out}(t) \cong \left| e^{j\phi(t)}e^{j\Delta\tau_{id}} + e^{j\Delta\tau_{ie}} \right|^2 \cong \cos^2 \frac{\phi(t)}{2}. \tag{6.31}$$

Since the phase $\phi(t)$ takes the value of either 0 or π, the photocurrent $I(t)$ becomes either 1 or 0 depending on the data. Thus, the data are recovered as an intensity modulation. When the delays of the decoder and encoder are unmatched by more than several coherence times, i.e., when $|\Delta\tau_{ie} - \Delta\tau_{kd}|$ $(i \neq k)$ is greater than a few times τ_c, the fields remain substantially uncorrelated and negligible coherent interference takes place. Signals from these unmatched encoders are, therefore, not recovered. The differences between all encoder delays must significantly exceed the source coherence time to prevent crosstalk between channels. It is desirable to use sources with broad linewidth since this allows short interferometer delays to be used and also reduces the detected optical beat noise. The beat noise will severely limit the number of subscribers.

6.4.3   Coherent versus incoherent OCDMA

OCDMA has its root in the above coherent multiplexing, but it is more practical because it gets rid of constraints on optical path differences, and it introduces an optical code sequence, instead of an optical path difference, as a signature to distinguish the desired signal from interference in a multiple access system. In Fig. 6.9 the basic architecture of

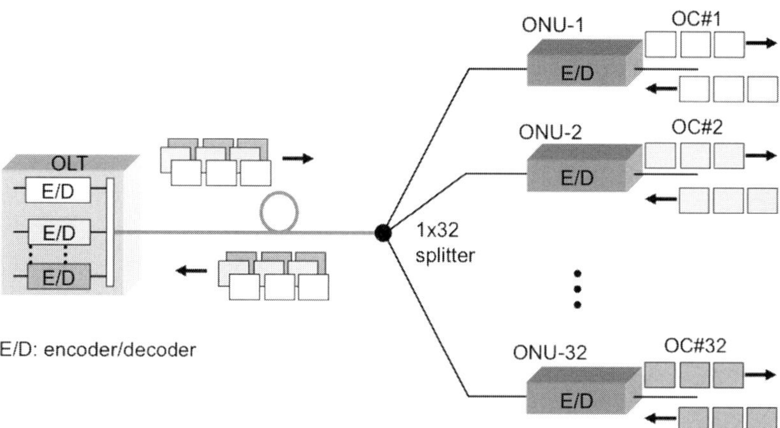

Figure 6.9 Basic architecture of an OCDMA access network.

an OCDMA network is shown. An OLT located in the central office is connected via a $1 \times N$ ($N = 32$) star coupler with N subscribers where the ONUs are installed. The optical encoder and decoder (E/D) are located at the transmitter and receiver at each ONU, respectively. In addition, each transmitter and receiver at the OLT has an optical encoder and decoder, respectively, and hence N optical encoders and decoders are located at the OLT. These are additional devices to the conventional PON setup, peculiar to OCDMA. OCDMA systems are categorized into two types, depending on the coherence of the light source: incoherent and coherent OCDMA systems. Incoherent OCDMA uses an incoherent light source with broadband emission spectrum such as a light emitting diode (LED), and it employs binary OOK optical code. One of the reasons why an incoherent light source was used in the early experiments in the 1980s is that coherent light sources such as laser diodes with a narrow linewidth were not available. In incoherent OCDMA the levels of the transmitted optical code sequence correspond to light "on" or light "off." In this correlation the peak intensity of the auto-correlation function equals the summation of the light intensity of the number of chips bearing "1" in the code sequence. In one of the pioneering incoherent OCDMA works [7], a 32-chip optical code based on a prime code sequence with prime number 5 was implemented using a fiber delay-line, and single channel transmission was conducted at the bit rate of 3 Mb/s.

By contrast, coherent OCDMA inherits the virtues of coherence multiplexing. The availability of coherent light sources and phase-controllable optical devices made coherent OCDMA possible in the 1990s. It uses a coherent light source with a long coherence time, and it employs PSK optical code. In Figs. 6.10(a) and (b), the correlation and cross-correlation of 8-chip OOK and bipolar PSK code sequences, respectively are compared. For OOK code, summation of the correlation in the discrete form of Eq. (6.18) is performed in the intensity and, for example, the auto-correlation peak is 4. For binary PSK code, summation of the correlation is performed in the amplitude, and as a consequence the auto-correlation intensity peak after square-law detection by the photodetector

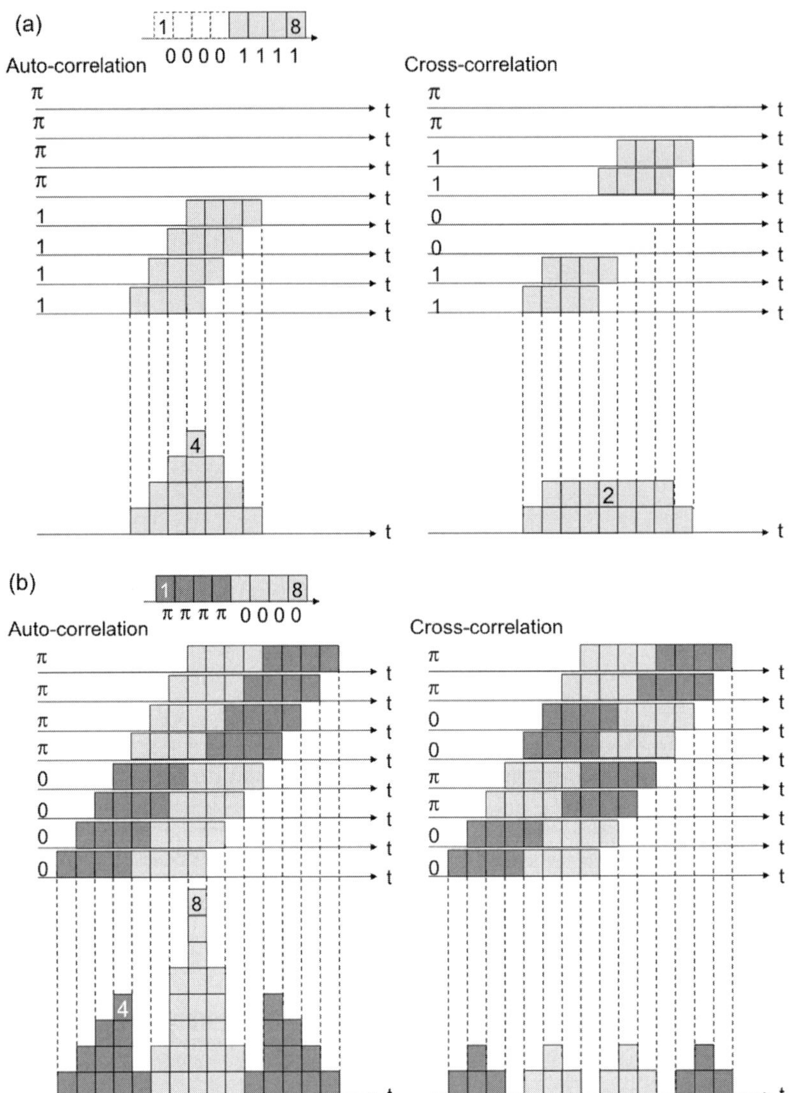

Figure 6.10 Correlation and cross-correlation of 8-chip (a) OOK and (b) bipolar PSK code sequences.

becomes 64. This is a sharp contrast between incoherent and coherent OCDMA, resulting in a remarkable difference in the multiple access interference (MAI) noise characteristics.

Despite the superior correlation properties of coherent OCDMA over incoherent OCDMA, however, the signal-interference beat noise unique to coherent OCDMA poses a big challenge. Square-law detection by a photodiode of the auto-correlation and cross-correlation induces the beat noise, and this is difficult to remove completely by optical time-gating and optical thresholding. The theoretical analysis of the beat noise will be described in Section 7.3.

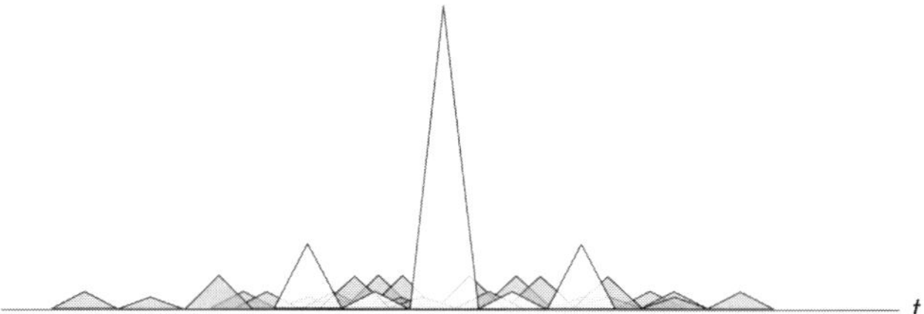

Figure 6.11 Asynchronous OCDMA.

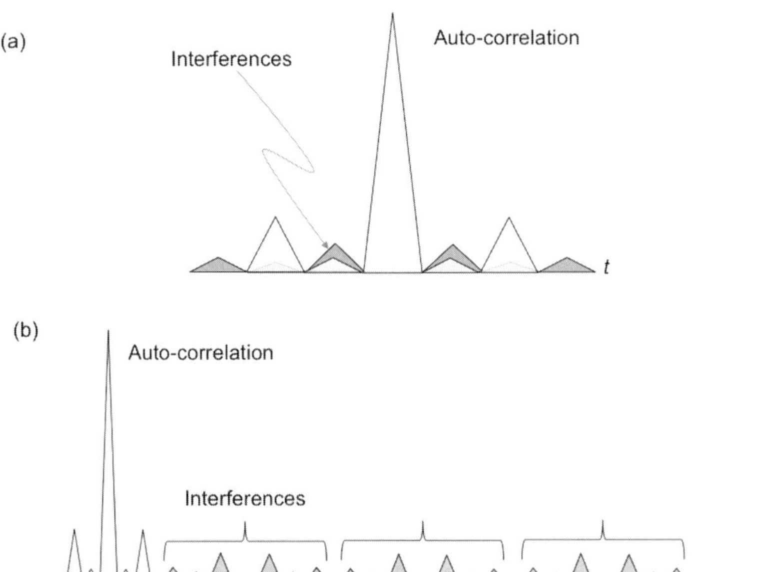

Figure 6.12 Synchronous OCDMA: (a) chip-synchronized OCDMA, and (b) slot-synchronized OCDMA.

6.4.4 Asynchronous versus synchronous OCDMA

One of the unique and most attractive properties of OCDMA-PON is its asynchronous nature. Unlike TDMA it does not require timing coordination between the OLT and ONUs. Asynchronous OCDMA-PON has "tell-and-go" multiple access capability and allows the subscribers to send their data signals at any time without the timing pre-coordination required by TDMA-PON. In asynchronous coherent OCDMA, the magnitude of interference noise, including MAI noise and beat noise between the desired signal and the interference signals, will vary randomly in a bit time duration, according to the relative positions of the cross-correlations with that of the desired auto-correlation peak as shown in Fig. 6.11. Therefore, the price is that the system margin has to be set assuming the worst scenario, i.e., the largest interference noise has to be taken into account. In synchronous OCDMA, on the other hand, there are two levels of timing synchronization: chip synchronization shown in Fig. 6.12(a) and slot synchronization shown in Fig. 6.12(b). Chip synchronization can realize the best scenario, exhibiting the

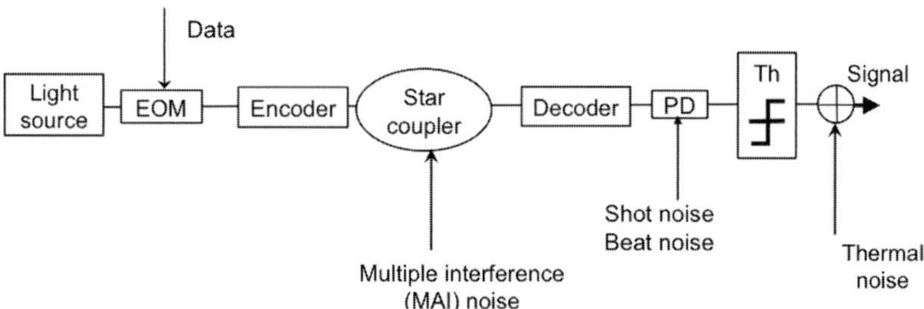

Figure 6.13 Model of OOK OCDMA-PON with bit rate detection.

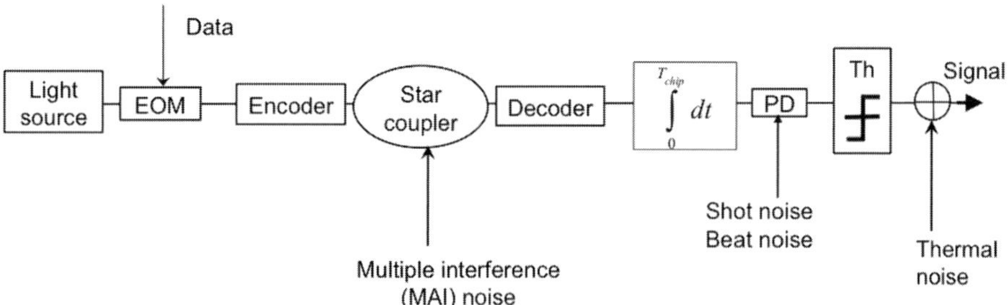

Figure 6.14 Model of OOK OCDMA-PON with chip rate detection.

least interference noise. It can be understood from Fig. 6.12(a) that the interference noise is minimized when the timing is aligned so that the centers of the auto-correlation and cross-correlation functions are aligned. In the slot synchronized OCDMA shown in Fig. 6.12(b) one bit duration is segmented into the number of active subscribers, and each slot is allocated to a different subscriber so that there is no overlap between the slots [8] as for TDMA. In this way, slot synchronized OCDMA is completely free from interference noise. However, time-slotted OCDMA has to pay the price for the synchronization, which limits the code length, resulting in a limited number of subscribers. Besides this limitation, the most significant disadvantage is the loss of "tell-and-go" multiple access capability without pre-coordination, and the system requires timing coordination as TDMA-PON does.

6.5   Signal impairment

6.5.1   Evaluation of system performance

We consider sources of noise in a generic OCDMA system in Fig. 6.13. The optical signal bearing digital information is encoded with an optical code and then combined with the other optical signals by the star coupler. At the receiver, the combined signals are decoded, and the desired signal along with interfering signals are converted to electrical signals after photodetection, followed by thresholding. This conventional detection scheme is referred to as bit rate detection. Chip rate detection uses time-gating before thresholding as shown in Fig. 6.14. The choice of the detection scheme depends on the temporal

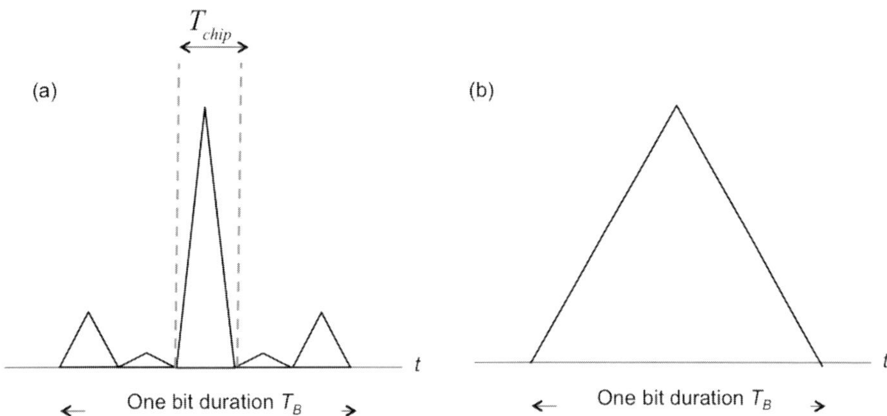

Figure 6.15 Auto-correlation waveforms: (a) with a sharp peak at the center and (b) extending over a bit time duration without peak.

waveform of the correlation after the optical decoding at the receiver as shown in Fig. 6.15. Chip rate detection is applied preferably to systems, for example, which use SSFBG en/decoders because the auto-correlation produces an impulse-like temporal waveform, shown in Fig. 6.15(a), and hence chip rate detection is applied in order to gate only the peak of the auto-correlation waveform, yielding a higher signal-to-noise ratio than bit rate detection. On the other hand, the auto-correlation waveform generated by the multiport en/decoder extends over one bit time duration, shown in Fig. 6.15(b), and therefore bit rate detection will be suitable.

In addition to the additive noise such as shot noise and thermal noise, other noise unique to coherent OCDMA systems, MAI noise and beat noise. The beat noise is inherent to the coherent nature, and is the beat between the desired signal and the interference from other subscribers generated by square-law detection using a photodetector. The bit error rate (BER) determines the signal quality of digital transmission systems and the signal-to-noise ratio determines the BER as described in Sections 4.5 and 4.6. In general, the additive noise described in Section 4.3 includes thermal noise, shot noise, and optical amplifier noise. In addition to these types of noise, the total noise of an OCDMA system is expressed as

$$\sigma_1^2 = \sigma_{th}^2 + \sigma_{sh}^2 + \sigma_{amp}^2 + \sigma_{MAI}^2 + \sigma_{beat}^2 \qquad (6.32)$$

where σ_{MAI}^2 and σ_{beat}^2, respectively, represent the variances of the MAI noise and the beat noise, peculiar to OCDMA systems. The beat noise is only inherent to coherent OCDMA, and incoherent OCDMA is free from beat noise. The definitions of "incoherent" and "coherent" OCDMA can be found in Section 6.4.3. From the viewpoint of beat noise, incoherent OCDMA has an edge over the coherent counterpart. Nevertheless, judging the overall system performance, coherent OCDMA is superior to incoherent OCDMA.

6.5.2   Chip rate detection

Chip rate detection is applied in order to gate only the peak of the auto-correlation waveform, yielding higher signal-to-noise ratio than bit rate detection. Chip rate detection is applied preferably to systems which use SSFBG en/decoders because the auto-correlation function produces an impulse-like temporal waveform as described in Section 7.3.1. Hereafter, the MAI noise and the beat noise are focused on. A model of OCDMA-PON with chip rate detection is shown in Fig. 6.14. It is assumed that at the receiver the decoder output is time-gated with the time window T_{chip} of the chip rate and detected with the photodetector, followed by thresholding in the electrical domain. Alternatively, the time-gated decoder output is thresholded in the optical domain, and detected with the photodetector. Let us assume that there are K active subscribers transmitting signals asynchronously in the network. If there are m ($\leq K - 1$) interference signals from undesired active subscribers at a given instant, the optical field of the signal illuminating the photodetector of the desired subscriber is given by [9]

$$E(t) = \sqrt{P_d}\exp\{j\omega_d t + \phi_d(t)\} + \sum_{i=1}^{m} \sqrt{P_i}\exp\{j\omega_i(t - \tau_i) + \phi_i(t - \tau_i)\} \quad (6.33)$$

where P_d and P_i are the optical intensities of the decoded signals of the desired and undesired subscribers, respectively, ω_d and ω_i are the optical frequencies, ϕ_d and ϕ_i are the phase noises, and τ_i is the relative propagation delay of the interferers to the desired subscriber. It is assumed that the phases ϕ_d and ϕ_i are of mutually independent Gaussian-distributed Wiener–Levy stochastic processes. The photocurrent of the detector at the receiver is given by

$$I_{ph} = \int_0^{T_{chip}} \Re(EE^*)dt + \int_0^{T_{chip}} n(t)dt$$

$$= T_{chip}\mathbf{R}P_d + T_{chip}\mathbf{R}\sum_{i=1}^{m} P_i + 2\mathbf{R}\sum_{i=1}^{m} \sqrt{P_i P_d} \times \int_0^{T_{chip}} \cos\{(\omega_i - \omega_d)t - \omega_i\tau_i$$

$$+ \phi_i(t - \tau_i) - \phi_d(t)\}dt$$

$$+ 2\mathbf{R}\sum_{i=1}^{m}\sum_{j=1}^{m} \sqrt{P_i P_j} \int_0^{T_{chip}} \cos\{(\omega_i - \omega_j)t - \omega_i\tau_i + \omega_j\tau_j$$

$$+ \phi_i(t - \tau_i) - \phi_j(t - \tau_j)\}dt + \int_0^{T_{chip}} n_0(t)dt \quad (6.34)$$

where $\mathbf{R}$ is the responsivity of the photodetector, and n_0 is the receiver noise current. It is assumed that all the signals have the same polarization. The first term on the right-hand side of Eq. (6.34) is the target signal power, the second term represents the MAI noise, and the third and fourth terms are the m primary signal–interference beat noise and

$m(m-1)/2$ secondary interference–interference beat noise, respectively. The final term represents the additive noise. We define the magnitude of interference as

$$\xi \equiv \frac{\langle P_i \rangle}{P_d}. \tag{6.35}$$

It can be assumed that $\xi = 1$. For example, in a coherent time-spread OCDMA-PON with the SSFBG encoder/decoder using a N_{chip}-chip long Gold code, $\xi \cong 1/N_{chip}$. The ratio of the variance of primary and secondary beat noise terms is about $2/(m-1)/\sqrt{\xi}$. If m is not very large so that $m\sqrt{\xi} = 1$, the secondary beat noise can be ignored. In the analysis hereafter, the secondary beat noise is neglected. We consider three cases, depending on the relation of the coherence of the light source and the chip rate.

Case I: Incoherent regime ($\tau_c = T_{chip}$)

In the incoherent regime the coherence time of the light source τ_c is much shorter than the chip time period, $\tau_c = T_{chip} \cdot \delta\phi_{id}$ are in random process and distribute uniformly over $[-\pi, +\pi]$ during the integral duration T_{chip}. Therefore, the integral of the cosine function gives zero, thus diminishing two beat noise terms. We can simplify Eq. (6.34) to

$$I_{ph} = T_c \mathbf{R} P_d + T_c \mathbf{R} \sum_{i=1}^{m} P_i + \int_0^{T_{chip}} n_0(t)dt. \tag{6.36}$$

Case II: Coherent regime ($\tau_c \geq T_{chip}$)

$\tau_c \geq T_{chip}$ is assumed in the coherent regime because the coherence of light should be maintained at least within each chip. Now our focus is on the third primary beat noise term in Eq. (6.34), having terms inside the cosine function. The first term $(\omega_i - \omega_d)T_{chip}$ is negligible with a typical value of $(\omega_i - \omega_d)/2\pi = 1 \sim 5\,\mathrm{GHz}$ and $T_{chip} \leq 10\,\mathrm{ps}$. $\omega_i \tau_i$ is approximated as a constant during the integral, and then it can also be neglected. $\phi_i(t - \tau_i) - \phi_d(t)$ depends on the coherent property of the optical pulse. As a consequence, the integral of the cosine function becomes zero and Eq. (6.34) is reduced to a simpler form:

$$I_{ph} = T_{chip}\mathbf{R} P_d + T_{chip}\mathbf{R} \sum_{i=1}^{m} P_i + 2T_{chip}\mathbf{R} \sum_{i=1}^{m} \sqrt{P_i P_d} \cos(\Delta\phi_i) + \int_0^{T_{chip}} n_0(t)dt,$$

$$\Delta\phi_i = (\omega_i - \omega_d)T_{chip} - \omega_i \tau_i + \delta\phi_{id}$$

$$\delta\phi_{id} = \phi_i(t - \tau_i) - \phi_d(t). \tag{6.37}$$

Here the secondary interference–interference best noise is ignored. $\Delta\phi_i$ represents the overall phase noise and is of random process that varies bit by bit over $[-\pi, \pi]$. This is the origin of the beat noise.

Case III: Partially coherent regime ($\tau_c \leq T_{chip}$)

In between the incoherent and coherent regimes there is a partially coherent regime characterized by $\tau_c \leq T_{chip}$. The coherence time is approximated as $\tau_c \cong 1/B_o$ where

B_o is the optical bandwidth of the system, and the typical value is 250 GHz (2–5 nm). The chip time is typically $T_{chip} \leq 10$ ps at the bit rate of 10 Gb/s. As a consequence, $T_{chip}/\tau_c \cong T_{chip}B_0$ is around 2.5, and therefore, practical PON systems in the real world will fall in this partially coherent regime. This model can be simplified if it is assumed that the relative phase is maintained as a constant within every time slot of coherent time τ_c, and they are mutually independent random processes distributed over $[-\pi, \pi]$ for different time slots. Under this assumption, the photocurrent of the detector at the receiver is given by

$$I_{ph} = T_c \mathbf{R} P_d + T_c \mathbf{R} \sum_{i=1}^{m} P_i + 2\tau_c \mathbf{R} \sum_{n}^{T_{chip}/\tau_c} \left[\sum_{i=1}^{m} \sqrt{P_i P_d} \cos(\Delta \phi_{in}) \right]$$

$$+ \int_0^{T_{chip}} n_0(t)dt$$

$$\Delta \phi_{in} = (\omega_i - \omega_d)\tau_c n - \omega_i \tau_i + \delta \phi_{id},$$

$$\delta \phi_{id} = \phi_i(t - \tau_i) - \phi_d(t). \tag{6.38}$$

Within the limit of $T_{chip}/\tau_c = 1$, Eq. (6.38) becomes that of the coherent regime Eq. (6.37), and within the limit of $T_{chip}/\tau_c \to \infty$, Eq. (6.38) becomes that of the incoherent regime Eq. (6.36).

The average BER of the system can be calculated as

$$\mathrm{BER} = \sum_{m=0}^{K-1} p(m)\mathrm{BER}(m). \tag{6.39}$$

where $p(m)$ is the probability that m of the $K - 1$ interfering subscribers are simultaneously "1"s and obeys the binomial distribution

$$p(m) = \frac{(K - 1)!}{(K - m - 1)!m!} 2^{-(K-1)} \tag{6.40}$$

and $\mathrm{BER}(m)$ is the BER with m interfering signals. With equal probability binary data, $\mathrm{BER}(m)$ can be expressed as

$$\mathrm{BER}(m) = \mathrm{P}(0)_{chip}\mathrm{Pe}(1|0)(m) + \mathrm{P}(1)_{chip}\mathrm{Pe}(0|1)(m)$$

$$= \frac{1}{2}\left[\left(2 - \frac{T_{chip}}{T_B}\right) \mathrm{Pe}(1|0)(m) + \frac{T_{chip}}{T_B}\mathrm{Pe}(0|1)(m) \right] \tag{6.41}$$

where $\mathrm{P}(0)_{chip}$ and $\mathrm{P}(1)_{chip}$ are the probabilities of chip mark "0" and "1", respectively, and $\mathrm{Pe}(1|0)(m)$ and $\mathrm{Pe}(0|1)(m)$ are the conditional error probabilities with density function with chip mark "0" and "1", respectively. They are given by

$$\mathrm{Pe}(0|1)(m) = \frac{1}{2}\mathrm{erfc}\left[\frac{P_d(1 + m\xi - D)}{\sqrt{2}\sigma_1} \right] \tag{6.42}$$

and

$$\mathrm{Pe}(1|0)(m) = \frac{1}{2}\mathrm{erfc}\left[\frac{P_d(D - m\xi)}{\sqrt{2}\sigma_0} \right]. \tag{6.43}$$

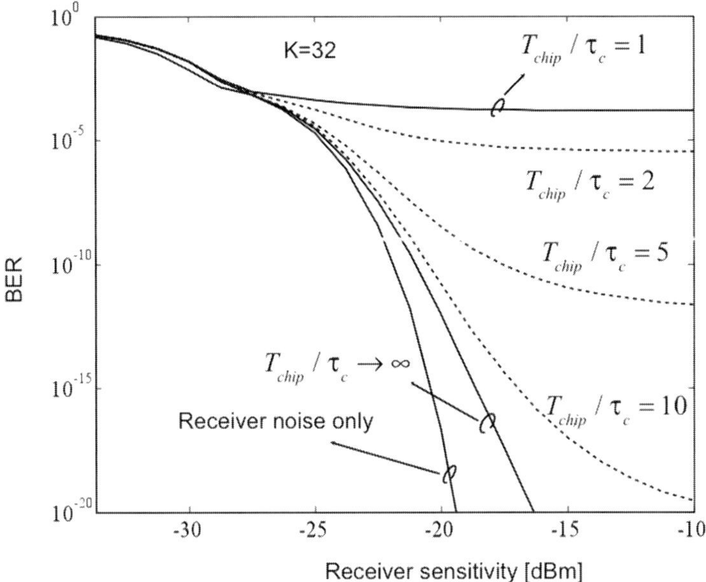

Figure 6.16 Numerical bit error rate plotted as a function of receiver sensitivity for various coherence ratios τ_c/T_{chip} [10]. © Reprinted by permission of IEEE.

where the decision threshold value D is given by

$$D = \frac{\sigma_0(1 + m\xi) + \sigma_1 m\xi}{\sigma_0 + \sigma_1} \tag{6.44}$$

and the noise variances for marks "1" and "0" for the incoherent regime are given by

$$\sigma_{1-in}^2 = \sigma_{MAI}^2 + \sigma_{th}^2 + \sigma_{1-shot}^2, \tag{6.45}$$

and

$$\sigma_{0-in}^2 = \sigma_{MAI}^2 + \sigma_{th}^2 + \sigma_{0-shot}^2. \tag{6.46}$$

In the coherent regime,

$$\sigma_{1-co}^2 = \sigma_{beat}^2 + \sigma_{MAI}^2 + \sigma_{th}^2 + \sigma_{1-shot}^2, \quad \text{and} \quad \sigma_{beat-1}^2 = 2m\xi P_d^2 \tag{6.47}$$

and

$$\sigma_{0-co}^2 = \sigma_{MAI}^2 + \sigma_{th}^2 + \sigma_{beat-0}^2, \ \sigma_{beat-0}^2 = m(m-1)\xi^2 P_d^2. \tag{6.48}$$

In Fig. 6.16 the numerical bit error rate is plotted as a function of receiver sensitivity for various coherence ratios T_{chip}/τ_c. As the coherent ratio T_{chip}/τ_c increases, the BER levels off, which means that the BER does not improve further with an increase in signal power.

6.5.3   Bit rate detection

Another detection scheme of OOK and IM-DD OCDMA systems is bit rate detection. A model of OCDMA-PON with bit rate detection is shown in Fig. 6.14. The difference from chip rate detection is that the photocurrent is integrated during one bit time duration T_B. Bit rate detection is preferable in a system which uses multiport en/decoders whose auto-correlation waveform extends over one bit time duration as shown in Fig. 6.4. The photocurrent at the receiver is written as

$$
\begin{aligned}
I_{ph} &= \int_0^{T_B} \mathbf{R}(EE^*)dt + \int_0^{T_B} n(t)dt \\
&= T_B \mathbf{R} P_d + T_B \mathbf{R} \sum_{i=1}^{m} P_i + 2\mathbf{R} \sum_{i=1}^{m} \sqrt{P_i P_d} \times \int_0^{T_B} \cos\{(\omega_i - \omega_d)t - \omega_i \tau_i \\
&\quad + \phi_i(t - \tau_i) - \phi_d(t)\}dt \\
&\quad + 2\mathbf{R} \sum_{i=1}^{m} \sum_{j=1}^{m} \sqrt{P_i P_j} \int_0^{T_B} \cos\{(\omega_i - \omega_j)t - \omega_i \tau_i + \omega_j \tau_j \\
&\quad + \phi_i(t - \tau_i) - \phi_j(t - \tau_j)\}dt + \int_0^{T_B} n_0(t)dt.
\end{aligned}
\tag{6.49}
$$

The most likely scenario for bit rate detection in a practical OCDMA system is in the partially coherent regime. The bit error rate is expressed as

$$
\begin{aligned}
\text{BER} &= \Pr(0)_{Bit}\text{Pe}(1|0) + \Pr(1)_{Bit}\text{Pe}(0|1) \\
&= \frac{1}{2}\left[\text{Pe}(1|0) + \text{Pe}(0|1)\right].
\end{aligned}
\tag{6.50}
$$

Here, the error probabilities of logical data "1" and "0" are expressed as

$$
\text{Pe}(0|1) = \frac{1}{2}\text{erfc}\left[\frac{P_d(1 + \sum_{i=2}^{m}\frac{P_i}{P_d} - D)}{\sqrt{2}\sigma_1}\right]
\tag{6.51}
$$

$$
\text{Pe}(1|0) = \frac{1}{2}\text{erfc}\left[\frac{P_d(D - \sum_{i=2}^{m}\frac{P_i}{P_d})}{\sqrt{2}\sigma_1}\right]
\tag{6.52}
$$

where

$$
P_d = \int_0^{T_B} |E_{ac}|^2 dt
$$

$$
P_i = \int_0^{T_B} |E_{cc}|^2 dt
\tag{6.53}
$$

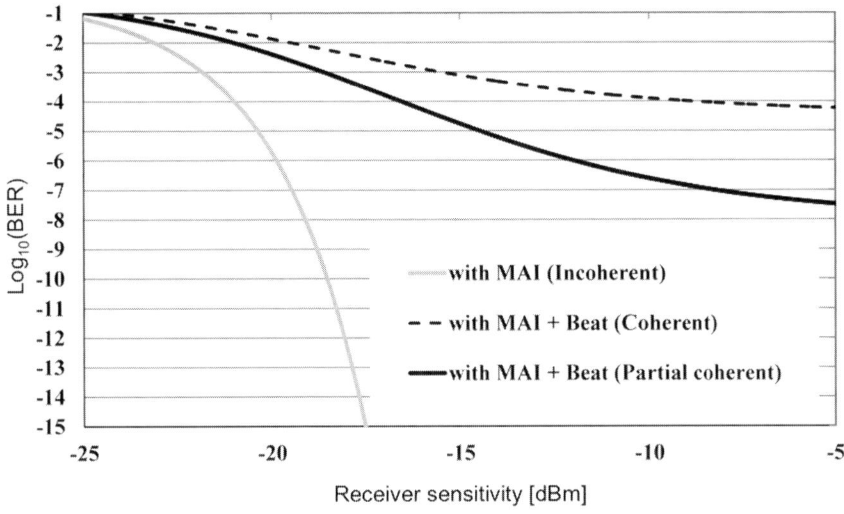

Figure 6.17 Numerical BERs plotted as a function of receiver sensitivity.

where E_{ac} and E_{cc} are the electric fields of auto-correlation and cross-correlation wave-forms, respectively. The decision threshold D is determined by

$$
D = \frac{\sigma_0 \left(1 + \sum_{i=2}^{m} \frac{P_i}{P_d}\right) + \sigma_1 \sum_{i=2}^{m} \frac{P_i}{P_d}}{\sigma_0 + \sigma_1}.
\tag{6.54}
$$

In Fig. 6.17 the numerical BERs of bit rate detection are plotted as a function of receiver sensitivity for a variety of coherence times. At the coherent limit, the beat noise shows error floor while, on the other hand, at the incoherent limit the BER shows no error floor because the beat noise completely disappears.

7   Optical encoding and decoding

7.1   Code sequences

For OCDMA applications, a type of code sequence which has maximum auto-correlation function and minimum cross-correlation function and at the same time proven large cardinality has to be selected from among a wide variety of code sequences. First, a brief review of code sequences used for wireless CDMA systems in the past is presented.

7.1.1   Prime code

The antecedent version of prime codes was first introduced for cellular mobile communication systems utilizing the FH-SS technique described in Section 6.1. The purpose of designing such a code was to support asynchronous transmissions from mobile units to base stations using signaling waveforms that had uniformly small cross-correlation functions for any relative time shift. The structure of the original prime code was based on the theory of linear congruence. Prime code was first introduced in OCDMA networks in the early 1980s [1]. There is a family of prime codes which includes the original prime code, extended prime code, synchronized prime code, $2n$ prime code, generalized prime code, carrier-hopping prime code, multi-length carrier-hopping prime code, concatenated prime code, and multicarrier prime code [2].

The original prime code is constructed using finite field arithmetic. Construction of the prime code with code length $N = p^2$ begins with the Galois field $GF(p) = \{0, 1, \ldots, p-1\}$ of a prime number p. A prime sequence $S_i = (s_{i,0}, s_{i,1}, \ldots, s_{i,j}, \ldots, s_{i,p-1})$ is constructed by multiplying every element j of $GF(p)$ by an element i under modulo-p as

$$s_{i,j} = i \cdot j \ (\mathrm{mod} \ p). \tag{7.1}$$

For example, the prime code sequence over $GF(5)$ is shown in Table 7.1. To construct the original prime code, each one of the prime sequences is mapped onto a binary $(0,1)$ code sequence $C_i = (c_{i,0}, c_{i,1}, \ldots, c_{i,k}, \ldots, c_{i,p^2-1})$, by assigning 1s in positions designated by k in Eq. (7.2) and 0s in all the other positions according to

$$c_{i,k} = \begin{cases} 1 \ \text{if} \ k = s_{i,j} + jp & \text{for} \ j = 0, 1, \ldots, p-1 \\ 0 & \text{otherwise.} \end{cases} \tag{7.2}$$

Table 7.1 Prime sequence $S_i = (s_{i,0}, s_{i,1}, s_{i,2}, s_{i,3}, s_{i,4})$ over $GF(5)$

i	$s_{i,0}$	$s_{i,1}$	$s_{i,2}$	$s_{i,3}$	$s_{i,4}$
0	0	0	0	0	0
1	0	1	2	3	4
2	0	2	4	1	3
3	0	3	1	4	2
4	0	4	3	2	1

Table 7.2 Prime code over $GF(5)$, C_0, C_1, C_2, C_3, and C_4

i	$c_{i,0}$	$c_{i,1}$	$c_{i,2}$	$c_{i,3}$	$c_{i,4}$
0	10000	10000	10000	10000	10000
1	10000	01000	00100	00010	00001
2	10000	00100	00001	01000	00010
3	10000	00010	01000	00001	00100
4	10000	00001	00010	00100	01000

The weight and cardinality of the prime code over $GF(p)$ are both equal to p. The length of the prime code sequence is p^2. The code sequences with $p = 5$ are shown in Table 7.2. The number of 1s per code sequence is equal to p.

The cross-correlation function between two distinct code sequences of the prime code is at most two. Figure 7.1(a) shows the auto-correlation function of the code sequence C_3 of the prime code over $GF(5)$ for the data sequence 1110010100. The peak of the auto-correlation is equal to five, as expected. The cross-correlation function of C_3 with C_2 for the same data stream, shown in Fig. 7.1(b), is at most two. This is because the number of coincidences of 1s for all shifted versions of any two code sequences is only one or two.

7.1.2   Gold code

The Gold code is attractive for CDMA because a large number of orthogonal sequences is available. Gold code sequences are generated by combining a pair of preferred maximal length sequences $\{a, b\}$ using modulo-2 addition in the pair of shift registers, shown in Fig. 7.2. The maximal length sequence has length $N = 2^n - 1$, which is generated using an n-stage binary linear feedback shift register. The Gold code is a set of $N + 2$ code sequences given by $\{a, b, a + b, a + \tau b, \ldots, a + \tau^{N-1}b\}$, where τ^i denotes an operator that shifts the sequence cyclically i chip intervals to the left. An important characteristic of a maximal length code is its two-valued discrete auto-correlation function. The auto-correlation function is given by

$$\psi(t) = \begin{cases} N & \text{for } t = 0 \\ -1 & \text{for } t = i\frac{T_B}{N}, \ i = \pm 1, \pm 2, \ldots \end{cases} \tag{7.3}$$

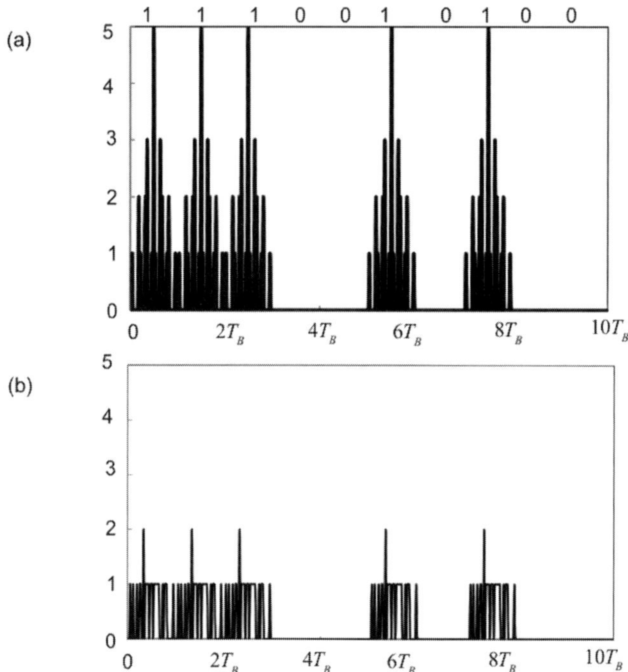

Figure 7.1 Prime code sequence C_3 of the prime code over $GF(7)$. (a) Auto-correlation function and (b) cross-correlation function with C_2 for the data sequence 1110010100.

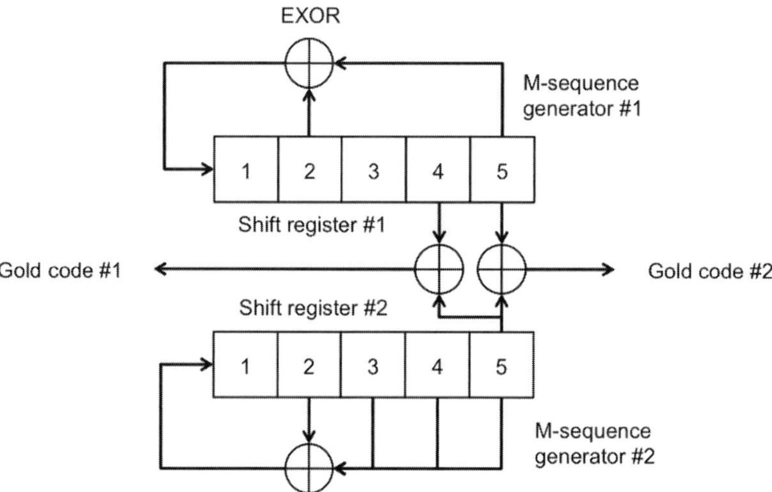

Figure 7.2 Configuration of a Gold code sequence generator using 5-stage shift register.

Arbitrary selection of maximal length code pairs can result in very poor correlation performance. However, the preferred pairs of maximal length codes, as selected by the Gold-derived algorithm, always exhibit the minimum possible cross-correlation peaks, equal to $2^{(n+1)/2} + 1$ (n odd) and $2^{(n+2)/2} - 1$ (n even). These correlation properties

Table 7.3 Performance of preferred pairs compared with worst case pairs [2],
© Reprinted by permission of Pearson

Degree n	Period	Undesired correlation		
		Worst case	Preferred pair	Difference (DB)
5	31	11	9	1.7
6	63	23	15	3.7
7	127	41	17	7.6
8	255	95	31	9.7
9	511	113	33	10.7
10	1023	383	63	15.7
11	2047	287	65	12.9
12	4095	1407	127	20.9
13	8191	703	127	14.9

Table 7.4 Preferred pairs for Gold code generation [2], © Reprinted by permission of Pearson

Degree	Period	Preferred pairs	Bound
5	31	[5,2] [5,4,3,2]	9
7	127	[7,3] [7,3,2,1] [7,3,2,1] [7,5,4,3,2,2]	15
9	511	[9,4] [9,6,4,3] [9,6,4,3] [9,8,4,1]	33

are summarized in Table 7.3 [2]. This implies that certain maximal length sequences will interfere very strongly with each other. Since the set of preferred maximal length sequences is few in number, the maximal length codes are not suitable for CDMA, where a large number of assignable addresses is required. In Table 7.4, some examples of preferred pairs for the generation of Gold codes are given.

For the Gold code, the number of 1s in each code sequence varies with the pair of code sequences, and so do the peaks of the auto-correlation function and the sidelobes. The number of coincidences of "1"s between shifted versions of two code sequences can be large and, therefore, so can the peak of the cross-correlation function. For example, two preferred pairs of maximal length codes having $N = 31 \ (=2^5 - 1)$ chips in Table 7.4, represented by a fifth-order polynomial of the forms, are plotted along with two generated Gold codes in Fig. 7.3. The amplitude takes the value either $+1$ or -1, depending on the code sequence:

$$[5, 2] = 1 + x^2 + x^5 \tag{7.4}$$

$$[5, 4, 3, 2] = 1 + x^2 + x^3 + +x^4 + x^5. \tag{7.5}$$

The auto-correlation and cross-correlation functions of Gold codes 1 and 2 are shown in Figs. 7.4(a)–(c). The transmitted data sequence is (1110010100). Each data bit "1" of duration T_B is encoded into a waveform, consisting of a Gold code sequence of $N = 31$ chips. Data bits "0" are not encoded. The auto-correlation peak is equal to 31.

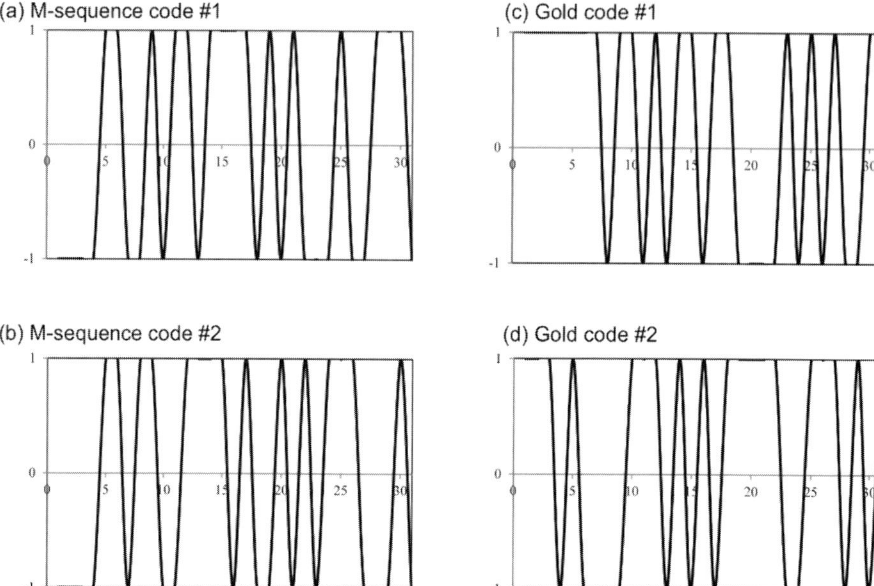

Figure 7.3 Gold code generation from preferred pairs of maximum-length sequence codes [5,2] and [5,4,3,2]: (a) M-sequence code 1, (b) M-sequence code 2, (c) Gold code 1, and (d) Gold code 2.

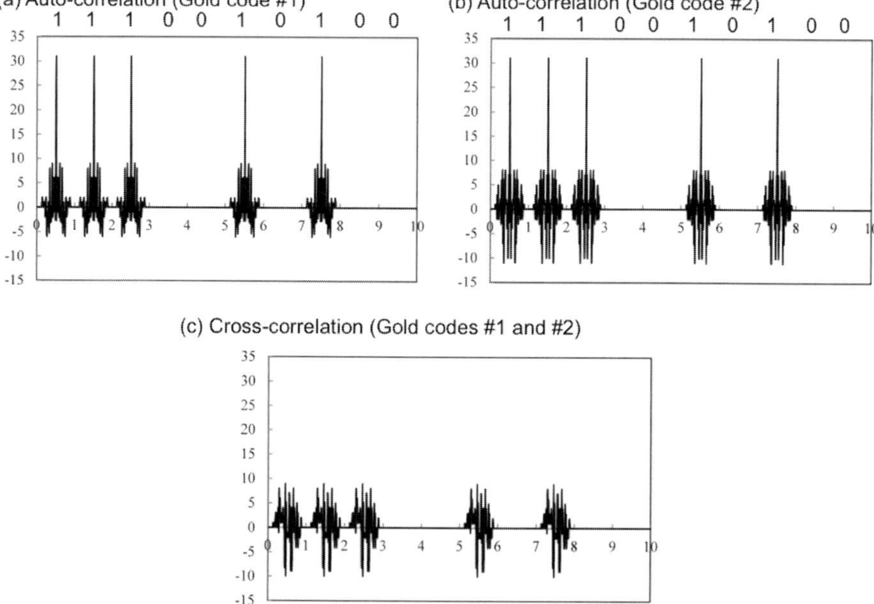

Figure 7.4 Auto-correlations and cross-correlations of Gold codes for the data sequence 1110010100. (a) Auto-correlation function of Gold code 1, (b) auto-correlation function of Gold code 2, and (c) cross-correlation of Gold codes 1 and 2.

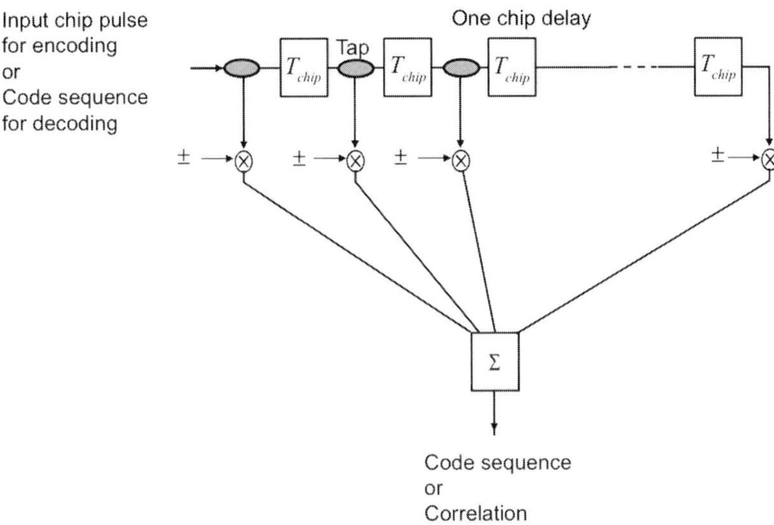

Figure 7.5 Tapped delay-line filter.

7.2 Optical code sequences

7.2.1 On-off-keying code

In Fig. 7.5 a generic configuration of tapped delay line filter consisting of taps and weighing components is illustrated. The input chip is delayed by the chip time T_{chip}, the time interval between the neighboring chips. The weight takes the sign either plus or minus, and the tap ratio is set so that the amplitude of the tapped chip pulse is equal to the other amplitudes. A single chip pulse is launched from the left and is split each time it passes through the tap, and the tapped pulse changes or does not change the sign of the amplitude. The encoding and decoding can both be performed with an identical device. There is a fundamental difference between wireless CDMA and OCDMA. The sign of the weight a_i ($i = 1, 2, \ldots, N$) is either positive or negative in wireless CDMA, but incoherent OCDMA uses OOK coding, and the sign of the weight is positive only [3]. The optical tapped delay line en/decoder is implemented by using an array of fiber delay lines, an optical power splitter, optical switches, and an optical combiner as shown in Fig. 7.6. The optical switch determines the code sequence by shutting or opening the gate. If the switch is in the off state, there is no output from the arm, representing zero in the code sequence. In the encoding, the input optical chip pulse is launched, split equally and guided to the optical fiber delay-lines. The length difference in the delay-line corresponds to the chip time T_{chip} or an integral multiple of T_{chip}. The tapped signals are combined incoherently after experiencing the delay on each arm, resulting in an optical code sequence. In the decoding, an optical code is launched, and the correlation is obtained from the output. It is noted that incoherent OCDMA uses an incoherent light source, characterized by nearly zero visibility V. The visibility is defined in Section 6.4.1. This means that there is no chance that two incoherent lightwaves will

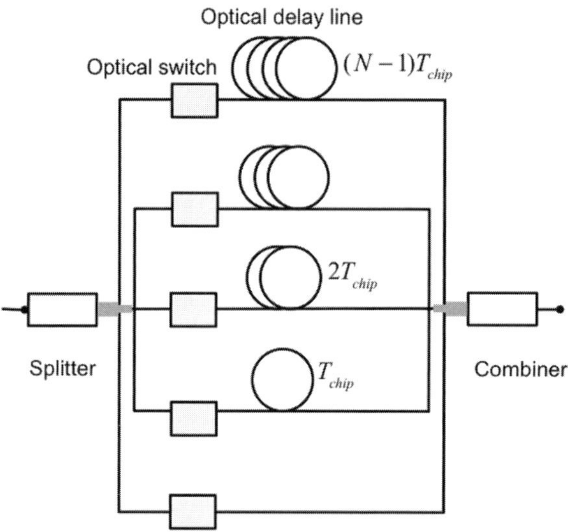

Figure 7.6 Fiber delay-line en/decoder.

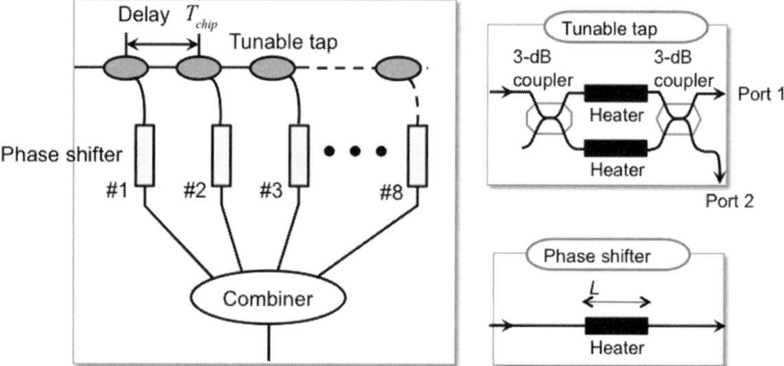

Figure 7.7 Tunable optical tapped delay-line en/decoder with phase shifters [5]. © IEEE by permission.

interfere with each other, and this is why combining the split signals simply sums their powers.

7.2.2 Phase-shift-keying code

Coherent OCDMA systems use optical PSK codes. In the optical PSK code the phase of the optical carrier shifts chip by chip, and the phase shift can be either bipolar or multi-level. The generic architecture of a coherent optical en/decoder consists of a tunable optical tapped delay with a phase shifter on each branch, as shown in Fig. 7.7. Both the tap ratio and the phase of each phase shifter can be tuned according to the code sequence. The input optical pulse is partially tapped at each splitter and experiences the designated phase shift on the branch. The tapped and phase-shifted chip pulses are coherently combined. The waveguide device is fabricated on a silicon substrate using PLC technology. Both the tap ratio of the splitter and the phase of the phase shifter can be tuned by changing the temperature. The tap consists of a Mach–Zehnder interferometer with a thin-film heater on each arm, shown in the inset of Fig. 7.7. The insertion loss is less than 1 dB. The extinction ratio is smaller than -30 dB, and the response time of the switch is around $1{\sim}2$ ms [4]. PLC is widely used in optical communication systems because it provides compactness and excellent stability as well as low loss. The fabrication process uses a silica glass layer of thickness 30 μm or more. GeO_2-doped silica glass is used for the core layer. The waveguide pattern is formed using photolithography and reactive ion etching. The output from Port 2, E_{out2}, for the input from Port 1, $E_{in1} = 1$, is obtained from Eq. (4.20) as

$$E_{out2} = \cos\frac{\phi}{2}\exp\left(j\frac{\phi}{2}\right)\exp(j\omega_c t).$$
(7.6)

Each tap ratio is tuned so that the amplitude of each chip is equal. The phase of the desired code can be properly set with the phase shifter by compensating for the phase shift suffered by the tap. For tuning the phase shifter by $\Delta\phi$, the change in the refractive index Δn on one arm is given by

$$\Delta\phi = \frac{2\pi\Delta n L}{\lambda}$$
(7.7)

where L is the length of the phase shifter, and λ is the wavelength of operation. The relation between the change in refractive index and the temperature change in the silica waveguide is given by

$$\frac{\Delta n}{\Delta T} = 2\times 10^{-8}/K.$$
(7.8)

Figure 7.8 shows the temporal waveforms of an optical encoded signal of code sequence $(\pi, 0, \pi, 0, \pi, 0, \pi, 0, \pi, 0)$, the auto-correlation $(0, \pi, 0, \pi, \pi, 0, \pi, 0)$ and cross-correlation $(\pi, 0, \pi, 0, \pi, 0, \pi, 0, \pi, 0)$ of a 200 Gchip/s, 8-chip optical tapped delay-line encoder [5]. Here, one important figure of merit of the optical en/decoder is the chip rate R_{chip} in units of chips per second [chip/s], defined by

$$R_{chip} = \frac{1}{T_{chip}}.$$
(7.9)

It represents the number of chips in a second. As the chip rate R_{chip} becomes higher, for a given bit rate the code length N_{chip} can become longer.

7.3 Optical encoders and decoders

7.3.1 Superstructured fiber Bragg grating encoder and decoder

A superstructured fiber Bragg grating (SSFBG) is defined as a number of sections of grating, imposed along a length of fiber, having a rapidly varying refractive index modulation of uniform amplitude and pitch, whose envelope is slowly varying in its amplitude and phase. One type of SSFBG has several discrete sections of grating of

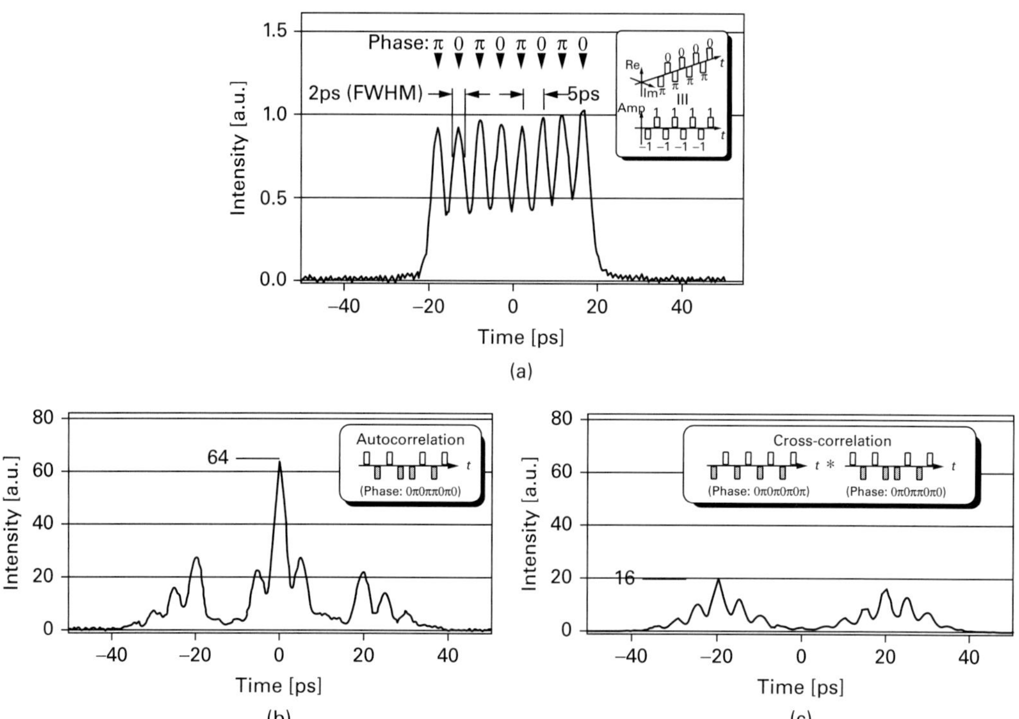

Figure 7.8 Temporal waveforms encoded with a 200 Gchip/s, 8-chip optical tapped delay-line encoder: (a) encoded signal, (b) auto-correlation, and (c) cross-correlation [5]. © IEEE by permission.

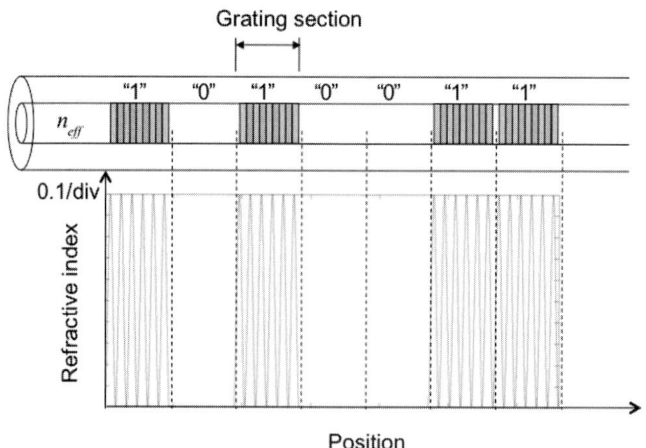

Figure 7.9 Schematic of superstructured fiber Bragg grating for OOK coding.

the same length, of which each grating is formed by modulating the refractive index with a uniform amplitude as shown in the lower trace of Fig. 7.9. When an optical pulse is launched in an SSFBG, the reflected pulse is split into chip pulses whose chip count is equal to the number of grating sections, and either full or zero refractive

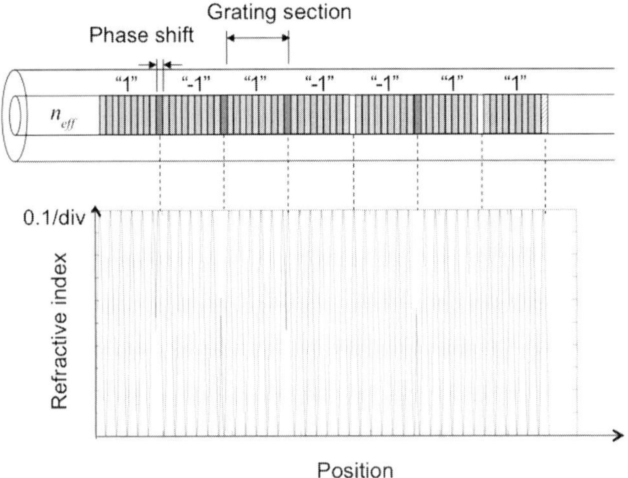

Figure 7.10 Schematic of superstructured fiber Bragg grating for binary PSK (BPSK) coding.

index modulation produces an OOK (unipolar) coded temporal waveform, for example, (010011) as shown in the upper trace. Another type of SSFBG generates PSK code sequences, and has discrete sections of uniform grating, and according to the code sequence the positions of the discrete sections shift by a quarter of a wavelength, $\lambda/4n$, as shown in the upper trace of Fig. 7.10. This gives a phase shift π to the reflected pulse, thus generating binary PSK (bipolar) code $(1 -11 -111)$ as shown in the lower trace.

As the operation mechanism of the FBG is based on deflection theory, we review briefly the deflection of a lightwave at the grating. Let us consider a lightwave incident on a periodic perturbation of the refractive index such as a grating with the period Λ at an angle of incidence θ_i which is deflected at an angle θ_r, as shown in Fig. 3.22. The corrugated refractive index of the grating section is given by cosinusoidal modulation as

$$n(z) = n_0 + \Delta n_B \cos\left(\frac{2\pi}{\Lambda}z\right) \tag{7.10}$$

where n_0 and Δn_B are the refractive index of nominal fiber and the modulation depth of the refractive index, respectively. The uniform rapid refractive index modulation simply defines the central frequency/wavelength of the grating's reflection band. The incidence and deflection angles are governed by the Bragg condition, and Eq. (3.64) is rewritten as

$$\sin\theta_r^{(l)} = \sin\theta_i + \frac{l\lambda}{n_{eff}\Lambda} \quad (l = 0, \pm 1, \pm 2, \ldots) \tag{7.11}$$

where n_{eff} is the refractive index of the medium. The reflection grating holds with the case of first-order deflection,

$$l = -1, \ \theta_i = -\theta_r = \pi/2. \tag{7.12}$$

From Eq. (7.12) the relation between the period of the grating and the Bragg wavelength λ_B is determined as

$$\Lambda = \frac{\lambda_B}{2n_{eff}}.$$
(7.13)

The power reflection coefficient R at the Bragg wavelength, which is derived in Section 3.5.2, is given by

$$R = \tanh^2 \kappa L.$$
(7.14)

For example, the reflectivity R goes up to 90% with $\kappa L = 1.84$. However, in the weakly coupled SSFBG limit the reflectivity is $<20\%$, i.e., where the grating depth is determined in such a way that light penetrates the full grating length to the other end, and the individual elements of the grating contribute more or less equally to the reflected response. The wave vector response $k_0 N(v)$ can be shown to be given simply by the Fourier transform of the spatial superstructured refractive index modulation profile $n(z)$ [6]:

$$N(v) = \int_{-\infty}^{\infty} n(z)e^{j2\pi v z} dz$$
(7.15)

where v is the spatial frequency of the superstructured modulation profile. Similarly, the impulse response of a fiber grating $h(t)$ is given by the inverse Fourier transform of its frequency response $H(f)$

$$H(f) = \int_{-\infty}^{\infty} h(t)e^{j2\pi f t} dt.$$
(7.16)

The scaling factor $t = 2nz/c$ is used to convert from the spatial to the temporal domain, where n is the refractive index, and c is the speed of light. This conversion implies that the spatial frequency u is proportional to the optical frequency f. The frequency response of the grating $N(u)$ can be rewritten as

$$N(u) = N \left(\frac{2n}{c} f \right).$$
(7.17)

The frequency spectrum of the reflected lightwave is given by the convolution of the Fourier transform of the incident signal $X(f)$ and the frequency response of the grating $N(f)$ as

$$Y(f) = X(f) * N(f).$$
(7.18)

A uniform FBG has a single spatial frequency component $u = 1/\Lambda$, specified by Eq. (7.13), and only the lightwave with the Bragg wavelength is deflected. By contrast, the SSFBG has a broadband spatial spectrum. This is why the optical code sequence can be formed by illuminating with light with a broad spectrum such as a short optical pulse. The impulse response of a weak grating has a temporal profile given by the complex form of the refractive index superstructure modulation profile of the grating. For example, in the case that the grating superstructure is simply amplitude modulated, i.e., the grating phase is uniform, the impulse response follows precisely the amplitude modulation profile used to write the grating. When a short but finite bandwidth pulse

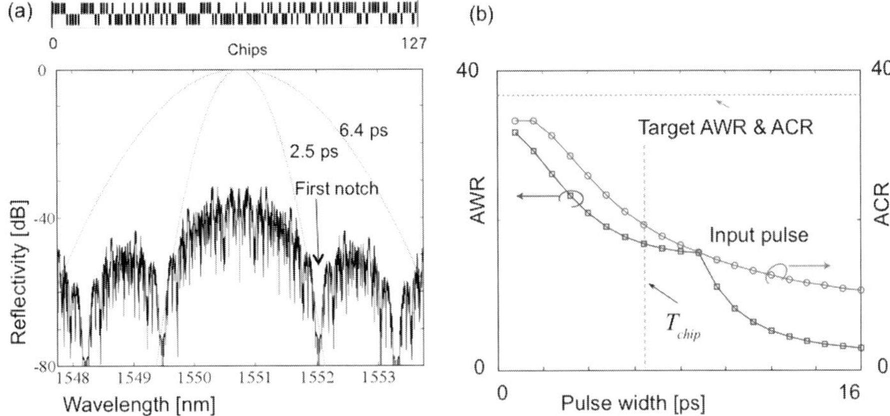

Figure 7.11 (a) Spectrum and (b) performance of 160 Gchip/s, 127-chip PSK code generated by SSFBG [7]. © Reprinted by permission of the Optical Society of America.

(i.e., not an impulse pulse) having temporal waveform $x(t)$ is reflected from an SSFBG, it is transformed into a pulse $y(t)$ with a temporal shape given by the convolution between the input pulse and the impulse response of the grating,

$$y(t) = x(t)^* h(t). \tag{7.19}$$

7.3.1.1   SSFBG encoder and decoder for PSK codes

We consider the SSFBG en/decoder which generates the PSK code sequences [7]. The modulation depth of the refractive index Δn_B is uniform along the z-axis, and only the positions of some discrete sections shift by a quarter of the wavelength, $\lambda/4n_{eff}$, as shown in the lower trace of Fig. 7.10. Assume that the input optical pulse is transform limited with Gaussian shape, for example, the SSFBG 127-chip Gold code encoder with index modulation $\Delta n_B = 1.0 \times 10^{-6}$ and corresponding chip duration $L_{chip} = 0.64$ [mm], as shown at the top of Fig. 7.11(a). The two phase values 0 and π correspond to the binary code 0 and 1. The chip interval is calculated as

$$T_{chip} = \frac{2n_{eff}L_{chip}}{c} \cong 6.4 \text{ [ps]}. \tag{7.20}$$

The chip rate R_{chip}, defined by Eq. (7.9), is calculated to be 160 Gchip/s. The spectrum profile is symmetric, and the first notch appears around 160 GHz apart from the center wavelength. Two parameters are introduced as criteria for the design of SSFBG en/decoders: the auto-correlation peak-to-cross-correlation ratio (ACR) and the auto-correlation peak-to-wing ratio (AWR), shown in Fig. 7.12 and defined as follows:

ACR = Peak intensity of auto-correlation/maximum intensity of cross-correlation,
AWR = Peak intensity of auto-correlation/maximum intensity of wing of auto-correlation.

Table 7.5 Code subsets selected from 127-chip and 511-chip Gold code sequences [7], © Reprinted by permission of the Optical Society of America

127-chip Gold code	ACR >			17.9				20.5	
(total 129 codes)	AWR >	17.9	30	35	38	42	35	38	42
	Number of codes	129	110	82	11	48	22	29	19
511-chip Gold code	ACR >			70			75	77	
(total 513 codes)	AWR >	70	110	120	135	140	130	130	140
	Number of codes	513	430	201	107	76	40	29	31

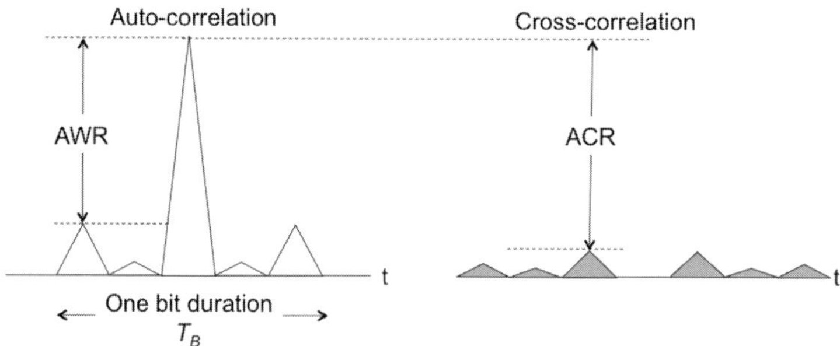

Figure 7.12 Design criteria: ACR, the ratio of the peak intensity over the maximum cross-correlation and AWR, the ratio of auto-correlation intensity peak over the maximum wing level.

The numerical ACR and AWR are plotted as a function of the input pulse width in Fig. 7.12(b). The ACR is the average for 10 codes. The target ratios ($\sim$37) are calculated using the OC sequences. As the pulsewidth becomes shorter, the AWR and ACR become closer to the target value, while an increase in the pulsewidth results in a decrease in both the ratios. The shorter optical pulse, therefore, is more suitable for obtaining better correlation properties.

Let us consider the modulation depth of the grating. It is recalled that the low reflectivity of the SSFBG causes high insertion loss of the en/decoder. This will be a critical issue in its application in practical systems, for instance in PON systems, because the power budget is always tight. A compromise has to be made between the insertion loss and the correlation properties. Then, the question arises of what is the optimum reflectivity, in other words, what is the modulation depth of the refractive index Δn_B for the SSFBG en/decoder. In Table 7.5 the subsets of PSK 127-chip and 511-chip Gold codes for several levels of correlation criteria are summarized, based upon calculations of all the values of ACR and AWR for all the code sequences. If the criteria for ACR and AWR are set too high, only a small number of codes is available. The setting of these criteria has to be done carefully, taking into account the requirements of the systems, and then a particular code subset will be selected.

To find the optimum refractive index depth Δn_B, the performance of SSFBG en/decoders in different Δn_B regimes is evaluated using ACR and AWR. Figure 7.13 shows temporal waveforms of the generated binary PSK Gold codes of 127-chip SSFBG

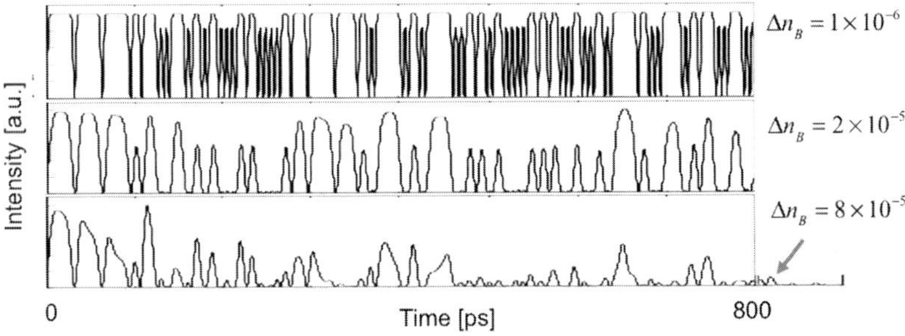

Figure 7.13 Temporal waveforms of the generated binary PSK Gold codes of 127-chip SSFBG encoders for various values of Δn_B [7]. © Reprinted by permission of the Optical Society of America.

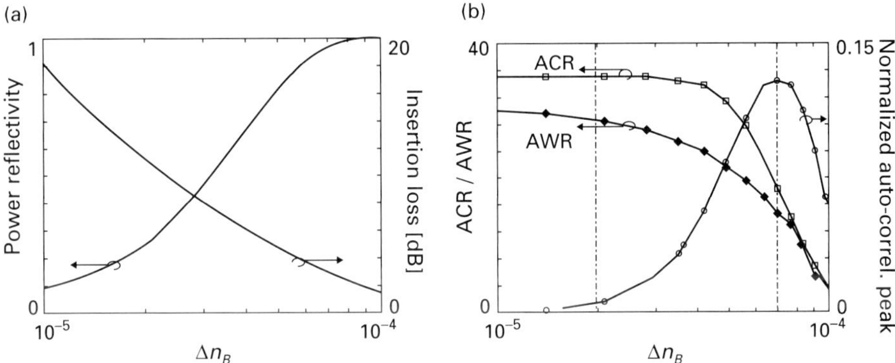

Figure 7.14 (a) Peak reflectivity and insertion loss versus Δn_B and (b) ACR and AWR along with the normalized peak intensity versus Δn_B [7]. © Reprinted by permission of the Optical Society of America.

encoders for various values of Δn_B. The intensity of the generated code is quite uniform for low Δn_B, while with increasing Δn_B the intensity becomes uneven and decays toward the tail as indicated by the arrow. These waveform degradations are due to two factors: one is that, due to the high reflection, only a small portion of the input light penetrates to the other end of the grating, and the other is the stronger multiple reflections between different grating sections. The calculated power reflectivity and the insertion loss are plotted as a function of Δn_B in Fig. 7.14(a). The insertion loss represents the overall optical power reflectivity of the SSFBG. With increasing Δn_B the reflectivity increases and levels off around $\Delta n_B = 7 \times 10^{-5}$, and the insertion loss decreases. The calculated ACR and AWR along with the normalized auto-correlation peak value are plotted against Δn_B in Fig. 7.14(b). It is noteworthy that both the ACR and AWR decrease monotonically with increasing Δn_B. Therefore, high ACR and AWR and low insertion loss cannot go together. A compromise solution giving a higher ACR and AWR and lower insertion loss will be found in the region of $\Delta n_B \leq 5 \times 10^{-5}$.

To date, the longest code length which has been realized with an SSFBG is 511 chips [8]. It can generate 640 Gchip/s binary PSK Gold code sequences. The modulation depth

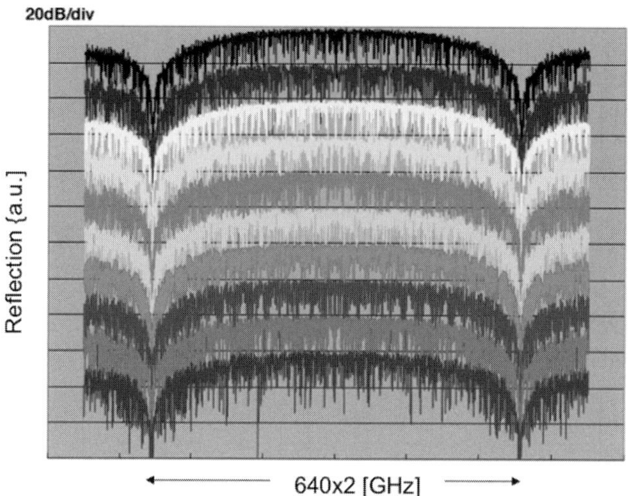

Figure 7.15 Reflection spectra of 511-chip PSK Gold codes generated by an SSFBG encoder [7]. © Reprinted by permission of the Optical Society of America.

of the refractive index is set at $\Delta n_B = 1.8 \times 10^{-4}$ by compromising between the ACR and AWR and the peak reflectivity. The chip length and the total length of the grating are 0.156 mm and 80 mm, respectively, which corresponds to a chip interval of 1.6 ps equivalent to a chip rate of 640 Gchip/s. The SSFBG is fabricated using the two-beam holographic technique described in Section 7.3.1.3. The reflection spectra of the test SSFBG are shown in Fig. 7.15. The profiles are symmetric with respect to the center wavelength of 1550 nm, and the first notches appear at 640 GHz on both sides of 1550 nm. The temporal waveforms of 511-chip PSK Gold code along with the auto-correlation and cross-correlation are shown in Fig. 7.16. The encoded waveform stretches to about 800 ps. Well-defined auto-correlation peaks and low-level cross-correlation are observed. The ACR and AWR of the test SSFBGs are estimated to be roughly 40~14 and 23~10, respectively, which are not as good as the theoretical predictions in Table 7.5. The Bragg wavelength can be tuned with temperature and/or stress in the fiber. It is of practical importance to provide some tolerance to the deviation of the Bragg wavelengths of the SSFBG encoder and decoder. Regarding the Bragg wavelength mismatch caused by the temperature variations, the auto-correlation and cross-correlation peaks are plotted as a function of the temperature difference between the encoder and decoder in Fig. 7.17. The temperature coefficient of a fiber grating is assumed to be 1.25×10^{-2} [nm/°C] due to the variation in fiber length. The experimental results agree well with the theoretical calculations. The temperature deviation tolerance will be ±0.3 °C.

7.3.1.2   Apodized profile

The light reflected from the far end of the grating section of the FBG suffers more loss after penetrating through the whole grating than that reflected from the near end.

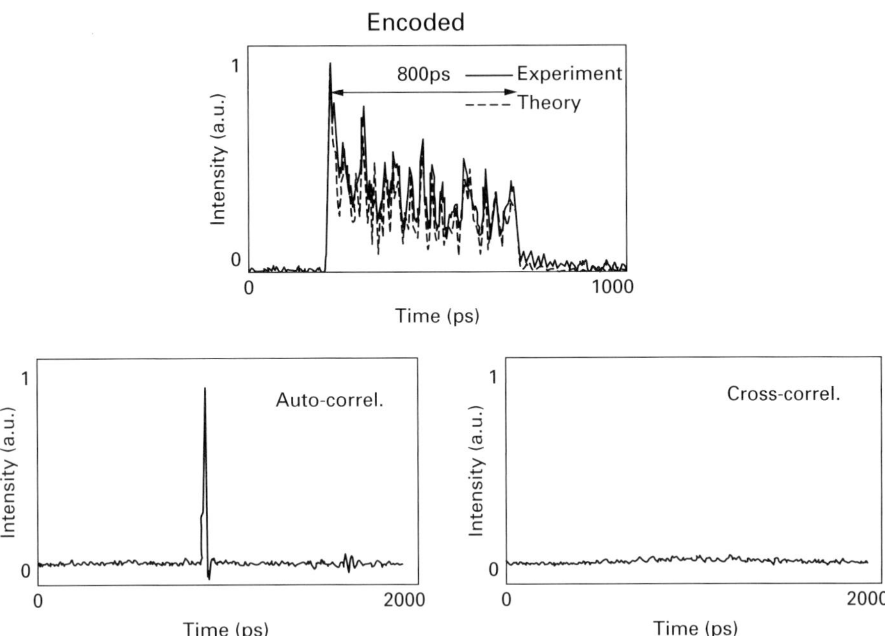

Figure 7.16 Temporal waveforms of the generated 511-chip binary PSK Gold codes [8]. © Reprinted by permission of the Optical Society of America.

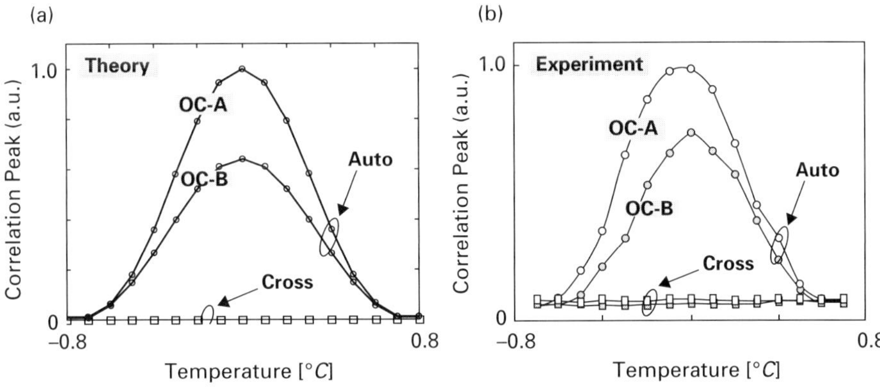

Figure 7.17 Auto-correlation and cross-correlation peaks versus temperature mismatch between the encoder and decoder: (a) theoretical and (b) experimental [8]. © Reprinted by permission of the Optical Society of America.

A straightforward method to compensate for the intensity difference in the reflected signals from the individual sections introduces the apodization technique. The apodized refractive index profile along the z-axis has a gradually increasing modulation depth of the refractive index from the input end along the fiber axis. This refers to the

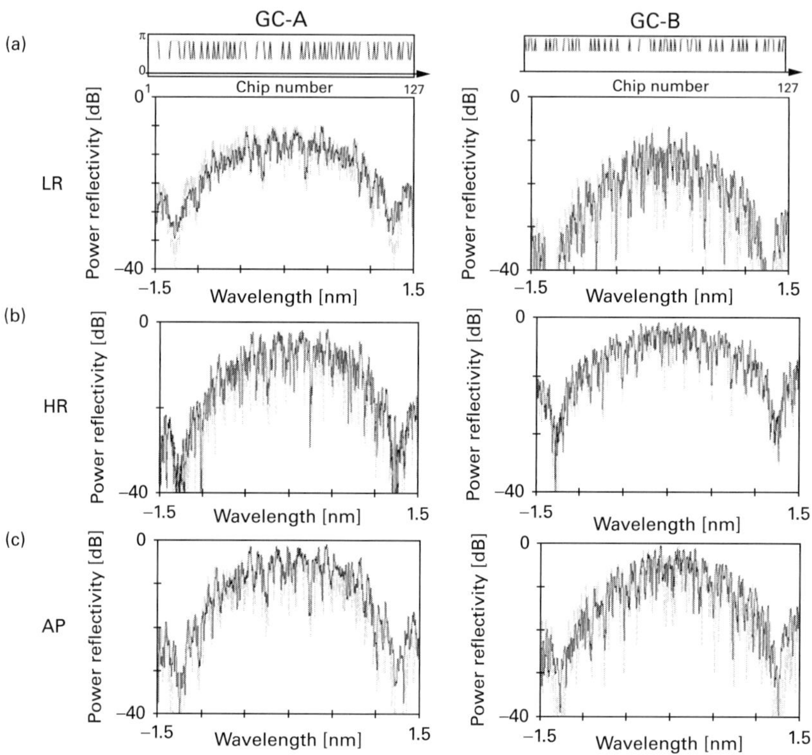

Figure 7.18 Test 127-chip, 160 Gchip/s, binary PSK Gold code SSFBG en/decoders. Two Gold code sequences on the top and the measured (solid lines) and calculated (dashed lines) reflection spectra: (a) uniform LR, (b) uniform HR, and (c) AP SSFBG en/decoders [7]. © Reprinted by permission of the Optical Society of America.

"apodization" of the grating, which has a non-uniform envelope of modulated depth of the refractive index. Complicated apodization profiles illustrated in Fig. 7.11 will be described in Section 9.3. This has been demonstrated using test SSFBG en/decoders [7]. Three types of test 127-chip, 160 Gchip/s, binary PSK Gold code SSFBG en/decoders were compared, including apodized (AP), uniform low-reflectivity (LR), and uniform high-reflectivity (HR) SSFBGs. The refractive index depths Δn_B of the LR and HR devices are 1.0×10^{-5} and 4.0×10^{-5}, respectively. In the AP device the modulation depth of the refractive index of the grating section increases monotonically from the input end with average value $\Delta n_B = 4.0 \times 10^{-5}$. From the spectra of encoded signals of the two Gold codes in Fig. 7.18, the first notch is located 160 GHz (1.28 nm) from the center. Apparently, there is not much difference in the profiles. The measured temporal waveforms of the input pulse and the reflected signals of LR, HR, and AP SSFBG encoders are shown in Fig. 7.19. There is a remarkable difference in the waveforms. The encoded signal from the LR encoder has a uniform waveform but the intensity is low. The waveform of the HR sample has a higher intensity but there is decay from the precursor to the tail. The waveform of the AP sample shows improved uniformity compared with

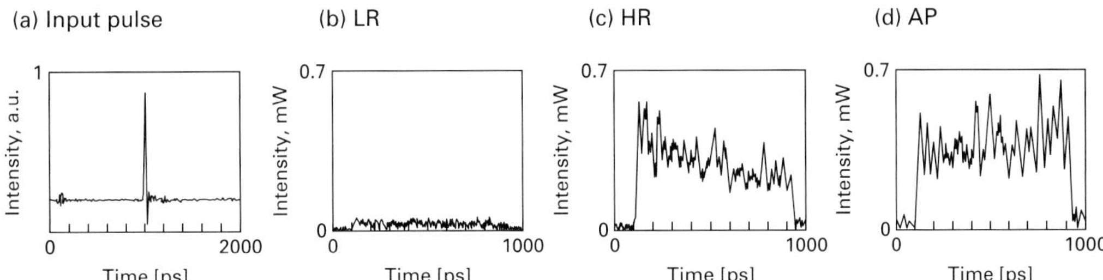

Figure 7.19 Test SSFBG en/decoders. Waveforms of (a) input pulse, and generated OC-A signals from (b) LR, (c) HR, and (d) AP samples [7].© Reprinted by permission of the Optical Society of America.

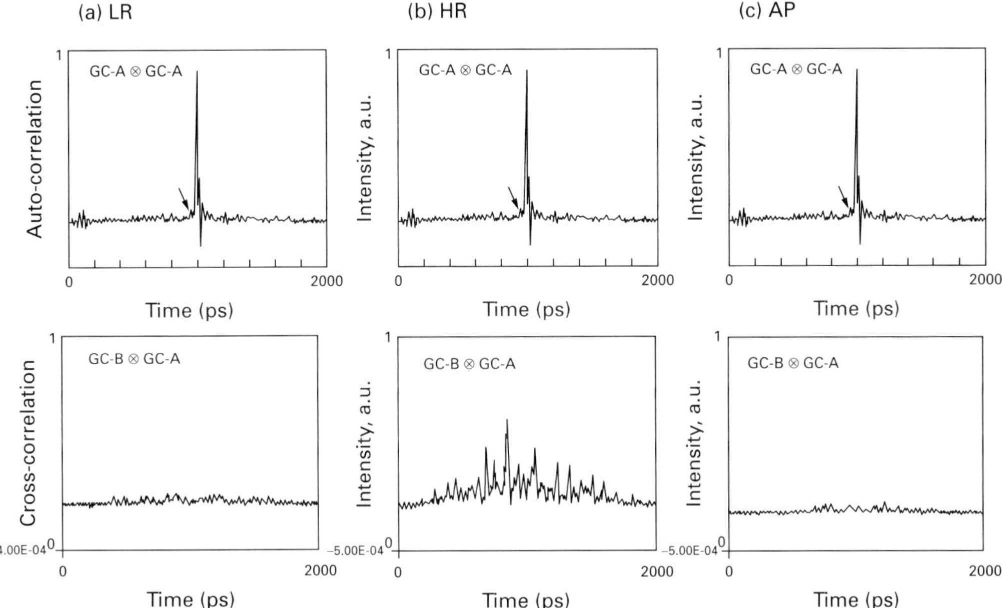

Figure 7.20 Test SSFBG en/decoders. Auto-correlation and cross-correlation waveforms of (a) LR, (b) HR and (c) AP samples. The intensity is normalized by the auto-correlation peaks [7]. © Reprinted by permission of the Optical Society of America.

the HR device, and has the highest intensity level. The measured auto-correlation and cross-correlation waveforms are shown in Fig. 7.20. The maximum wings in the auto-correlation waveforms, indicated by an arrow in the figures, are located in front of the auto-correlation peaks. The ACR and AWR are lower in the case of the HR sample than in the case of the LR sample. The AP sample shows the best ACR and AWR. This is confirmed by the BER measurements of OCDMA system experiments using the test SSFBG en/decoders. The bit error rates measured for the three test SSFBG en/decoders with and without multiplexing are shown in Fig. 7.21. In the case of two-code multiplexing (filled marks in the figure), only the AP SSFBG en/decoder can achieve error-free transmission.

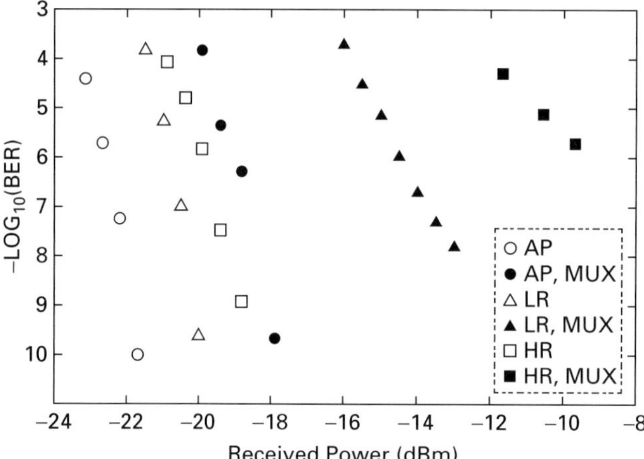

Figure 7.21 BER performance of LR, HR, and AP SSFBG en/decoders for single code and two-code multiplexing [7]. © Reprinted by permission of the Optical Society of America.

Consequently, the AP SSFBG en/decoder has better overall BER performance, compared to the LR and HR samples.

7.3.1.3 Fabrication methods

In silica fibers the photosensitive effect is induced, and their optical properties are changed permanently (a lifetime of twenty-five years is predicted) when they are exposed to intense radiation of a blue or ultraviolet laser [9]. A conventional silica fiber doped with germanium is extremely photosensitive. The first grating formation was discovered accidentally when an intense argon-ion laser radiation at a wavelength of 514.5 nm was launched into a germanium-doped fiber. After several minutes the reflected light intensity started to increase, and eventually almost all the light was reflected from the fiber. It was confirmed that a very narrowband Bragg grating filter had been formed over the entire 1 m length of the fiber. Later, this was developed as the single-beam internal technique to fabricate fiber gratings. A shortcoming of this technique is that the grating can only be formed near the wavelength of the laser used to illuminate it. As germanium-doped fibers exhibit little photosensitivity in the spectral region above 500 nm, this technique is not effective in the regions of 1300 nm and 1550 nm.

There are other techniques for fabricating fiber gratings, including continuous grating writing using a phase mask and holographic techniques. The phase mask is made from a flat slab of silica glass which is transparent to ultraviolet light. On one of the flat surfaces, a one-dimensional periodic surface relief structure is etched using a photolithographic technique. The shape of the periodic pattern approximates a square wave in profile. The optical fiber is placed almost in contact with the corrugations of the phase mask as shown in Fig. 7.22. An ultraviolet exposure beam passes through a narrow slit and is incident normal to the phase mask. It is scanned along the mask, and is diffracted by

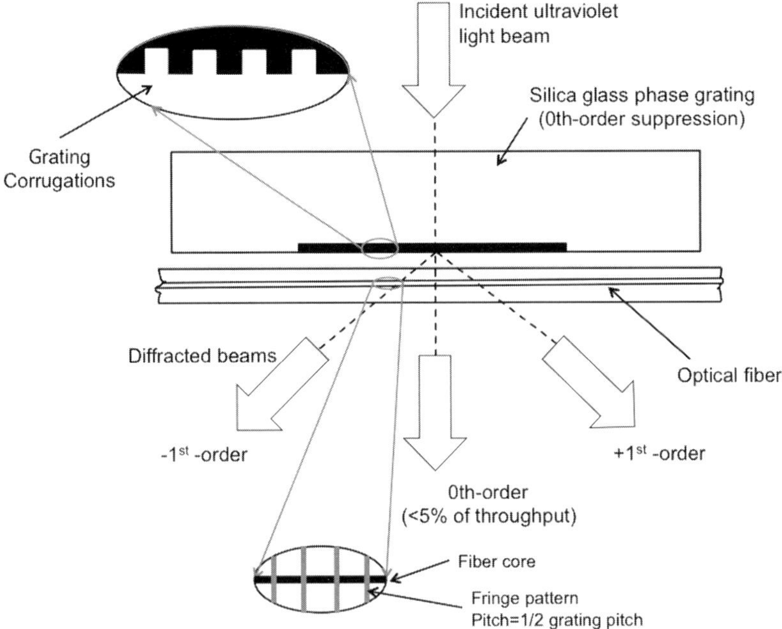

Figure 7.22 Phase mask technique without zero-order diffraction [9]. © Reprinted by permission of IEEE.

the periodic corrugations of the phase mask. For writing PSK code, the phase difference is introduced chip by chip by controlling the relative position between the fiber and the mask during exposure. The amplitude of the phase mask grooves is chosen to reduce the light transmitted in the zero-order beam to less than 5% of the total throughput. These choices result typically in more than 80% of the throughput being in the ± 1 diffracted beams. This technique has the advantage that the fiber cladding is transparent to the ultraviolet beam whereas the fiber core is highly absorptive to the ultraviolet beam. The phase mask technique greatly simplifies the manufacturing process for Bragg gratings, yet still yields gratings with high performance. A drawback of the phase mask technique is obviously that a separate phase mask has to be prepared for each different Bragg wavelength. The change in refractive index Δn_B in germanium-doped single-mode fiber ranges from 10^{-5} to 10^{-3}. Using hydrogen loading Δn_B can reach as high as 10^{-2}.

The two-beam holographic technique solves the problems of the single-beam internal technique and the phase mask technique. As shown in Fig. 7.23, the fiber is irradiated from the side with two intersecting coherent ultraviolet light beams. The two overlapping beams interfere, producing a periodic interference pattern that writes a corresponding periodic refractive index grating in the core of the fiber. The holographic technique for grating fabrication has two principal advantages. Bragg gratings can be printed in the fiber core without removing the glass cladding. Furthermore, the period of the photo-induced grating Λ depends on the angle 2θ between the two interfering coherent

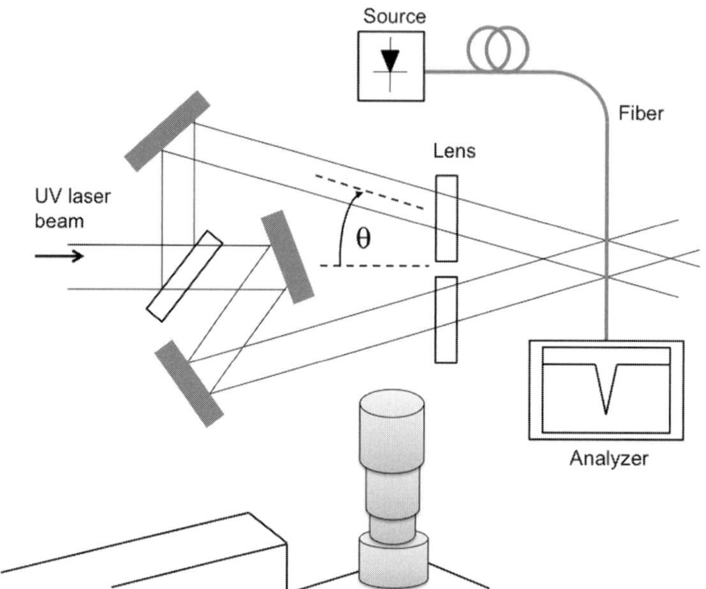

Figure 7.23 Two-beam holographic technique [9]. © IEEE by permission.

ultraviolet light beams:

$$|\mathbf{k}_B| = 2\,|\mathbf{k}_{UV}|\sin\theta$$
$$\Lambda = \frac{\lambda_{UV}}{2\sin\theta} \tag{7.21}$$

where the wave vectors of the Bragg grating and the ultraviolet beam are denoted by $\mathbf{k}_B$ and $\mathbf{k}_{UV}$. Thus, even though ultraviolet light is used to fabricate the grating, Bragg gratings can be formed at much longer wavelengths of interest for devices which have applications in fiber optic communications and optical sensors.

7.3.2   Multiport encoder and decoder in arrayed waveguide grating configuration

This device is a silica-based passive PLC, with N inputs and N outputs. The device has an arrayed waveguide grating (AWG) configuration, but the corresponding design rules are different from those of a wavelength multiplexer/demultiplexer [10]. N different PSK optical codes are generated simultaneously at N different outputs as shown in Fig. 7.24. It looks like a conventional AWG filter. The optical code generated is N-chip, N phase-level PSK code sequences. It acts as a decoder as well. When an optical code is fed into one input, all the correlation signals between the input code and the N codes are generated at the different device outputs. Due to the unique cyclic property like an AWG, the combination of numbers at the input port of the device at which an optical code is launched and at the output port number at which the auto-correlation peak appears immediately tells what code is detected. The cyclic property of a five-port en/decoder is illustrated in Fig. 7.25. The superscripts and subscripts of OC denote the port number and optical

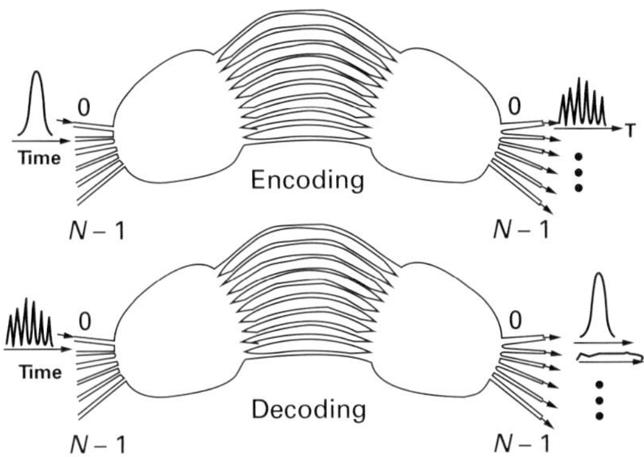

Figure 7.24 Encoding and decoding in a multiport en/decoder [10]. © Reprinted by permission of IEEE.

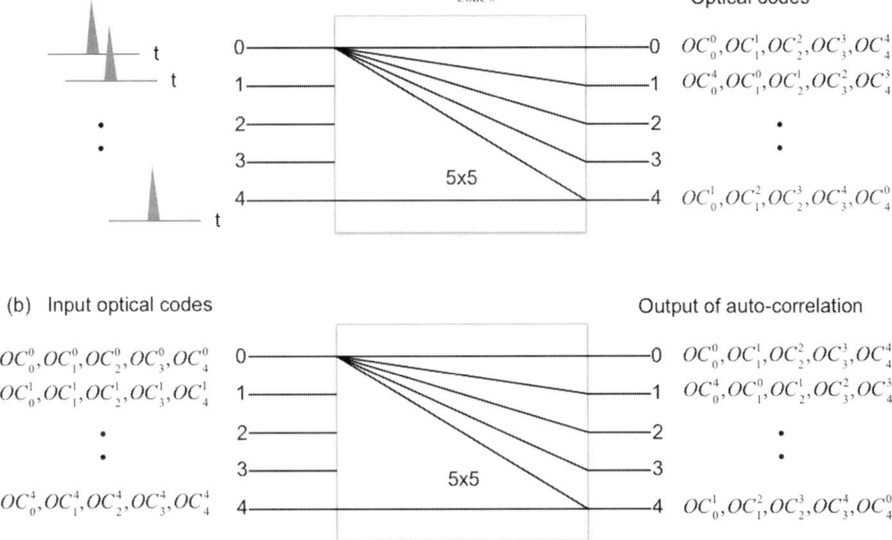

Figure 7.25 Cyclic property of a 5×5 multiport en/decoder. The superscripts and subscripts denote the port number and optical code number, respectively. (a) Encoding and (b) decoding.

code number, respectively. For example, five codes OC_0^0, OC_1^0, OC_2^0, OC_3^0, OC_4^0 are launched in input Port 0, the correlations come out from different output ports: OC_0^0 from Port 0, OC_1^0 from Port 1, ..., OC_4^0 from Port 4.

7.3.2.1   Principle of design

The mechanism used to build a set of optical codes can be described easily by analyzing the AWG in the time domain. If a short optical pulse is driven into one of the input ports

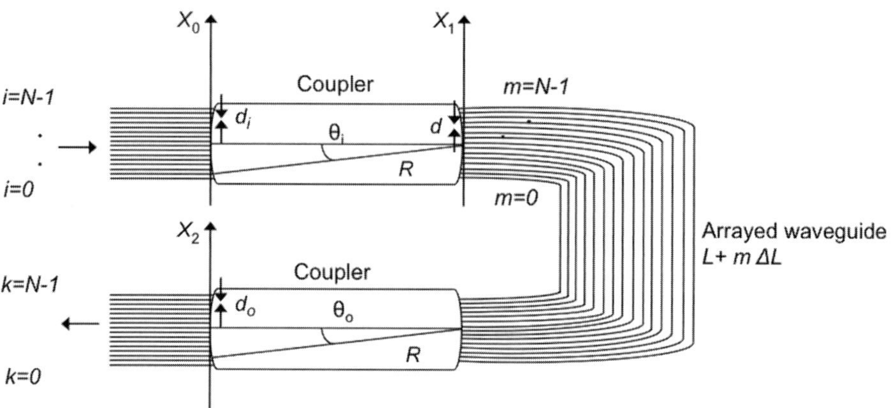

Figure 7.26 Multiport encoder in AWG configuration [10]. © Reprinted by permission of IEEE.

shown in Fig. 7.26, N copies of the pulse are generated by the input slab coupler, with phase given by the Rowland circle configuration where the positions of all the input and output ports are located on the circles. The optical pulses from different input ports are incident on the arrayed waveguides, travel different paths in the arrayed waveguides, and the output slab coupler recombines the pulses to build N codes at the outputs. Each PSK code is composed of N optical chips, and the differential path delay $\Delta \tau = n_{eff} \Delta L / c$ in the grating is chosen to be larger than the input pulsewidth, so that the chips in the optical code do not overlap. The impulse response between the input port i and the output port k is given by

$$h_{ik}(t) = \sum_{m=0}^{N-1} \exp\left[-j \frac{\pi n_{eff} d}{\lambda}(2m - N + 1)(\sin \theta_i + \sin \theta_0) \right]$$
$$\times \delta\left(t - n_{eff} \frac{L + m \Delta L}{c} \right), \quad i, k = 0, 1, \ldots, N - 1 \qquad (7.22)$$

where $\delta(\ )$ is the Dirac delta, n_{eff} is the effective refractive index, L is the shortest length of the arrayed waveguides, and θ_i and θ_o are the diffraction angles in the input and output slab couplers, respectively:

$$\sin \theta_i \cong (2i - N + 1)\frac{d_i}{2R}$$
$$\sin \theta_o \cong (2k - N + 1)\frac{d_o}{2R}. \qquad (7.23)$$

where $d_i d_i = d_0$ Assume that all the spacings of the input and output ports are equal, $d_i = d_o$. The parameters are set so that the port count N satisfies

$$\frac{\lambda R}{n_{eff} d d_o} = N. \qquad (7.24)$$

The above constraint on the parameters is peculiar to the en/decoder, and this differs from the conventional wavelength mux/demux in AWG configuration. It is noteworthy that the number of arrayed waveguide gratings is the same as the number of input/output ports N, in the en/decoder but this is not always the case in the wavelength mux/demux. Equation (7.22) is reduced to

$$h_{ik}(t) = \sum_{m=0}^{N-1} \exp\left[-j\frac{\pi}{N}(2m - N + 1)(i + k - N + 1)\right] \times \delta(t - m\Delta\tau), \quad i, k$$
$$= 0, 1, \ldots, N - 1. \tag{7.25}$$

For the sake of simplicity, $L = 0$ is assumed as its value does not affect the code generation and decoding, but only corresponds to a constant time delay. The code generated is a time-spread N-ary PSK code, consisting of N chips with time interval $\Delta\tau$. The number N of the device inputs and outputs coincides with both the number of codes having N chips. In particular, for a given input port i, it is obvious from Eq. (7.25) that the code generated at the output $k = N - i - 1$ has all the chips with identical phase zero. Otherwise, each chip incurs a different phase shift. A more accurate model of the multiport en/decoder in the AWG configuration, based on diffraction theory, has been developed, which takes into account the spatial profile of the waveguide mode and the loss of non-uniformity due to the position of the port with respect to the central input and output ports.

The transfer function $H_{ik}(f)$ from the input i to the output k in the frequency domain is obtained by Fourier transforming Eq. (7.25) as

$$H_{ik}(f) = \sum_{m=0}^{N-1} \exp\left\{-j\pi(2m - N + 1)\left(\frac{i + k + 1}{N} + \Delta\tau f\right)\right\}$$
$$= \exp\{-j\pi(N - 1)\Delta\tau f\}\frac{\sin\{\pi(i + k + 1 + N\Delta\tau f)\}}{\sin\left\{\pi\left(\frac{i+k+1}{N} + \Delta\tau f\right)\right\}}, \quad i, k$$
$$= 0, 1, \ldots, N - 1. \tag{7.26}$$

It is noteworthy that in the frequency domain each code is equispaced with the frequency spacing of

$$\Delta f = \frac{\text{FSR}}{N} \tag{7.27}$$

where the inverse of the time interval $\Delta\tau$ corresponds to the free spectral range (FSR) of the en/decoder

$$\text{FSR} = \frac{1}{\Delta\tau} = \frac{c}{n_{\text{eff}}\Delta L}. \tag{7.28}$$

The power spectra of 16-chip optical code sequences calculated from Eq. (7.26) are plotted in Fig. 7.27. The waveguide parameters are summarized in Table 7.6. The frequency interval between the peaks of adjacent codes is 12.5 GHz with FSR $= 200$ GHz. This contrasts with the code sequence generated by an SSFBG in that the center wavelengths of all the code sequences are the same, the Bragg wavelength. The spectrum of the code partly overlaps with those of neighboring codes. If there is no overlap in

Table 7.6 Layout parameters [10], © Reprinted by permission of IEEE

Symbol	Description	Value
f_0	Carrier frequency at $\lambda = 1550$ nm	193.292 THz
N	Number of ports	16
R	Radius of grating circle	20.85 mm
d	Spacing of arrayed waveguide	24.6 μm
w_g	Width of arrayed waveguide	7 μm
d_i	Spacing of input waveguide	56.47 μm
d_o	Spacing of output waveguide	56.47 μm
w	Width of input/output waveguides	50 μm
ΔL	Path difference of arrayed waveguide	1.0316 mm
n_{eff}	Effective refractive index	1.468

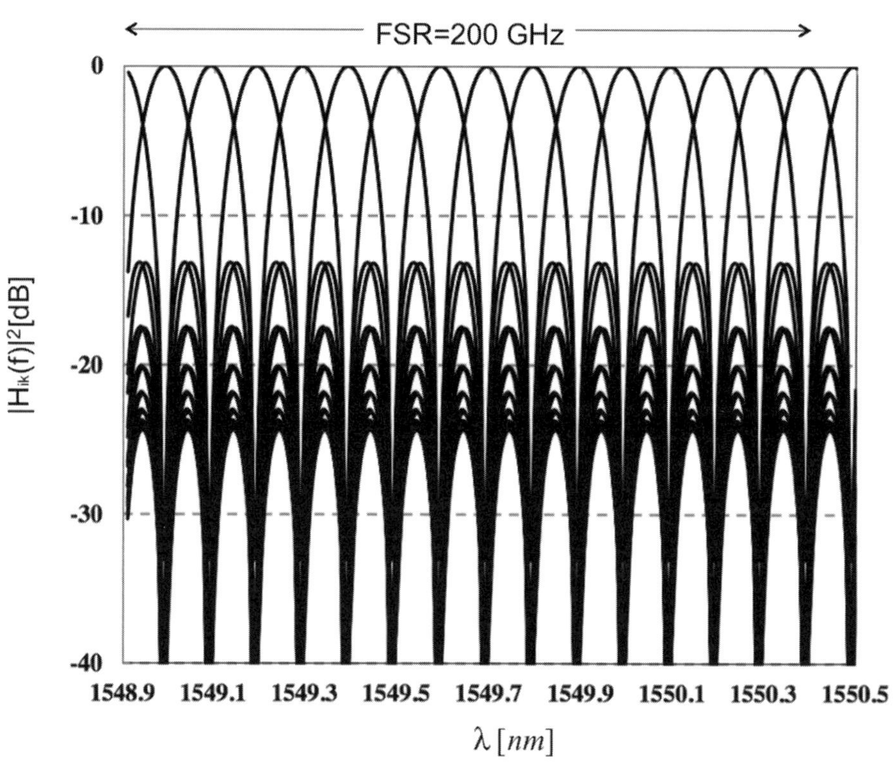

Figure 7.27 Power spectra of 16 optical code sequences [10]. © Reprinted by permission of IEEE.

the spectra between two codes, the codes are orthogonal. But this is not the case, and hence the code sequences are "pseudo-orthogonal." The correlation properties will be described hereafter. The phases of 16 codes obtained from the output ports of the encoder are shown in Fig. 7.28.

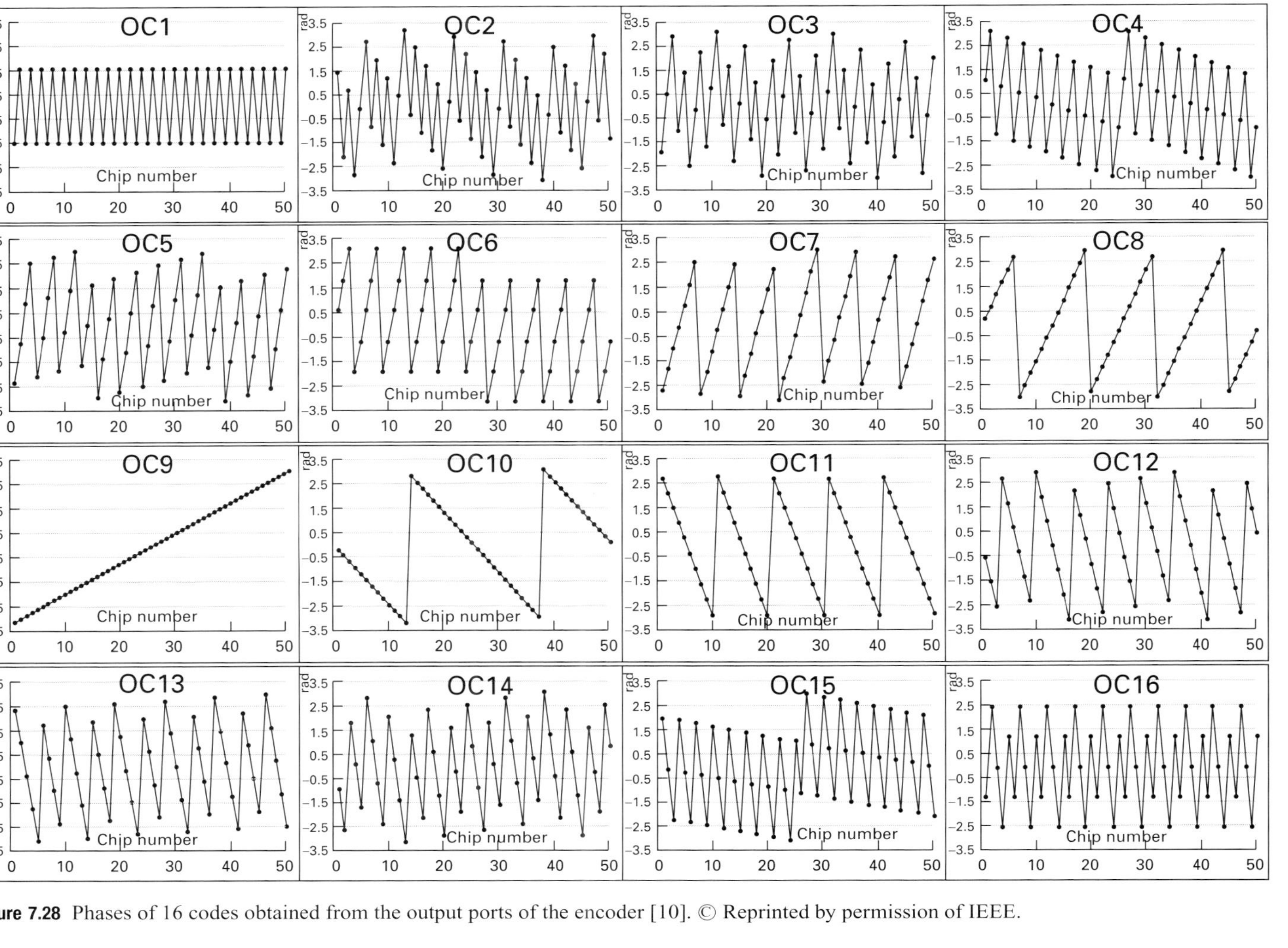

Figure 7.28 Phases of 16 codes obtained from the output ports of the encoder [10]. © Reprinted by permission of IEEE.

7.3.2.2   Correlation properties

The correlation between the codes is given using Eq. (7.25) as

$$
h_{ik}(t) * h_{ik'}(t)
$$

$$
= \sum_{m=0}^{N-1} \sum_{m'=0}^{N-1} \exp\left[-j\frac{\pi}{N}(2m - N + 1)(i + k - N + 1)\right]
$$

$$
\times \exp\left[-j\frac{\pi}{N}(2m' - N + 1)(i + k - N + 1)\right] \delta\{t - (m + m')\Delta\tau\}
$$

$$
= \sum_{m=0}^{N-1} \exp\left[-j\frac{2\pi(m + 1)}{N}(i + k + 1)\right] \times \sum_{m'=0}^{m} \exp\left[-j\frac{2\pi m'}{N}(k' - k)\right]\delta(t - m\Delta\tau)
$$

$$
+ \sum_{m=N}^{2N-2} \exp\left[-j\frac{2\pi(m + 1)}{N}(i + k + 1)\right]
$$

$$
\times \sum_{m'=M+1}^{2N-1} \exp\left[-j\frac{2\pi m'}{N}(k' - k)\right]\delta(t - m\Delta\tau), \; i, k, \; k' = 0, 1, \ldots, N - 1.
$$

$$
(7.29)
$$

The correlation waveform extends over $(2N - 1)\Delta\tau$. The auto-correlation function is obtained by putting $m = m'$ in Eq. (7.29)

$$
h_{ik}(t) * h_{ik}(t) = \sum_{m=0}^{N-1} \exp\left[-j\frac{2\pi(m + 1)}{N}(i + k - N + 1)\right](m + 1)\delta(t - m\Delta\tau)
$$

$$
+ \sum_{m=N}^{2N-2} \exp\left[-j\frac{2\pi(m + 1)}{N}(i + k + 1)\right](2N - m - 1)\delta(t - m\Delta\tau), i, k
$$

$$
= 0, 1, \ldots, N - 1 \tag{7.30}
$$

At the center of the waveform $t = (N - 1)\Delta\tau$, the auto-correlation peak appears as

$$
\text{Auto-correlation peak} = N.
$$

Furthermore, the maximum sidelobe of the auto-correlation function is $N - 1$. Therefore, the ratio of the peak intensity to the maximum wing of auto-correlation, AWR, defined in Section 7.3.1.2 is obtained as

$$
\text{AWR} = \frac{N - 1}{N}. \tag{7.31}
$$

Compared with that of code generated by the SSFBG, for example, $1/N$ for Gold code, the wing of auto-correlation is higher and closer to its peak. A closed form of the total integrated power of the auto-correlation signal can be evaluated as

$$
A_{ac} = \sum_{j=0}^{N-1} \int_{0}^{T_B} \left|(j + 1)\exp\left[-\frac{(t - j\Delta\tau)^2}{2\sigma}\right]\right|^2 dt
$$

$$
+ \sum_{j=N}^{2N-2} \int_{0}^{T_B} \left|(2N - j - 1)\exp\left[-\frac{(t - j\Delta\tau)^2}{2\sigma}\right]\right|^2 dt. \tag{7.32}
$$

The cross-correlation function is obtained from Eq. (7.29) by putting $m \neq m'$ as

$$
h_{ik}(t) * h_{ik'}(t) = \exp\left\{-j\frac{\pi(k-k')}{2N}\right\} \sum_{m=0}^{2N-2} \exp\left\{-j\frac{2\pi(m+1)}{N}\left(i+1+\frac{k-k'}{2}\right)\right\}
$$

$$
\times \frac{\sin\left\{\frac{2\pi(m+1)(k-k')}{N}\right\}}{\sin\left\{\frac{\pi(k-k')}{N}\right\}}\delta(t-m\Delta\tau),\ i,k,\ k'
$$

$$
= 0, 1, \ldots, N-1. \tag{7.33}
$$

The maximum cross-correlation peak occurs at $t = \{(2q+1)/2(k-k')-1\}\Delta\tau$ with $q = 0,\ 1,\ \ldots,\ \lfloor 2(2N-1)(k-k')/N\rfloor$ and is obtained as

$$
\text{Cross-correlation peak} = \frac{1}{\sin\left\{\frac{\pi(k-k')}{N}\right\}},\ i,k,=k'=0,1,\ldots,N-1. \tag{7.34}
$$

The ratio of the peak intensity of auto-correlation to the maximum intensity of cross-correlation, ACR, defined in Section 7.3.1.1, is obtained as

$$
ACR = N^2 \sin^2\left\{\frac{\pi(k-k')}{N}\right\},\ i,k=k'=0,1,\ldots,N-1 \tag{7.35}
$$

The 16-chip PSK code generated by the device for an input 2 ps Gaussian pulse and the auto-correlation function are plotted in Figs. 7.29(a) and (b), respectively. Figures 7.29(c) and (d) show the cross-correlation functions between codes generated at two adjacent output ports and at two far apart ports, respectively. In this case, ACP = 16, and the maximum cross-correlation peak (CCP), for two codes generated at adjacent ports, is 5.13, so that the corresponding code-detection parameter is $ACR = (ACP/CCP)^2 = 9.74$; on the other hand, the maximum CCP for two outputs k and $k' = k + N/2$ is 1, and the corresponding code-detection parameter is ACR = 256. The parameter ACR is plotted in Fig. 7.30 versus the number of labels N, for different values of k and k': we observe that r increases with N and that it tends asymptotically to the upper-bound value $\pi^2(k-k')^2$. Therefore, if we increase N to increase the code cardinality, the code-detection parameter remains practically unchanged. Figures 7.31(a)–(p) show the measured OCs generated at output Ports 1 to 16 when an input pulse with 2.5 ps full width at half-maximum (FWHM) from a mode-locked laser diode is fed into input Port 1 [11]. The temporal waveform is measured using a streak camera. Each code is a 16-chip PSK code and is composed of 16 chip pulses with a different phase. The time interval between two consecutive chips is $\Delta\tau = 5$ ps, so that the code chip rate is $1/\Delta\tau = 200$ Gchip/s. The measured temporal waveforms of the correlation are shown in Figs. 7.32(a)–(p). An OC 1 is launched into the encoder input Port 1. The auto-correlation appears from output Port 1, whereas all cross-correlation signals are measured at the other Ports 2–15.

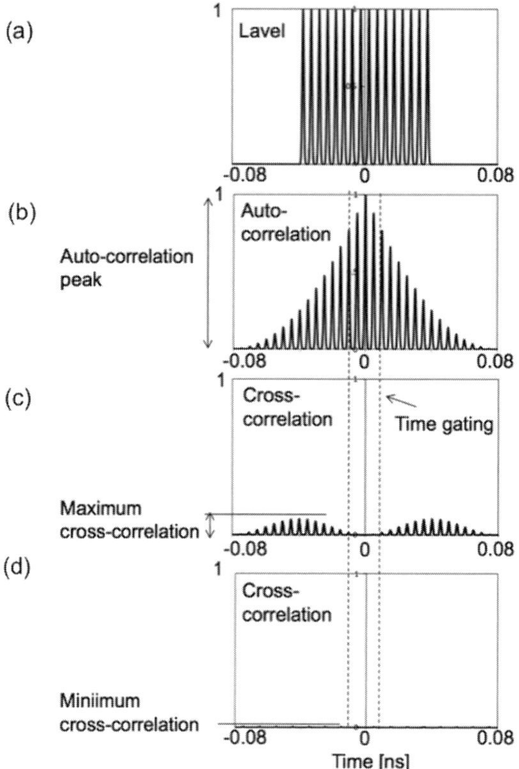

Figure 7.29 (a) PSK code generated by the device for an input 2 ps Gaussian pulse. (b) Auto-correlation function, (c) maximum and (d) minimum cross-correlation functions between two codes generated at adjacent output ports.

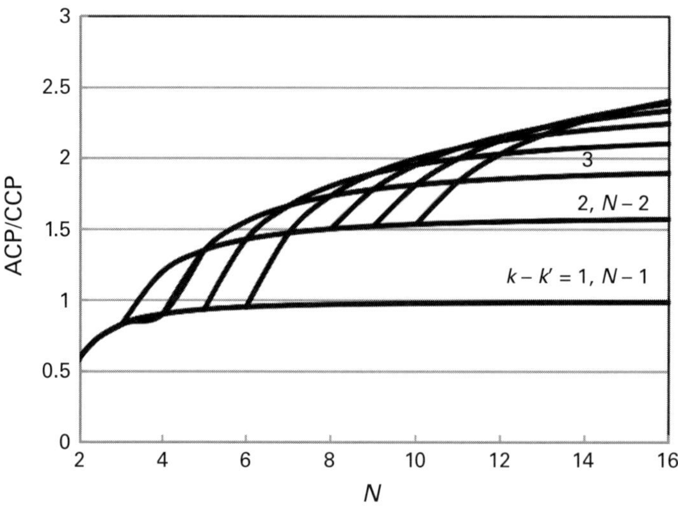

Figure 7.30 Code-detection parameter ACR versus the number of codes N; codes are generated at outputs k and k' [10]. © Reprinted by permission of IEEE.

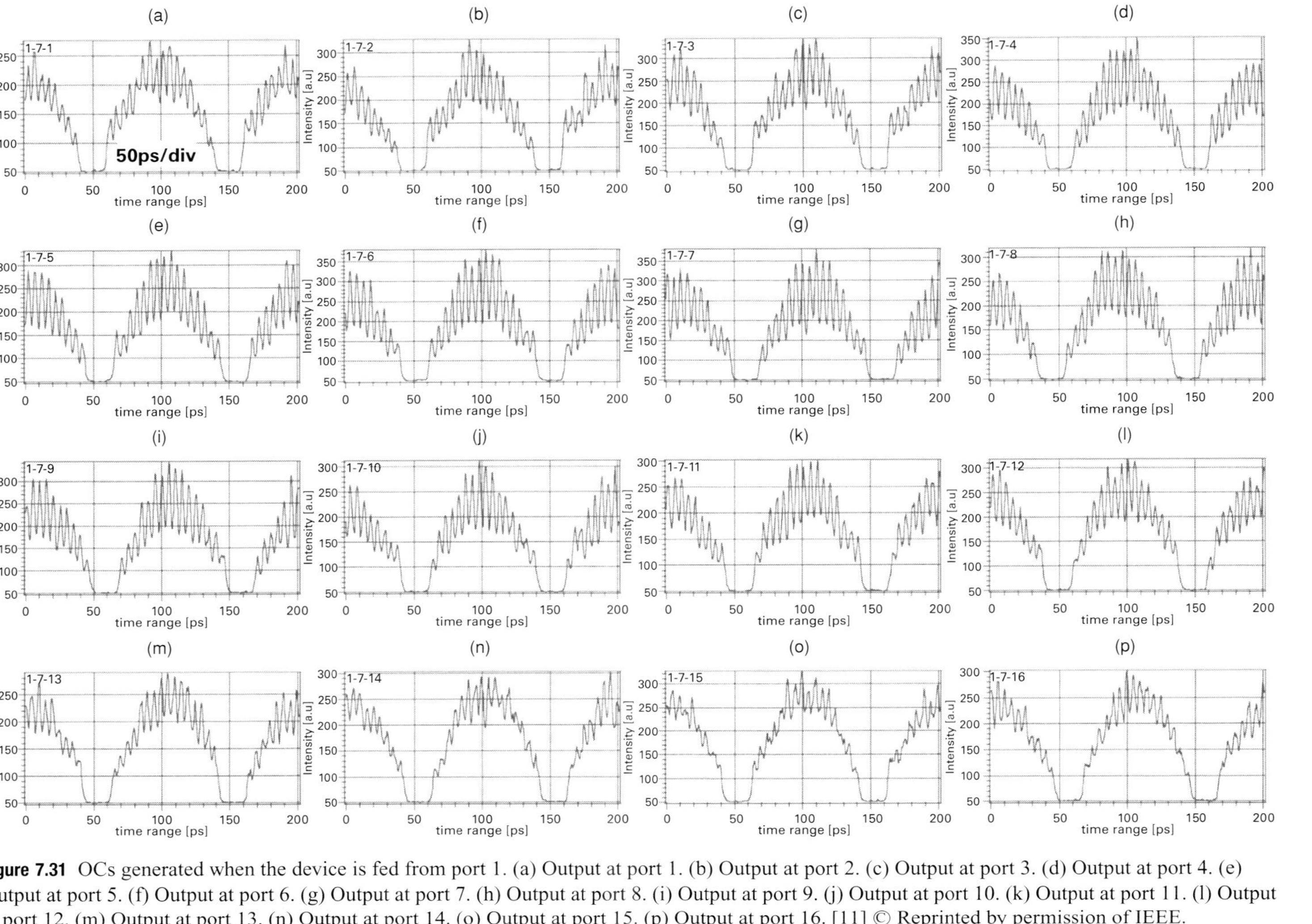

Figure 7.31 OCs generated when the device is fed from port 1. (a) Output at port 1. (b) Output at port 2. (c) Output at port 3. (d) Output at port 4. (e) Output at port 5. (f) Output at port 6. (g) Output at port 7. (h) Output at port 8. (i) Output at port 9. (j) Output at port 10. (k) Output at port 11. (l) Output at port 12. (m) Output at port 13. (n) Output at port 14. (o) Output at port 15. (p) Output at port 16. [11] © Reprinted by permission of IEEE.

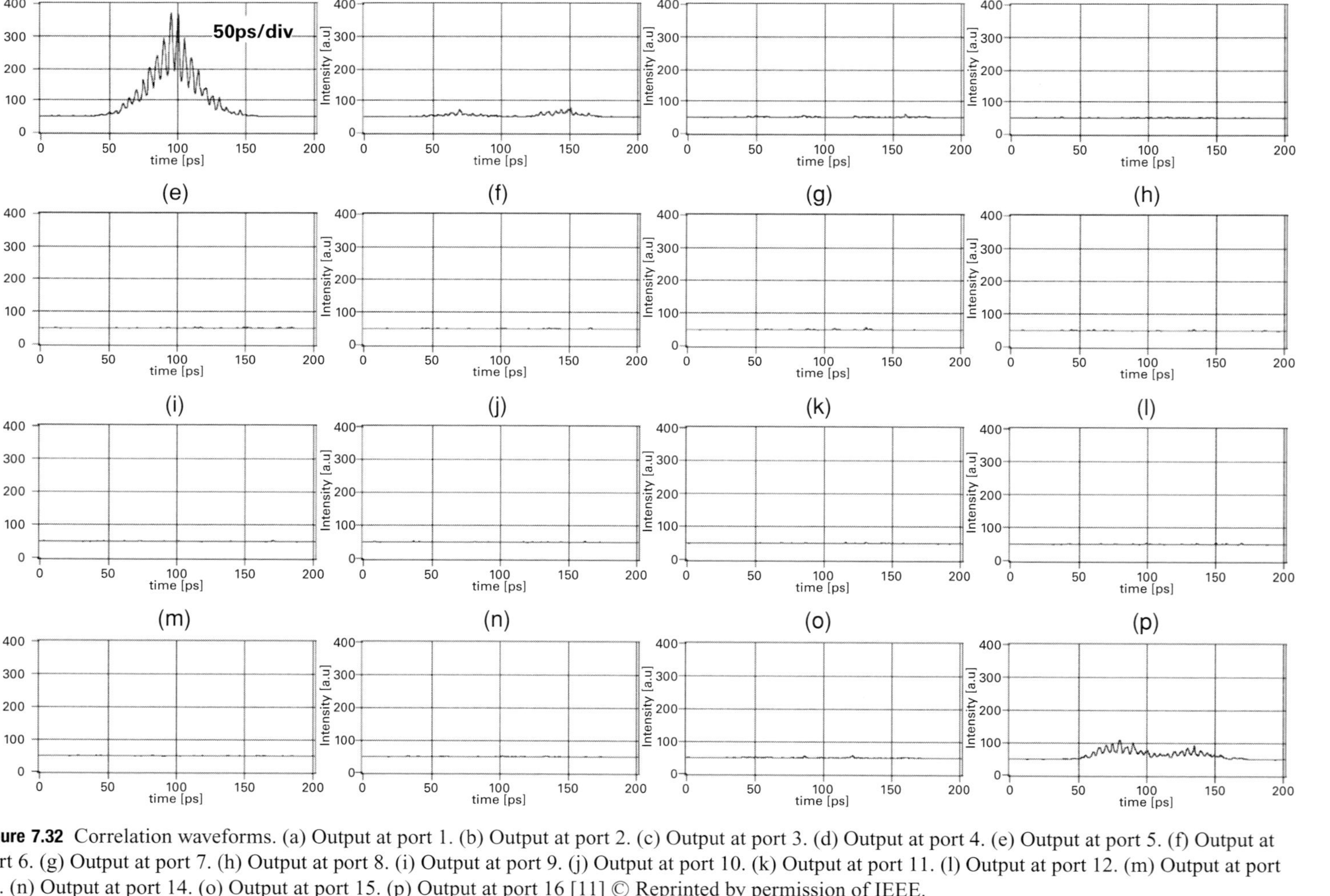

Figure 7.32 Correlation waveforms. (a) Output at port 1. (b) Output at port 2. (c) Output at port 3. (d) Output at port 4. (e) Output at port 5. (f) Output at port 6. (g) Output at port 7. (h) Output at port 8. (i) Output at port 9. (j) Output at port 10. (k) Output at port 11. (l) Output at port 12. (m) Output at port 13. (n) Output at port 14. (o) Output at port 15. (p) Output at port 16 [11] © Reprinted by permission of IEEE.

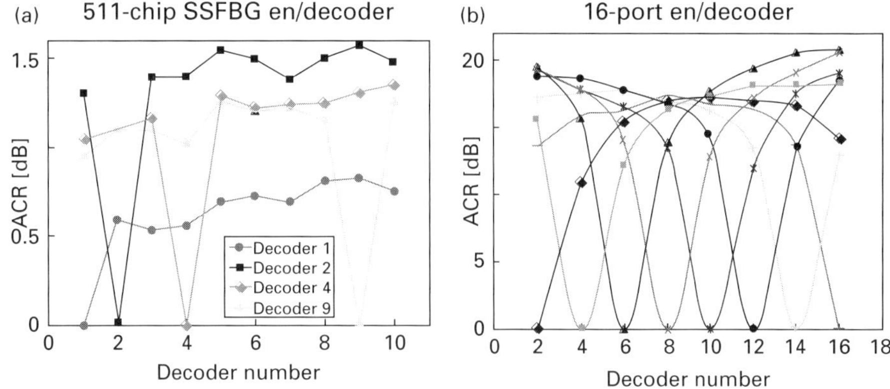

Figure 7.33 ACRs of (a) 511-chip SSFBG and (b) 16-port en/decoders plotted as a function of code number [12]. © Reprinted by permission of IEEE.

7.3.3 Correlation between a heterogeneous combination of multiport and SSFBG encoder and decoder

Two different types of optical en/decoder were introduced in Sections 7.3.1 and 7.3.2. It is interesting to compare their correlation characteristics. The calculated ACRs of SSFBG and multiport en/decoders are plotted as a function of code number in Fig. 7.33(a) and (b), respectively [12]. Here ACR becomes unity (0 dB) for the case of auto-correlation. The lowest value of ACR of the multiport decoder occurs between two codes generated at two adjacent output ports (i.e., $k = k' + 1$). The ACR increases with the port count N, and it tends asymptotically to the upper-bound value $\pi^2(k - k')^2$. Compared with the ACR of the 511-chip bipolar code generated by the SSFBG encoder shown in Fig. 7.33(a), the ACR of the 16-port encoder shown in Fig. 7.33(b) is higher by an order of magnitude.

Heterogeneous usage of different types of optical en/decoder in an OCDMA-PON significantly improves the system flexibility as well as the system cost. An experimental demonstration is described in Section 9.1. There is a discrepancy in the code property between SSFBG and multiport en/decoders. The SSFBG encoder generates a bipolar code sequence, and it adopts time-gating in the decoding to curve out the impulse-like auto-correlation peak. By contrast, the multiport encoder generates a multi-level PSK code sequence, and bit rate detection without time-gating is adopted. Obviously, the bipolar code of the SSFBG en/decoder and the multi-level PSK code of the multiport en/decoder show poor correlation with each other. To realize the heterogeneous usage of either the multiport encoder and SSFBG decoder or the SSFBG encoder and multiport decoder, the gap in the code sequence between the bipolar code of the SSFBG and the multi-level PSK code of the multiport en/decoder has to be filled.

The auto-correlation waveforms of bipolar code generated by a 63-chip, 640 Gchip/s SSFBG en/decoder are compared with the auto-correlation waveform generated by a 16-chip, 200 Gchip/s multiport en/decoder, respectively, in Figs. 7.34(a) and (b) [13]. The impulse-like waveform of the bipolar code contrasts with the triangular temporal profile of the 16-level phase-shifted code generated by the multiport encoder, as

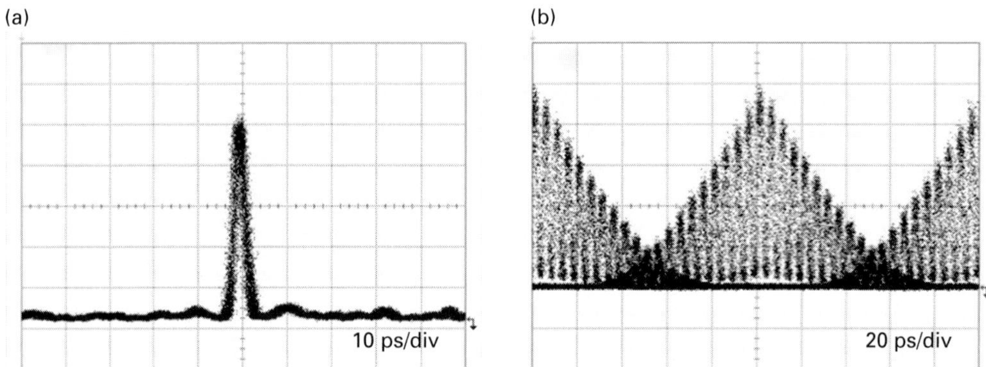

Figure 7.34 Auto-correlation waveforms of (a) a 63-chip, 640 Gchip/s bipolar code generated by an SSSFBG encoder and (b) a 16-level phase-shifted 16-port encoder [13]. © Reprinted by permission of IEEE.

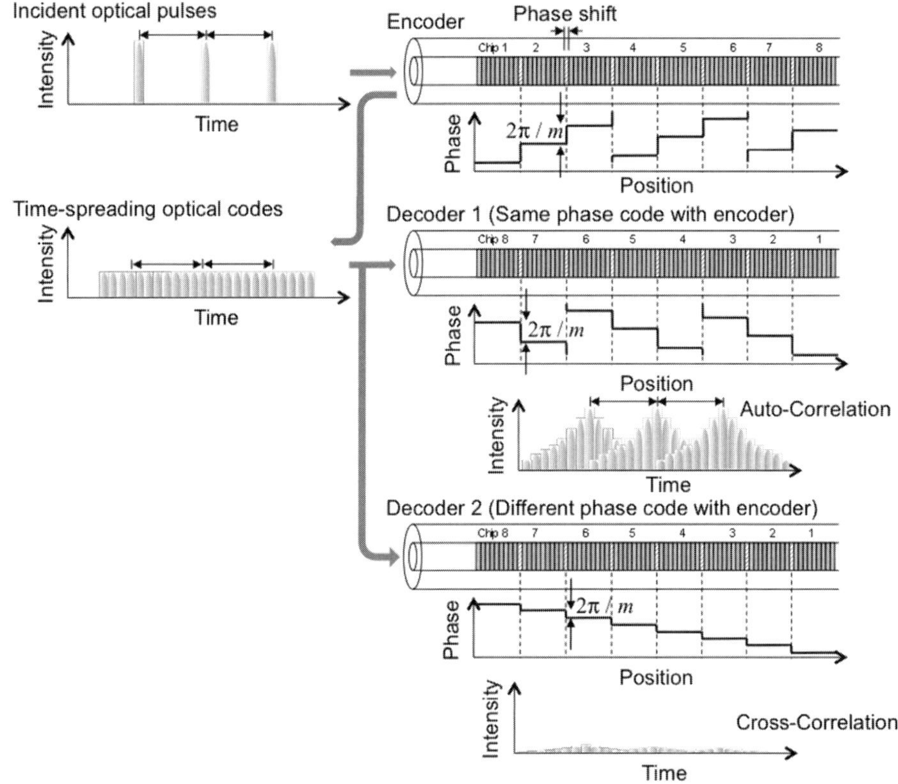

Figure 7.35 Generation and correlation of multi-level PSK code using an SSFBG en/decoder [14]. © OFS 2012, SPIE by permission.

described in Section 7.3.2. To solve the above matching problem with the multi-port en/decoder, an SSFBG en/decoder which generates a multi-level PSK code has been developed. Figure 7.35 illustrates schematically the generation and correlation of a multi-level PSK en/decoder using an SSFBG en/decoder [14]. In the case with m-level PSK code, the position of each grating section shifts in proportion to the phase

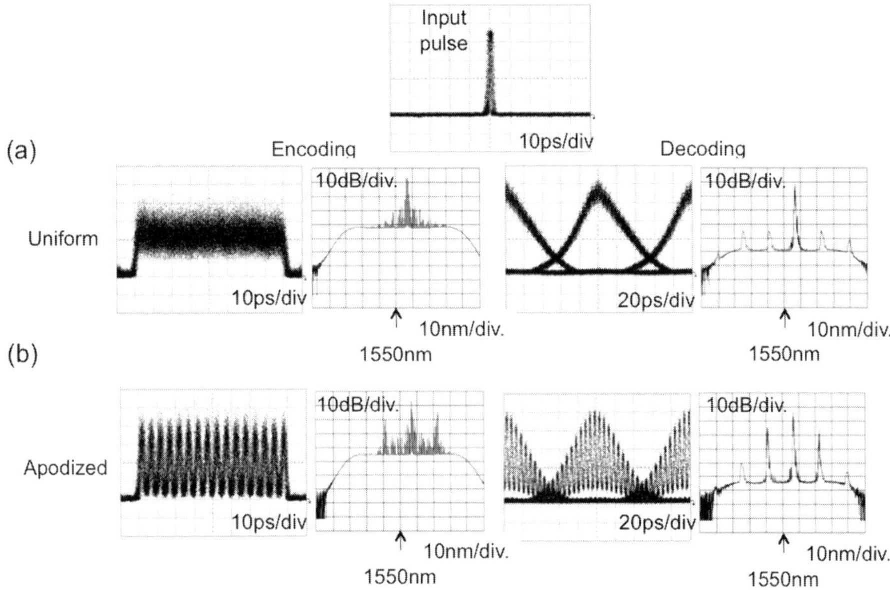

Figure 7.36 Temporal waveforms of a 16-chip, 200 Gchip/s optical code and its auto-correlation waveform generated by an SSFBG encoder: (a) without and (b) with apodization [15]. © Reprinted by permission of ECOC 2009.

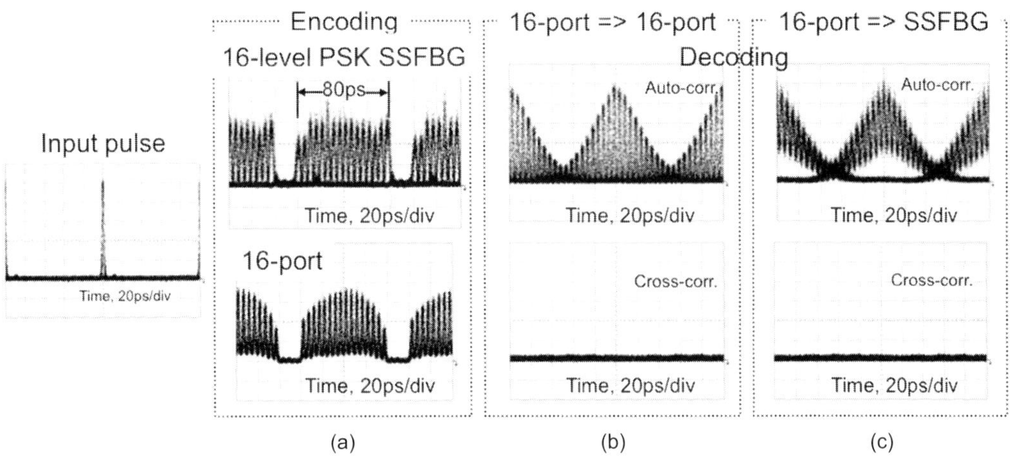

Figure 7.37 (a) Encoded waveforms of 16-level PSK codes of 16-level PSK SSFBG and 16×16 multiport encoder, correlation waveforms of (b) the multiport–multiport en/decoder, and (c) the SSFBG–multiport–multiport en/decoder [16]. © Reprinted by permission of IEEE.

shift, denoted by $2\pi/m$, $(m = 1, 2, \ldots)$. For $m = 8$ as an example, the shift of the grating position is $\lambda/16$ with respect to the position of zero phase shift when the round trip is taken into account. In Figs. 7.36(a) and (b) 16-level PSK code generated by a 16-chip, 200 Gchip/s SSFBG en/decoder having uniform and apodized profiles, respectively, is shown [15]. The design parameters of the SSFBG encoder are center wavelength 1546 nm, chip length 0.52 mm, total length of grating 8.32 mm, and the 16 phase levels are generated by shifting the chip grating by a step of $\pm\lambda/8$. The peaks of each individual chip of encoding/decoding waveforms generated from the uniform SSFBG are not well defined. On the other hand, the apodized SSFBG generates encoding/decoding waveforms, showing well-defined chip structure. The correlation properties between a 16-port multiport and 16-level PSK SSFBG were investigated experimentally. Figure 7.37 shows that the auto-correlation waveforms resemble each other well, comparing the hybrid en/decoder and the SSFBG–SSFBG and multiport–multiport en/decoders [16]. This confirms the practical viability of using both types of en/decoder in a single PON, for example, a single multiport en/decoder at the OLT and SSFBG en/decoders at each ONU.

8 Data confidentiality

8.1 Security architecture for systems provisioning end-to-end communications

Any entity or organization requires an Information Security Management System (ISMS). The transfer of sensitive information, such as financial transactions, medical records, and intellectual property, currently relies on the Internet via high-speed, large-capacity optical networks, thanks to the cost effectiveness of IP networks. Loss of data confidentiality of the Internet would have a tremendous impact on society as a whole, and hence the security of information and communication systems has become a primary concern. Here, we summarize briefly the security architecture for systems provisioning end-to-end communications. The International Standards Organization (ISO) maintains generic security standards. The main objective is the protection of confidentiality, integrity, and availability (known as the CIA triad) [1].

The International Telecommunication Union Telecommunication Standardization Sector (ITU-T) recommendation X.800 has set standards for telecommunication systems and networks. The security is hierarchical, according to the equipment and facilities groups. Three security layers are identified: security of the infrastructure, services, and applications as described in Fig. 8.1 [2]. Each security layer has unique vulnerabilities, threats, and mitigations. The three security layers are linked to each other in such a way that the infrastructure security layer enables applications security via service security function. The ITU-T X.800 specifies a threat model, shown in Fig. 8.2, including five threats such as destruction, corruption, removal, disclosure, and interruption. Eight security dimensions, including the above-mentioned CIA, address a variety of network vulnerabilities, as shown in Fig. 8.3. The data confidentiality security dimension protects data from unauthorized disclosure by means of encryption. This is probably the best known and most commonly sought form of security in communications. Hereafter, we focus on the data confidentiality of OCDMA. The data integrity security dimension ensures the correctness or accuracy of data. The availability security dimension ensures that there is no denial of authorized access to network elements, stored information, information flow, services and applications due to events impacting the network. Disaster recovery solutions are included in this category. The types of actions to protect security are represented by three security planes: an end-user security plane, a control/signaling security plane, and a management security plane, shown in Fig. 8.4.

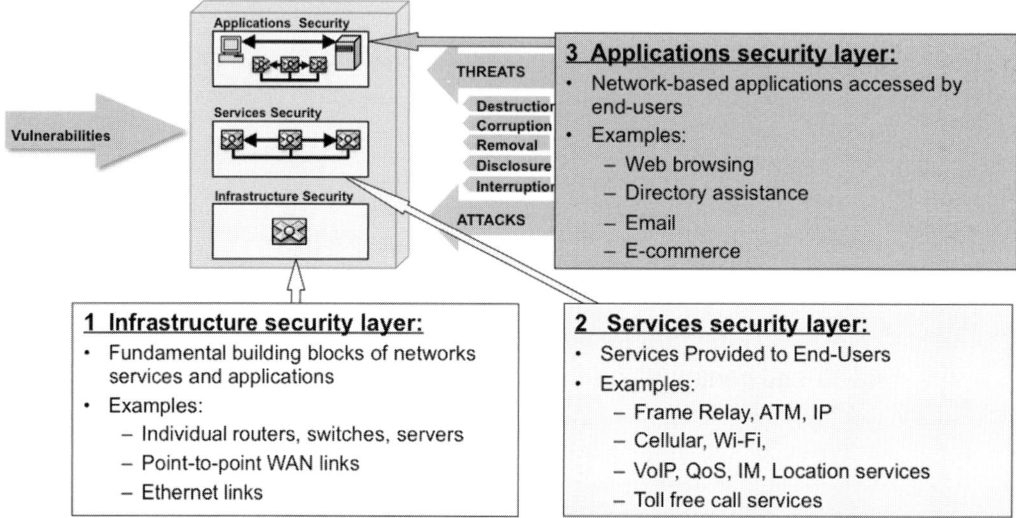

Figure 8.1 Three security layers described in ITU-T X.800 [2]. © Reprinted by permission of ITU-T and by courtesy of Z. Zeltan.

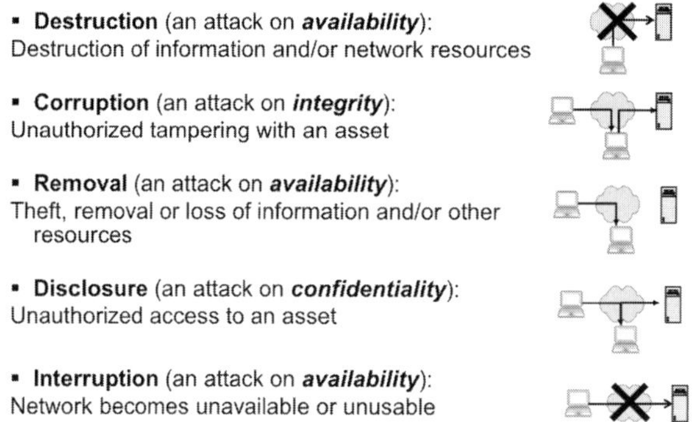

Figure 8.2 Five threats described in ITU-T X.800 [2]. © Reprinted by permission of ITU-T and by courtesy of Z. Zeltan.

Each security plane is applied to every security layer in Fig. 8.1 to yield nine (3×3) security perspectives.

In IP networks, for example, the infrastructure security layer is applied to individual routers and servers, and the service security layer is applied to basic IP transport and value-added services such as virtual private networks (VPNs), while the application security layer is applied to fundamental applications, for example, email and high-end applications such as e-commerce. Internet Protocol security (IPsec) is applied to the end-user security plane, and generalized multiprotocol label switching (GMPLS) is applied to the control/signaling security plane.

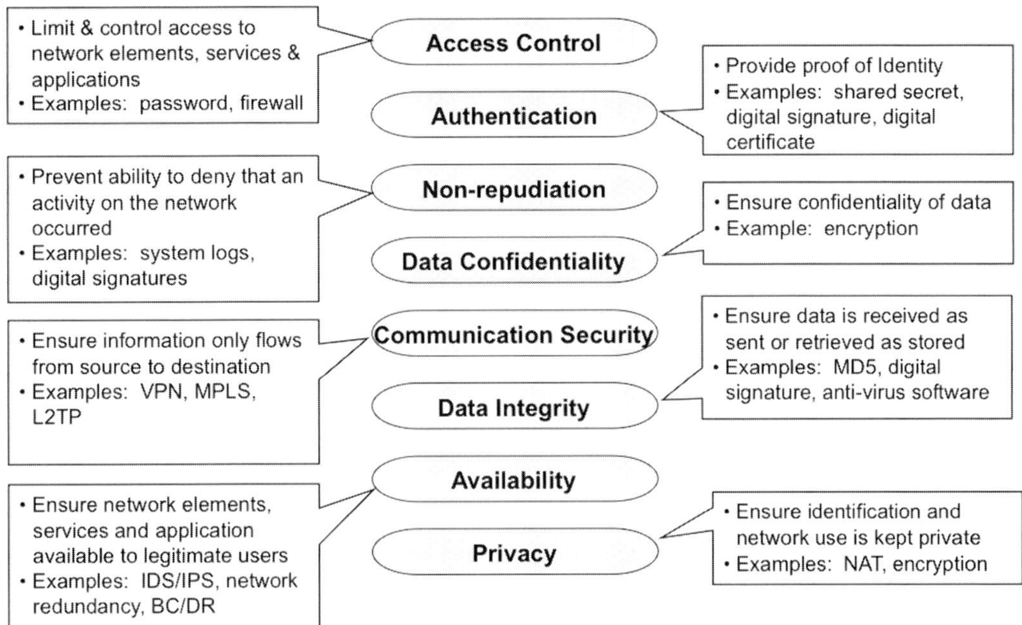

Figure 8.3 Eight security dimensions described in ITU-T X.800 [2]. © Reprinted by permission of ITU-T and by courtesy of Z. Zeltan.

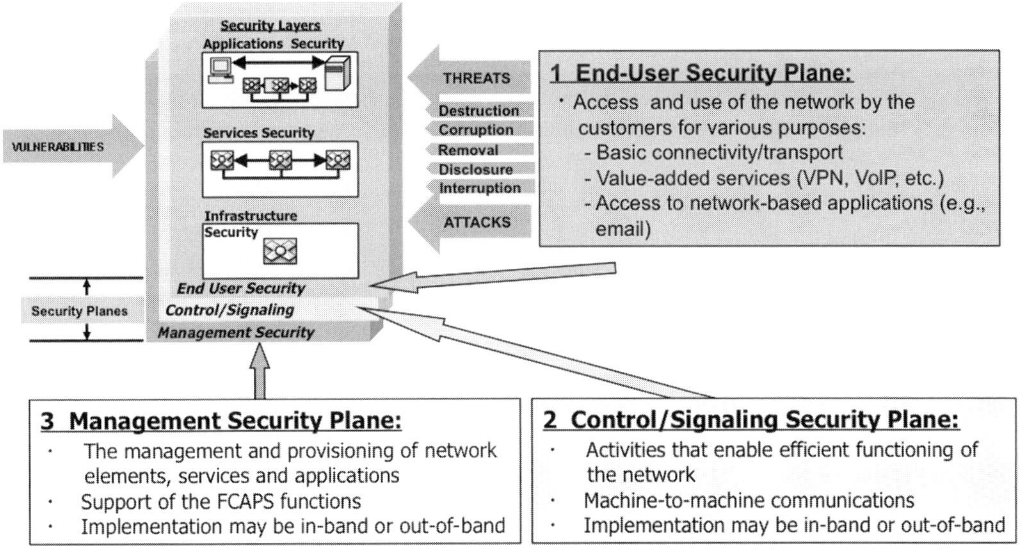

Figure 8.4 Three security planes described in ITU-T X.800 [2]. © Reprinted by permission of ITU-T and by courtesy of Z. Zeltan.

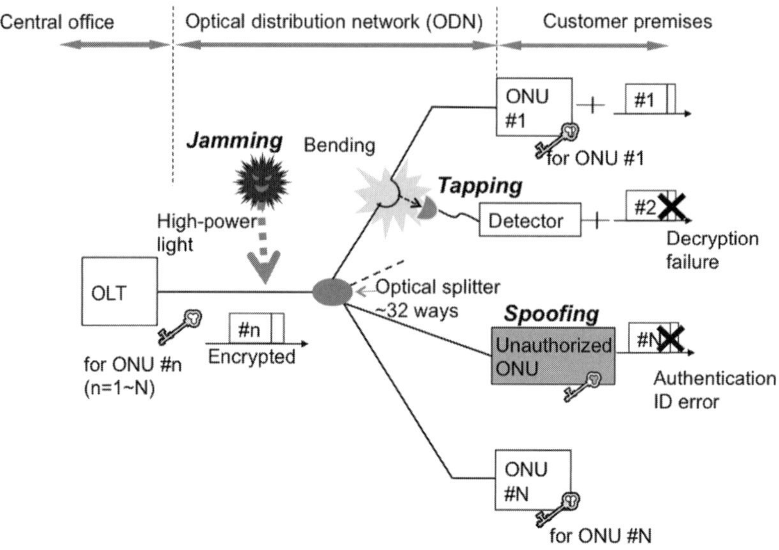

Figure 8.5 Security threats in an optical access network [1]. © Reprinted by permission of IEEE.

8.2   Security threats to a PON

There are two distinct classes of data confidentiality. The most confidential communication systems are called "unconditionally secure" if they are theoretically unbreakable even with infinite computational resources. If a system requires a sufficiently large amount of computational resource over a sufficiently long time to break, it is called "computationally secure." OCDMA is categorized as computationally secure.

All secure protocols are supported by modern cryptographies such as secure Hash algorithm 1 for integrity protection and authenticity and advanced encryption standard (AES) for confidentiality. In contrast to the service security layer and the application security layer shown in Fig. 8.1, the infrastructure security layer has not attracted much attention. The importance of the infrastructure security layer, however, should be stressed, because once a security breakdown occurs, a quick stopgap measure will not be easily implemented. It takes a painfully long time to repair a physically damaged photonic layer. This is in sharp contrast to the vulnerability of the upper layers in which security can be restored in a relatively short time by patching software or releasing new codes online. It should be remembered that the security is a "chain of trust," and the weakest part determines the security level of the whole system.

The architecture of current PON systems is a point-to-multipoint network topology and is inherently prone to security threats. In Fig. 8.5 potential security threats to a PON such as tapping, spoofing, and high-power jamming are illustrated. As shown in Fig. 1.30, optical fiber cables in the distribution cable section of the outside plant are laid on electric poles, and an eavesdropper would have the opportunity to tap leaked light at the bent portion of the fiber. An adversary could conduct spoofing by connecting an unauthorized

ONU to the PON. To gauge the vulnerability to security attack, "Kerckhoffs' principle," used in the analysis of cryptographic systems, is assumed. This states that a potential adversary is technologically sophisticated, has significant resources and knows what types of OCDMA signals are being sent: the data rate, the type of encoding, and the structure of the codes, but he or she does not know the particular code that an individual user employs. A malicious attacker could also inject laser light with intent to jam the communication. When high-power light is injected into a PON, most likely from the distribution cable link, photodetectors at the OLT and/or ONUs will be destroyed, and eventually the entire PON will malfunction.

IEEE 802.3ah standard specifies the authentication and encryption mechanism. ITU-T G.984 GPON recommends use of the AES for downstream transmission. The AES is a specification for the encryption of electronic data, which was announced by the National Institute of Standards and Technology (NIST), USA in 2001. AES uses block ciphering, and the algorithm described by AES is a symmetric-key algorithm, meaning the same key is used for both encrypting and decrypting the data. Block ciphering is detailed in Section 8.4. In commercially available EPON systems and upcoming 10G-EPON systems, layer 2 security functions are implemented, for example, using AES128. IEEE 802.1ae defines the encryption protocol to provide secure communication functions to the network, and IEEE 802.1af defines the control of communication connections by means of authentication and the key exchange protocol for encryption. Thanks to these security functions, the downstream data are not available to anyone and privacy is guaranteed in many commercially available EPON systems and upcoming 10G-EPON systems.

8.3   Bit ciphering versus block ciphering

8.3.1   Bit ciphering

A bit-cipher cryptographic system creates a one-to-one correspondence between each bit from each user and an optical code. Three schemes of data modulation in an OCDMA system, including OOK, differential binay PSK (DBPSK), and code-shift-keying (CSK), are considered in Fig. 8.6 [3]. Obviously, as the code cardinality increases, the data confidentiality increases in the sense of computational security. OOK-OCDMA has security vulnerability because it encodes only the logical mark with an optical code and the logical space is left un-encoded. As shown in Fig. 8.7 the encoded signal is noise-like, and an eavesdropper cannot decipher the signal without knowledge of the optical code [4]. However, if the eavesdropper can tap the signal before multiplexing, even though the data are encoded with a designated optical code, the eavesdropper could easily break the security by simple data-rate energy detection without any information about the optical code. Therefore, in OOK-OCDM data-rate energy detection can be used to detect the logical mark, otherwise the receiver knows the logical space is sent.

Next we discuss the data confidentiality of binary PSK-OCDMA or DPSK-OCDMA using the balanced detection described in Section 4.7.3. A simplified model of OCDMA using DBPSK as data format (DBPSK-OCDMA) is shown in Fig. 8.8 [5].

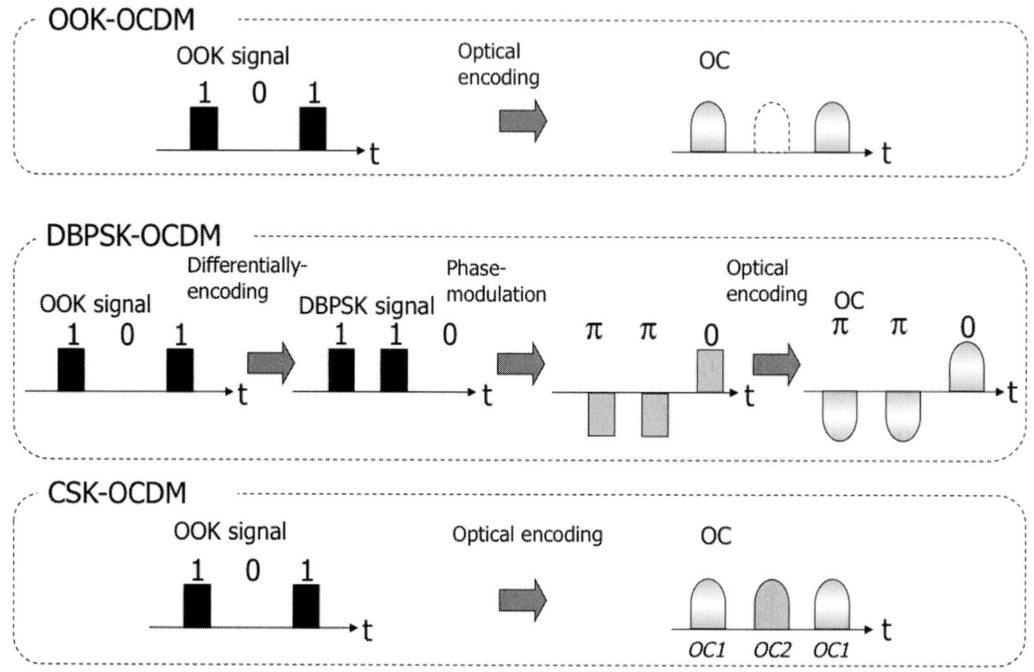

Figure 8.6 Three schemes of data modulation in OCDMA.

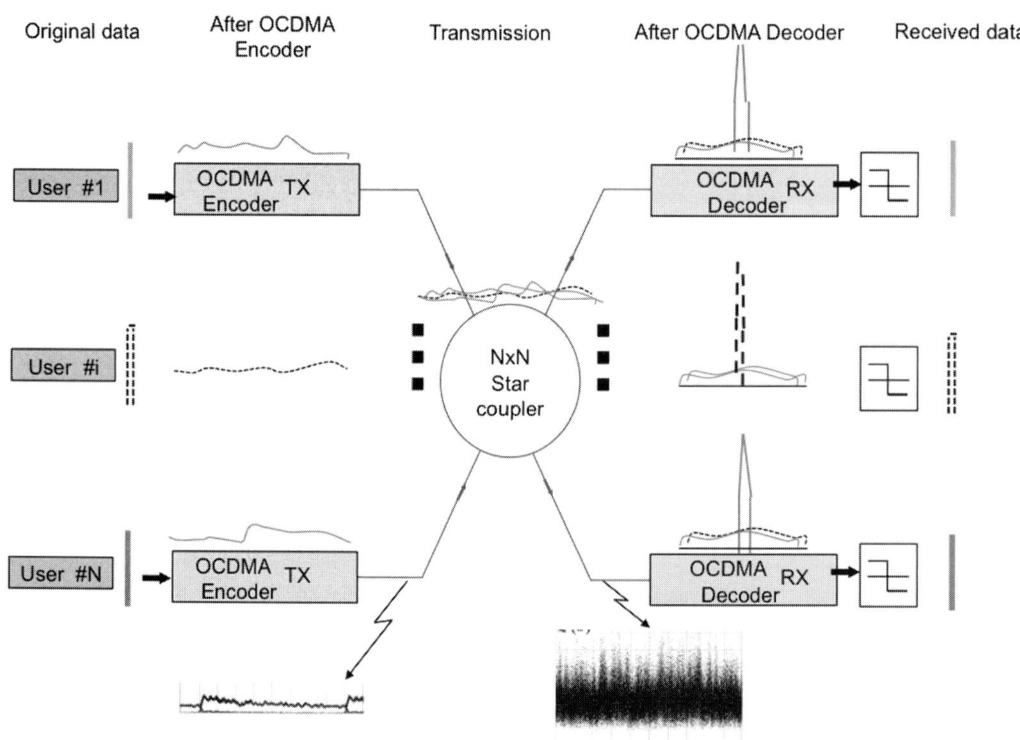

Figure 8.7 Noise-like signals in OCDMA-PON [4]. © Reprinted by permission of IEEE.

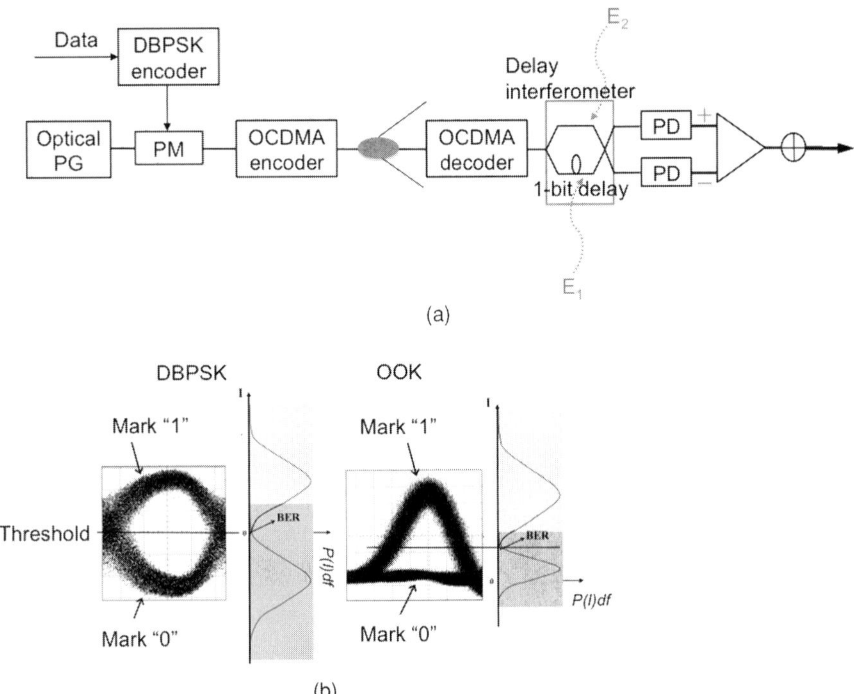

Figure 8.8 OCDMA using DBPSK modulation format. (a) System model and (b) eye diagrams of DBPSK-OCDMA and OOK-OCDMA [5]. © IEEE by permission.

Unlike OOK-OCDMA, both the logical mark and space are encoded with the same optical code. The data are first encoded by the DBPSK encoder, and then the electrical data signals are sent to the phase modulator, followed by optical encoding. At the receiver, the balanced photodetector is placed after the one-bit delay interferometer. In the inset of Fig. 8.8 eye diagrams and the noise probability density functions of the received DBPSK-OCDMA and OOK-OCDMA signals are compared.

The bit error rate of DBPSK-OCDMA can be evaluated theoretically. In Section 4.7.3 the basics of BPSK and DBPSK signals were described. Here, we focus on the performance of the DBPSK-OCDMA system. The electric fields of the signals for the mark and space at the detector are given by

$$E_1(t) = \sqrt{P_d}\exp\{j\omega_d t + \phi_d(t) + d(t)\pi\}$$
$$+ \sum_{i=1}^{m} \sqrt{P_i}\exp\{j\omega_i(t - \tau_i) + \phi_i(t - \tau_i)\}$$
$$E_0(t) = \sqrt{P_d}\exp\{j\omega_d t + \phi_d(t) + d(t - \tau_0)\pi\}$$
$$+ \sum_{i=1}^{m} \sqrt{P_i}\exp\{j\omega_i(t - \tau_i - \tau_0) + \phi_i(t - \tau_i - \tau_0)\} \tag{8.1}$$

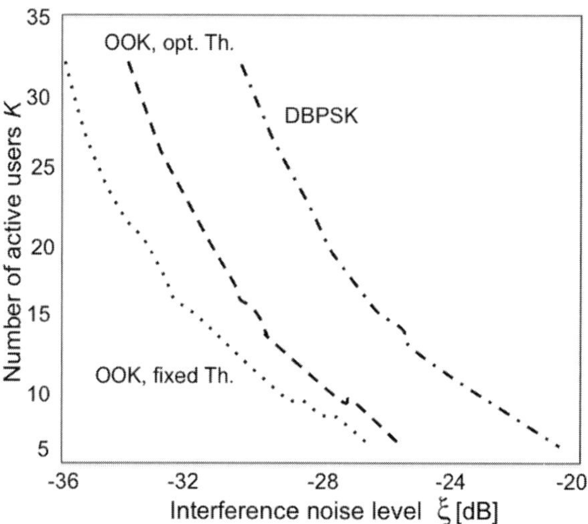

Figure 8.9 Number of active users against the interference noise level in OOK and DBPSK OCDMA. Fixed Th and Opt Th, respectively, represent the fixed and optimum threshold levels of OOK-OCDMA [5]. © Reprinted by permission of IEEE.

where $d(t)$ is encoded binary data and takes the value of either zero or unity, P_d and P_i are the optical intensities of the decoded signals of the desired and undesired subscribers, respectively. ω_d and ω_i are the optical frequencies, ϕ_d and ϕ_i are the phase noise, and τ_i is the relative propagation delay of the interferers. From Eq. (6.34) the photocurrent at the receiver for the case of chip rate detection is given by

$$
I_{ph} = \int_0^{T_{chip}} \Re(E_1 E_1^* - E_0 E_0^*)dt + \int_0^{T_{chip}} n(t)dt \propto T_{chip} P_d \cos\{\Delta d(t, T_B)\pi\}
$$

$$
+ T_{chip} \sum_{i=1}^{m} P_i + 2 \int_0^{T_{chip}} \cos\{\Delta\omega_{i,d}t - \Delta(\omega\tau)_{i,d} + \Delta\phi_{i,d}(t, \tau, T_B)\}dt
$$

$$
+ \sum_{i \neq j}^{\frac{m(m-1)}{2}} \sqrt{P_i P_j} \int_0^{T_{chip}} \cos\{\Delta\omega_{i,j}t - \Delta(\omega\tau)_{i,j} + \Delta\phi_{i,j}(t, \tau, T_B)\}dt + \int_0^{T_c} n(t)dt
$$

$$
(8.2)
$$

where the decoded binary data $\Delta d(t, \tau_0)$ take the value either zero or unity. The five terms in Eq. (8.2) represent data, MAI, primary signal–interference beat noise, secondary interference–interference beat noise, and receiver noise, respectively. Compared to coherent OOK-OCDMA in Eq. (6.34), the noise terms have the same distributions, while the data term changes from "0" to "$-T_{chip}P_d$" for marks "0," due to the fact that the threshold level is also changed to zero in DBPSK-OCDMA. In Fig. 8.9 the calculated numbers of active subscribers K of OOK-OCDMA and DBPSK-OCDMA for the bit error rate of 6×10^{-5} are plotted against the interference noise level ξ [4]. ξ is

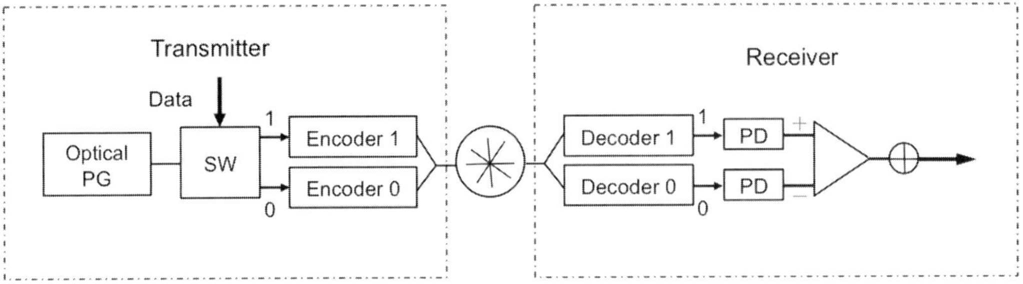

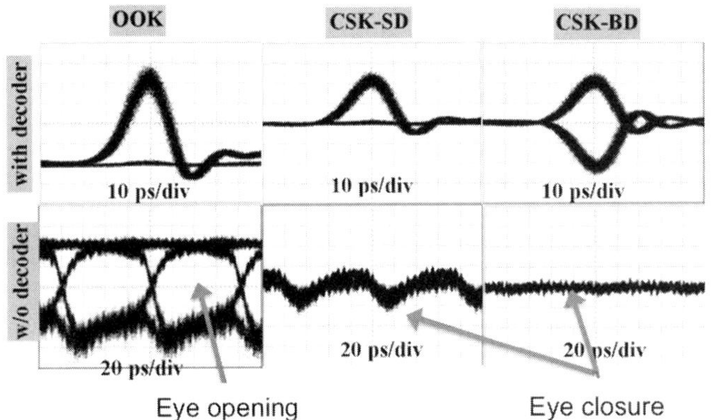

Figure 8.10 Code-shift-keying OCDMA [4]. © Reprinted by permission of IEEE.

the interference level defined in Eq. (6.35) which is approximately equal to the inverse of the code length. The data rate is 10 Gb/s, and the 511-chip long Gold code is used. For OOK-OCDMA, the results are shown for both the optimal threshold (opt Th) and fixed threshold (fixed Th). For a given number of active subscribers K, the interference level ξ of DBPSK-OCDMA can tolerate about 4 dB higher noise than OOK-DBPSK with the optimum threshold level (Th) and 5~6 dB higher than OOK-OCDMA with fixed threshold. Therefore, DBPSK-OCDMA can tolerate a higher level of MAI noise and beat noise compared to OOK-OCDMA, which allows it to accommodate more active users. From the security viewpoint, however, DBPSK-OCDMA-PON is still not sufficiently secure to guarantee data confidentiality if the transmitted signal is tapped before multiplexing. This is because an eavesdropper could decipher the transmitted data without any knowledge of the optical code, assuming that the adversary has a DBPSK decoder without having a proper optical decoder and a data-rate energy detector.

A scheme of code-shift-keying (CSK) or code-switching can prevent energy detection in that both the logical mark and space are encoded with different optical codes [4]. At the transmitter in the CSK scheme shown in Fig. 8.10, a 1×2 optical switch (SW) is driven by a binary data signal, so that the optical pulse is guided to one of the two output ports according to whether it represents a logical mark or space, followed by optical encoding with different optical codes. The encoded optical signals are combined again after the encoders, thus generating the CSK-OCDMA signal. At the receiver, the transmitted signals are split into two and detected with different optical decoders. The electric fields of the signals for the mark and space at the detector are

$$E_1(t) = d(t)\sqrt{P_d}\exp\{j\omega_d t + \phi_d(t)\} + \sum_{i=1}^{m}\sqrt{P_i}\exp\{j\omega_i(t - \tau_i) + \phi_i(t - \tau_i)\}$$

$$E_0(t) = \{1 - d(t)\}\sqrt{P_d}\exp\{j\omega_d t + \phi_d(t)\} + \sum_{i=1}^{m}\sqrt{P_i}\exp\{j\omega_i(t - \tau_i) + \phi_i(t - \tau_i)\}$$

$$(8.3)$$

where the notation defined in Eq. (8.1) is used. From Eq. (6.34) the photocurrent at the receiver for the case of chip rate detection is given by

$$I_{ph} = \int_0^{T_c}\Re(E_1 E_1^* - E_0 E_0^*)dt + \int_0^{T_c} n(t)dt$$

$$= \{2d(t) - 1\}T_c\Re P_d + (\text{MAI}_1 - \text{MAI}_0) + (\text{SIBN}_1 - \text{SIBN}_0) + \int_0^{T_c} n(t)dt.$$

$$(8.4)$$

Here, only the signal-interference beat noise is taken into account, and the secondary interference–interference beat noise is neglected. As the transmitted signal is split in half and guided to the photodetector, 3 dB loss occurs in the splitting, but the balanced detection will compensate for the splitting loss. Another penalty for the enhanced data confidentiality is that the code cardinality has to double compared with conventional OOK-OCDMA or DBPSK-OCDMA.

Figure 8.11 compares the calculated numbers of active subscribers for the bit error rate of 6×10^{-5} as a function of the single interference for CSK-OCDMA, OOK-OCDMA, and DBPSK-OCDMA. CSK-OCDMA can tolerate an interference noise level ζ more than 3 dB higher than OOK-OCDMA for a given value of K, whereas DBPSK-OCDMA can tolerate the ζ up to 1 dB higher than CSK-OCDMA. Therefore, the multi-user capability in CSK-OCDMA is improved with respect to OOK-OCDMA, and is slightly decreased with respect to DBPSK-OCDMA. More importantly, security can be significantly tightened in the CSK-OCDMA system because an eavesdropper cannot decipher the signal without knowing the optical codes.

8.3.2 Block ciphering

Block ciphering is widely used in electronic cryptography. As shown in Fig. 8.12, data are divided into sequences of m-bit blocks ($m = 8$) and encoded with a code Ci with at least $M = 2^m$ determinations. In M-ary optical code division multiplexing (OCDM) the secret key is not the optical code itself, but the correspondence between an optical code and a bit sequence. In this way, data confidentiality is protected by two levels of security. An adversary has first to detect the optical code, and then he or she has to

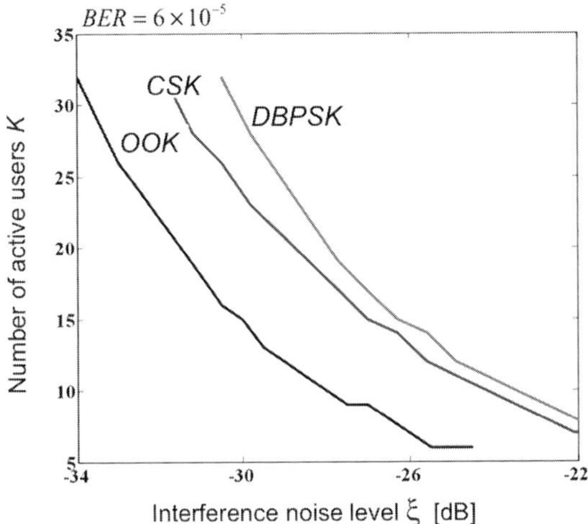

Figure 8.11 Number of active users against the interference noise level in OOK, DPSK, and CSK OCDMAs [4]. © Reprinted by permission of IEEE.

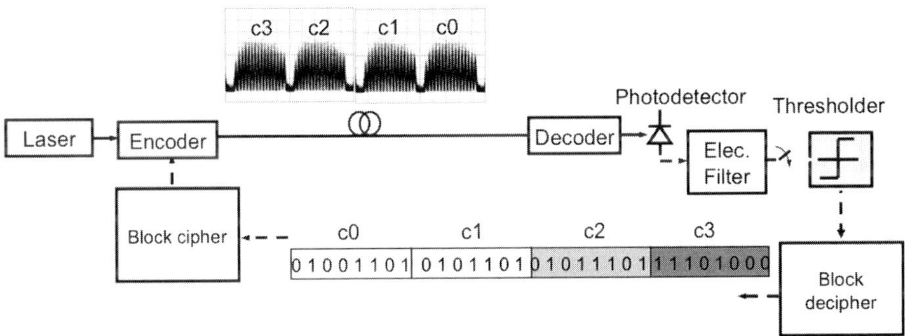

Figure 8.12 Scheme of block ciphering.

find the correspondence between the code and the bit sequence. Here, we start with a point-to-point link of M-ary OCDM and later consider multiple access based on M-ary OCDMA.

When N-chip long, N-level PSK codes generated by N-port optical encoders in AWG configuration, as described in Section 7.3.2, are applied to block ciphering, the maximum length of the bit sequence that we can encrypt is $M = \log_2 N$, and all the possible correspondences between the optical code and the bit sequence are $N!$. To increase the code cardinality further, it is possible to generate a "multi-dimensional" encoding [6]. If there are N PSK codes available for multi-dimensional encoding, one can stack codes up to n ($n < N$), and thus the overall n-dimensional codewords that can be generated is given by

$$\sum_{i=1}^{N} {_n}C_i = 2^N - 1. \tag{8.5}$$

Table 8.1 Three-bit block ciphering using 2-dimensional code

| | | 2-D | | | |
Three-bit block	1-D	OC1	OC2	OC3	OC4
000	OC1	✔			
001	OC2		✔		
010	OC3			✔	
001	OC4				✔
100	OC5	✔	✔		
101	OC6		✔	✔	
110	OC7			✔	✔
111	OC8	✔			✔

Table 8.2 System confidentiality of bit and block ciphering using multi-dimensional codewords

| | Bit cipher | | Block cipher | |
	1-D	n-D	1-D	n-D
Number of keys	N	$2^N - 1$	$N!$	$(2^N - 1)!$
Number of trials against COA	$\frac{N}{2}$	$\frac{2^N - 1}{2}$	$\frac{N!}{2}$	$\frac{(2^N - 1)!}{2}$
Number of bits against CPA	1	1	$(N - 1)\log_2 N$	$N(2^N - 1)$

All the possible correspondences between optical codes and bit sequences are $2^N!$. In Table 8.1, for example, all eight 3-bit blocks are assigned with 2-dimensional codes ($n = 2$) using only four codes. Note that 3-bit block ciphering cannot be done with three codes even if up to 3-dimensional code is introduced because the number of generated codewords is 7, from Eq. (8.5). The optical implementation to generate an n-dimensional codeword using a multiport encoder is described in Section 8.6.2.

The results of quantitative analysis of system confidentiality of OCDMA using an N-port optical encoder for 1-dimensional and multi-dimensional codewords are summarized in Table 8.2. Two sorts of attacks, exhaustive key search attack, which is the simplest cipher-text only attack (COA), or known plaintext attack (KPA) and chosen plaintext attack (CPA) are considered. COA is an attack model where the attacker is assumed to have access only to a set of cipher-texts. In a CPA, the adversary has access to the encryption function and can encrypt any plaintext message, trying to determine the secret key. For instance, knowledge that a common message (like a "Hello" packet) is transmitted can be used to break the network security using this attack. In this case, the lower bound security parameter of modern cryptanalysis is the number of plaintext bits that an attacker needs to know to detect the code. In a bit-ciphering scheme, just a single bit allows the adversary to intercept the data. In contrast, in a block-ciphering scheme, the number of message bits required is larger. In Fig. 8.13, the number of trials needed to break the security is plotted as a function of the number of ports N of the multiport

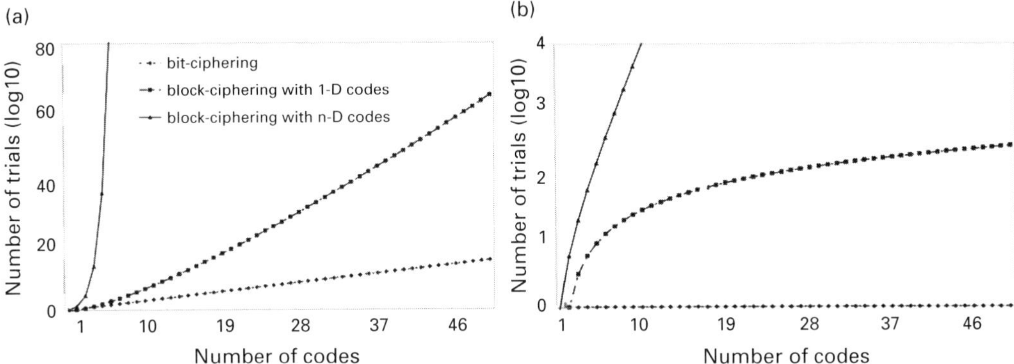

Figure 8.13 (a) Number of trials necessary to break confidentiality in a COA attack. (b) Number of trials necessary to break confidentiality in a CPA attack [7]. © Reprinted by permission of ICTON 2008.

encoder [7, 8] when an exhaustive key search attack, that is, the simplest COA and CPA, is assumed. The average number of trials required to break a code is $K/2$, where K is the overall number of secret keys. In the case of bit ciphering using n-dimensional codes, it is $K = 2^N - 1$, and for block ciphering, we have $K = (2^N - 1)!$.

8.4 Data confidentiality of OCDMA-PON

One of the advantages of OCDMA over other multiple access schemes such as TDMA and WDMA is its inherent data confidentiality, because messages are encoded at the transmitter and can be recovered only by the authorized user, who knows the optical code. MAI noise prevents an eavesdropper from intercepting a message without knowing the proper code. This assumption is regarded as the same as Kerckhoffs' principle introduced in Section 8.2, used in the analysis of cryptographic systems, which states that a potential adversary is technologically sophisticated, has significant resources, and knows what types of OCDMA signals are being sent: the data rate, the type of encoding, and the structure of the codes, but he or she does not know the particular code that an individual user employs. We consider that the adversary makes either COA or KPA or chosen plaintext attack CPA.

Theoretical analyses have shown that standard OCDMA encoding is not as secure as current cryptography such as AES. The conclusion drawn from [9] is that cryptography provides a much greater degree of confidentiality than does OCDMA encoding. However, an intelligently encoded OCDMA signal can force a potential eavesdropper to use a sophisticated and possibly expensive detector in order to be able to break the user's confidentiality. Rapid reconfiguration of codes can also increase the difficulty of interception. These factors can provide significant security advantages compared with standard optical communication technologies such as WDM, where a commercial off-the-shelf detector can be purchased to read the data.

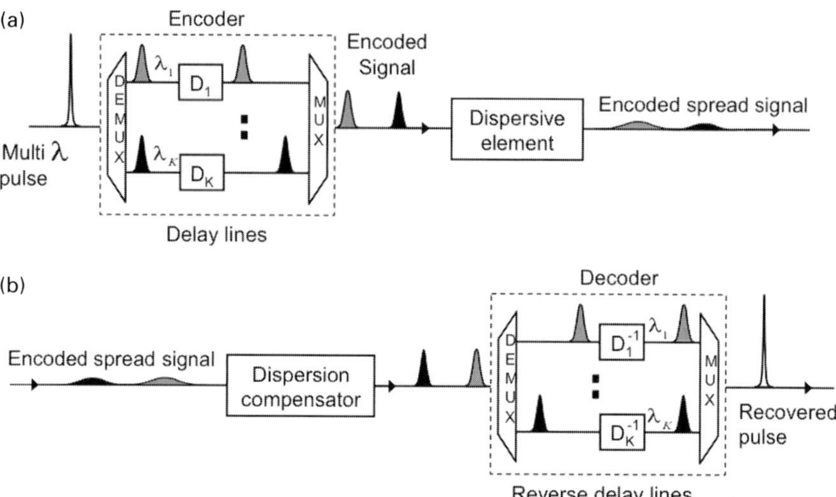

Figure 8.14 (a) Secure transmission over existing WDM channels and (b) incoherent OCDMA en/decoder using frequency hopping and time-spreading [10]. © Reprinted by permission of IEEE.

8.5   Steganographic security enhancement

Steganography hides messages so that no one, apart from the sender and intended recipient, even realizes there is a hidden message. The cryptographic security of OCDMA can be enhanced by steganographic security [10]. Figure 8.14(a) illustrates a system configuration which consists of public WDM channels using OOK modulation and a secure channel. As shown in Fig. 8.14(b) the optical encoding employs CSK for the secure channel where two different codes are assigned for the encoding of the logical mark and space, as described in Section 8.3.1. Encoding is performed by a frequency hopping technique using tunable delay-lines $D_i(i = 1, ..., K)$ for various wavelengths from a multi-wavelength light source. A random frequency hopping assignment is preferred in order to maximise the security of the system. The encoded time-delayed signal is subsequently time spread by a dispersive element. The signals of both the WDM channels and the secure channel are coupled independently into the optical fiber link, and the ASE noise from EDFA is added to the composite signal, resulting in the time-spread signal being hidden under the ASE noise floor during propagation. At the receiver, the WDM signals are wavelength demultiplexed, and each signal is detected by a photodetector, while the signal of the secure channel is dispersion compensated with complementary prescribed dispersion, followed by decoding with reverse delay-lines. It has been demonstrated experimentally that with the appropriate choice of parameters such as time spreading, intensity, and bit rate for the secure user, low BER can be achieved. It is noteworthy that the presence of the WDM channels provides an ad hoc security enhancement for the secure channel, increasing the difficulty for an eavesdropper to intercept and decode the secure signal.

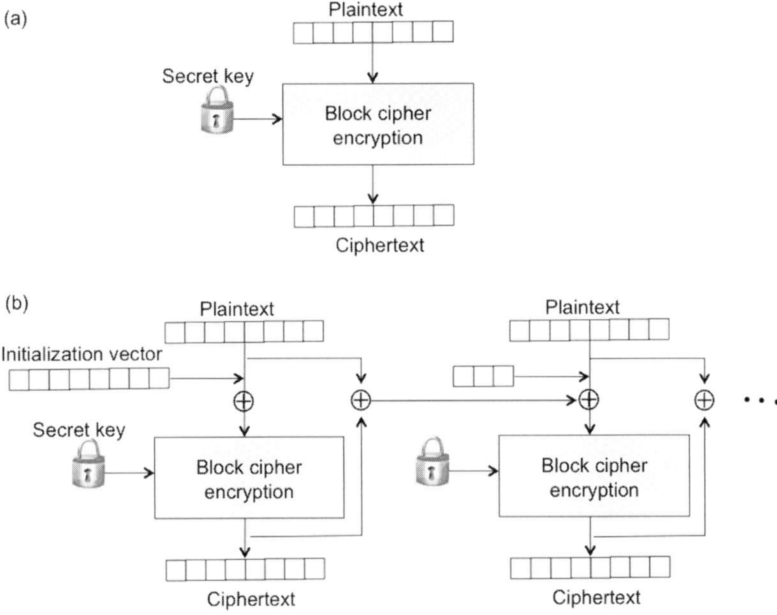

Figure 8.15 (a) Electronic codebook (ECB) mode. (b) Cipher block chaining (CBC) mode.

8.6   M-ary OCDMA

8.6.1   Principle of operation

There are several modes of operation in encryption based upon block ciphering. The simplest of the encryption modes is the electronic codebook (ECB) mode. The message is divided into bit blocks, and each block is encrypted separately as shown in Fig. 8.15(a). Since an identical bit block is always encrypted into an identical codeword, ECB does not provide stringent message confidentiality. By contrast, in the cipher block chaining (CBC) mode the codeword is determined depending not only on the individual bit block but also on all the bit blocks of the message in the past. Figure 8.15(b) shows the chain logical operation of exclusive OR (XOR) between the original bit block and the output of XOR operation at the transmitter. For encryption of the first bit block of the message, the initial vector is used. In this way the correspondence between the original bit block before the XOR operation and the assigned codeword varies with time, and therefore the data confidentiality is improved compared with the ECB mode.

M-ary OCDMA is a special class of OCDMA which adopts block ciphering [11]. In the optical implementation, M-ary OCDMA uses either the ECB mode or the CBC mode. Compared to conventional OCDMA systems, it presents two levels of data confidentiality: the adversary has first to detect the optical code, and then he or she has to find the correspondence between the code and the bit sequence. For starters, consider a scheme of point-to-point M-ary ($M = 16$) OCDM. At the transmitter the data must first be partitioned into separate 4-bit blocks, and according to the code lookup table each bit

Table 8.3 Code lookup table for 16-ary OCDMA

Optical code	Four-bit block			
	MSB			LSB
OC1	0	0	0	0
OC2	1	0	0	0
OC3	0	1	0	0
OC4	1	1	0	0
OC5	0	0	1	0
OC6	1	0	1	0
OC7	0	1	1	0
OC8	1	1	1	0
OC9	0	0	0	1
OC10	1	0	0	1
OC11	0	1	0	1
OC12	1	1	0	1
OC13	0	0	1	1
OC14	1	0	1	1
OC15	0	1	1	1
OC16	1	1	1	1

block is mapped onto a specific codeword. The data block is encoded with the assigned codeword and transmitted. At the receiver the transmitted encoded signal is decoded, and the correlation can identify the codeword. Thus, the original bit block is recovered based upon one-to-one correspondence of the codeword and the data block. In Table 8.3, the code lookup table for 16-ary OCDM is shown [12] as an example. 16 optical codes are used, and each 4-bit block is assigned with a different optical code. The point-to-point M-ary OCDM scheme can be extended straightforwardly to M-ary OCDMA.

8.6.2   Implementation of 16-ary OCDMA

Let us consider a 16-ary OCDM-based block-ciphering system [12]. It operates in CBC mode using the XOR operation at the transmitter/receiver. At the transmitter, shown in Fig. 8.16, a serial data bit stream is segmented every four bits by the serial-to-parallel (SP) converter. The logic operation XOR between the 4-bit data block and the previous data block that is stored in the memory is operated on-line and the resultant 4-bit block is mapped onto a codeword as shown in Table 8.3. The code lookup table is implemented by a 4-to-16 line coder with 16 output ports. Each output of this line coder generates the corresponding optical code by driving the 16-channel optical gate switch array using an LiNbO$_3$ intensity modulator (LN-IM) array. Only the optical pulse train passing through the optical gate (colored in gray) is forwarded to a designated input port of the 50-port optical encoder in AWG configuration, and one of 50 optical codes is generated. The optical code is 50-chip, 500 Gchip/s, 50-level PSK code. Since only 16 codes are needed in this experiment, 16 codes are selected every three output ports, in order to minimize

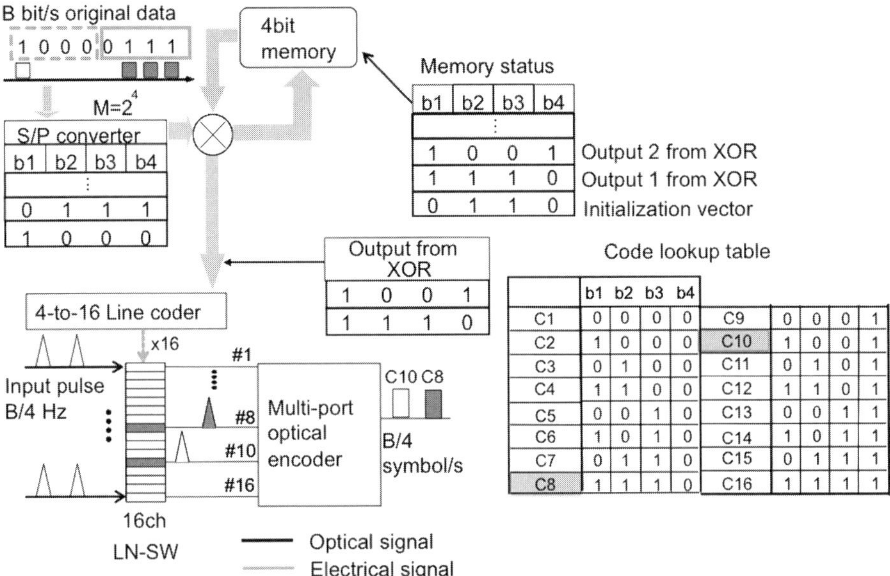

Figure 8.16 Transmitter of a 16-ary OCDM system [12]. © Reprinted by permission of IEEE.

the crosstalk between the codes. As an example, the incoming bit block (1, 0, 0, 0) is XORed with the initial bit set (initialization vector) (0, 1, 1, 0) stored in the 4-bit memory, resulting in Output1, (1, 1, 1, 0). Output1 is encoded into the optical code OC8, according to the code lookup table, and it is also stored in the memory. At the next step, the incoming bit block (0, 1, 1, 1) is XORed with Output1, from the memory, generating Output2, (1, 0, 0, 1), which in turn is optically encoded into OC10.

All the codes are generated at the same output port, so that the selection of the input port of the encoder determines which optical code is generated. The pulse repetition rate equates to the symbol rate, i.e., the bit rate divided by 4 ($= \log_2 16$). At the receiver shown in Fig. 8.17, the received optical codes are sent to the 16-port optical decoder, identical to the encoder, which has the same configuration as the encoder. An autocorrelation waveform appears only at one of the 16 output ports of the optical decoder, and the output port number indicates which is the received optical code. For the input of OC8 the optical pulse appears from the output Port 8 of the decoder and is converted into an electrical signal by the 16-channel O/E array. The output signal is launched into the 16-to-4 line decoder, and the resultant output is (1, 1, 1, 0). After the XOR operation with the same initialization vector (0, 1, 1, 0) stored in the 4-bit memory, the original 4-bit data sequence (1, 0, 0, 0) is recovered using the code lookup table.

Experiments using 2.488 Gb/s or 622 MSymbol/s 16-OCDM have been conducted. In the experimental setup shown in Fig. 8.18, a key component is the 500 Gchip/s 50×50 optical en/decoder, which generates 50-chip PSK optical codes with a single encoder. 16 code sequences are generated by using 16 ports, every 3 ports out of 50 ports. The short optical pulse of a mode-locked laser diode (MLLD) at 1565 nm with repetition rate of 9.95328 GHz in (ii) is downconverted to the 622 MHz pulse stream in (v), and

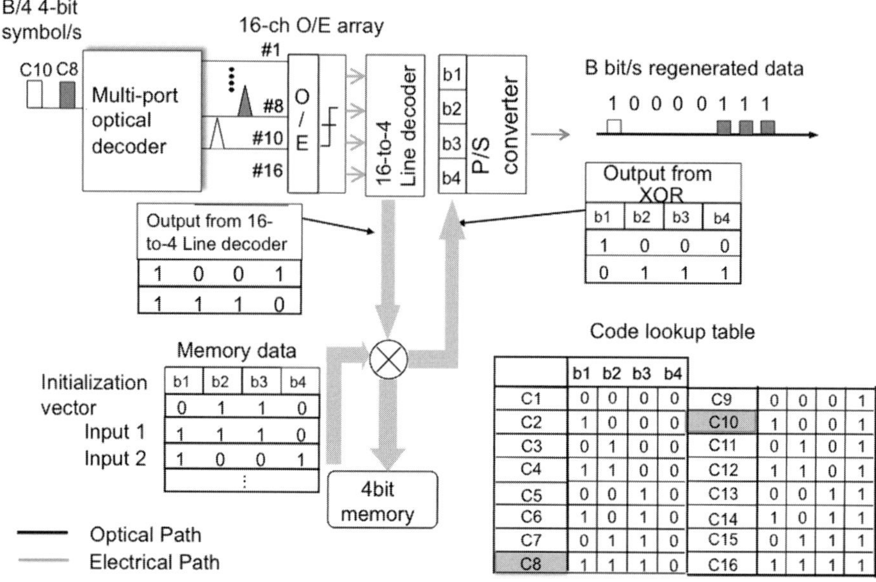

Figure 8.17 Receiver of a 16-ary OCDM system [12]. © Reprinted by permission of IEEE.

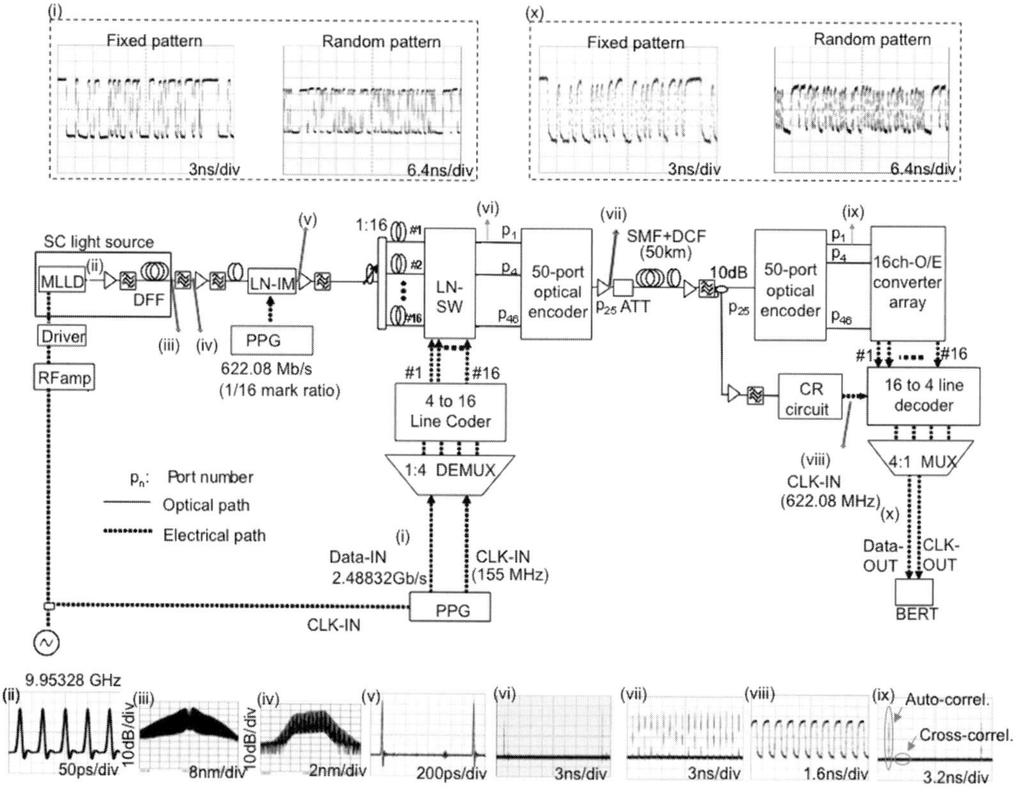

Figure 8.18 Experimental setup of 622 MSymbol/s 16-OCDM [12]. © Reprinted by permission of IEEE.

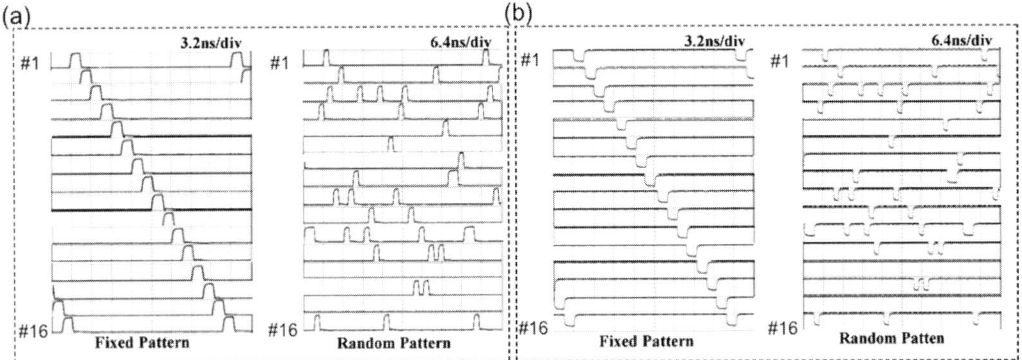

Figure 8.19 (a) Electric gate signals at each output of the 4-to-16 line coder for fixed pattern and random patterns and (b) electric gate signals at each output of the 16-to-4 line coder for fixed pattern and random patterns [12]. © Reprinted by permission of IEEE.

the SC with the center wavelength of 1550 nm in (iii) is generated using a 2 km long DFF and sliced out into 7.5 nm to cover the spectrum of the code in (iv). The details of SC are described in Section 5.2.2. The pulse stream is split into 16 arms by an optical coupler, and each arm is connected to the optical gate array composed of LN-SWs. The pulse can pass through the gate only when the gate is opened by the gate signal from the 4-to-16 line coder. In Fig. 8.19 the electric gate signals at the 4-to-16 line coder for fixed and random patterns at the transmitter and those at the 16-to-4 line coder at the receiver are shown. The output optical pulse of an LN-SW in (vi) is guided to one of the input ports of the optical encoder, and the optical code is generated in (vii). The fine structure of the 16-chip is not clearly seen.

After propagating in the 50 km long DCF the optical code is fed into the 50-port optical decoder, and the auto-correlation and cross-correlation outputs are observed in (ix) as the bit pattern of the input 4-bit block varies. As shown in Fig. 8.19, the gate signals at the 16-to-4 line coder for the fixed and random patterns confirm that the original fixed and random bit patterns are properly decoded. Finally, the measured bit error rates of the received serial data for fixed and random patterns without/with on-line XOR are shown in Fig. 8.20. The power penalties between B-to-B and after 50 km transmission for original random and fixed patterns are about 1.2 and 0.4 dB, respectively. On the other hand, in the XORed case, the power penalties between B-to-B and after 50 km transmission for random and fixed patterns are about 0.6 and 1.2 dB, respectively.

8.6.3 Scaling of the M-ary count

As the number of the ary count M increases, the data confidentiality of M-ary OCDMA becomes tighter. This is because the number of optical codes assigned to the bit blocks increases, and this imposes on the adversary more trials to find the correspondence between the code and bit sequence. Two approaches to enlarge the ary count are introduced here.

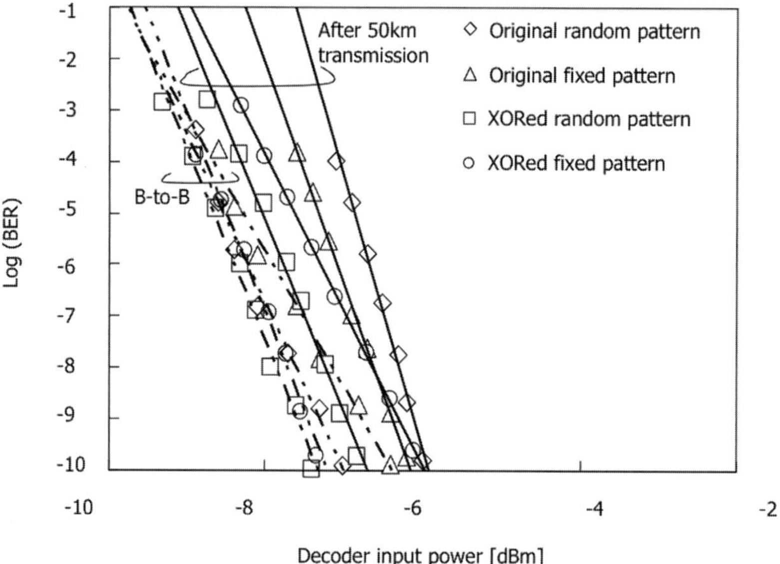

Figure 8.20 Measured bit error rates [12]. © Reprinted by permission of IEEE.

8.6.3.1 Polarization division multiplexing

One approach is to utilize polarization division multiplexing (PDM) or daul polarization (DP). This allows us to use the same code sequences on both vertical and horizontal polarizations because there is no interference between the two orthogonal polarizations in an optical fiber in the linear regime. In Fig. 8.21 the architecture of PDM $2^8(=256)$-ary OCDM transmission system using multiport en/decoder is shown. The operation principle is the same as for conventional M-ary OCDM, except for the PDM [13]. A notable difference is that a single-polarization M-ary OCDM, as in Section 8.6.2, would require $M(M = 2^m)$ codes while, on the other hand, PDM OCDM can do with $\sqrt{M}(\sqrt{M} = 2^{m/2})$. Therefore, for 265-ary the number of codes required is reduced to only 16. At the transmitter, a serial data bit stream at B bit/s is segmented every 8 bits by a SP converter, and each 8-bit block is sent to an 8-to-32 line coder. The former 4-bit block (higher-order bits, HX) and the latter 4-bit block (lower-order bits, LX) are mapped onto two codewords, according to Table 8.4. For example, the incoming 8-bit block (1, 0, 0, 0, 0, 1, 1, 1) is divided into the HX 4-bit block (1, 0, 0, 0) and the LX 4-bit block (0, 1, 1, 1) which are encoded into the OC2 and OC15 codes, respectively. 32 outputs of the line coder are time-interleaved into 16 lines by an electronic 32:16 multiplexer (Mux), and each output is guided to one of a 16-port optical gate switch array using LN-IMs, to generate a gate signal that selects an optical seed pulse corresponding to the OC. It would be possible to encode the LX and HX blocks onto two orthogonal polarizations using two identical en/decoders.

The optical codes for the HX and LX bit blocks come from two different output Ports 1 and 8 of the multiport encoder, respectively. The switch (SW) at these two outputs

Table 8.4 Code lookup table for PDM 256-ary OCDMA [13], © Reprinted by permission of the Optical Society of America

	OC1	OC2	OC3	OC4	...	OC15	OC16
OC1	00000000	00001000	00000100	00001100		00000111	00001111
OC2	10000000	10001000	10000100	10001100		10000111	10001111
OC3	01000000	01001000	01000100	01001100		01000111	01001111
OC4	11000000	11001000	11000100	11001100		11000111	11001111
.	.	.	.	.	.	.	.
.	.	.	.	.	.	.	.
OC15	01110000	01111000	01110100	01111100	...	01110111	01111111
OC16	11110000	11111000	11110100	11111100	...	11110111	11111111

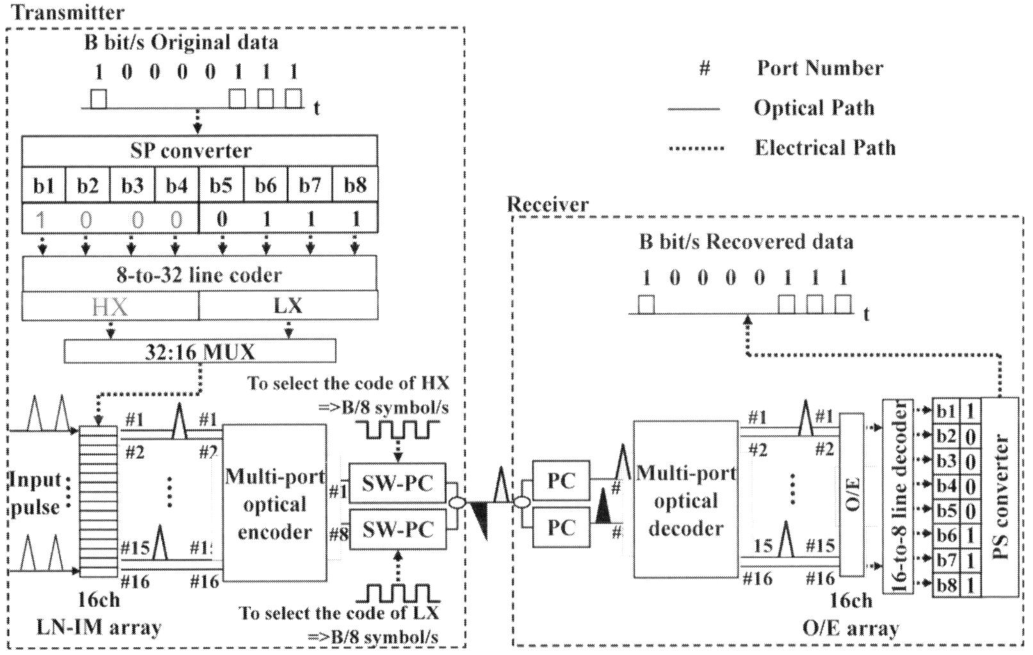

Figure 8.21 Architecture of PDM 256-ary OCDM system [13]. © Reprinted by permission of the Optical Society of America.

selects the HX and LX codes and the polarization controller (PC) rotates their polarization 90° and 0°, respectively. Therefore, the code repetition rate at each polarization state is not equal to $B/16$ but it is equal to the symbol rate at $B/8$ Symbol/s.

At the receiver, the 256-ary OCDM signal is split into two encoded signals with orthogonal polarization states by the PCs, and the HX and LX codes are launched to different input Ports 1 and 8 of the multiport optical decoder, which has the same configuration as the encoder. An auto-correlation waveform appears only at one of the 16 output ports of the optical decoder, and the output port number unequivocally identifies the received optical code. The output optical pulse from the decoder is converted into an electrical signal by the 16-channel optical-to-electrical (OE) array, and it is launched into the 16-to-8 line decoder, so that the original 8-bit data sequence is recovered via the parallel-to-serial (PS) converter, using the same code lookup table, see Table 8.4. The experimental demonstration has been conducted for 256-ary OCDM at 311 MSymbol/s or bit rate of 2.488 Gb/s.

Let us compare the data confidentiality of a PDM M-ary OCDM system with that of a conventional M-ary OCDM system. M-ary OCDM systems furnish both "optical" and "electrical" confidentiality, since an eavesdropper has first to decrypt the optical code, and then he or she has to find the correspondence with a sequence of bits. In PDM M-ary OCDM, the optical confidentiality of the two systems of orthogonal polarization is identical, since they use the same number of optical codes. Therefore, the number of codes is reduced from M to $\sqrt{M}$. Next, the electrical confidentiality is considered by evaluating the average number of trials that an adversary has to make to decrypt a message. To give a quantitative evaluation of the confidentiality of conventional and PDM systems, we consider an exhaustive key search attack, or brute force attack, which is the simplest COA attack. In this case, an eavesdropper is able to intercept only the cipher-text, i.e., the optical codes, and to break the system security an eavesdropper has to determine the correspondence between the optical code and the sequence of m bits. In a conventional M-ary OCDM system, M is equal to the number of optical codes, and the average number of trials needed to break the system security is equal to half of all the possible combinations, that is $M!/2$ as shown in Table 8.2. In a PDM OCDM, only $\sqrt{M}$ optical codes are used, and the confidentiality can be evaluated in the following way. First, the message of m bits is split in two parts, which can be chosen in a completely arbitrary way. Since the eavesdropper cannot know which $m/2$ bits have been selected to be encoded on the same polarization, he or she has to make some guesses and the only way to tell if his/her guess is right is by looking at the deciphered output to see if it is meaningful. The number of possible $m/2$-combinations of m elements, i.e., the number of sequences of $m/2$ bits taken over a set of m is

$$m(m-1)\cdots(m-m/2+1)/(m/2)! = m!/[(m/2)!]^2 \tag{8.6}$$

and is $8!/(4!)^2 = 70$ in the case of $m = 8$. Later, the two groups of $m/2$ bits are encoded separately onto $\sqrt{M}$ optical codes, using two independent lookup tables for the two polarizations, and the total number of possible choices is $(\sqrt{M}!\,)^2$. Therefore, the average number of trials that the eavesdropper has to make is

$$(\sqrt{M}!\,)^2 m!/[(m/2)!]^2. \tag{8.7}$$

The number of trials is plotted as a function of the number of optical codes in Figs. 8.22(a) and (b) for COA and CPA, respectively. We observe that the "electrical" confidentiality of a PDM M-ary system is enhanced with respect to that of a conventional system with the same number of optical codes.

Using 16 OCs, the system confidentiality against a COA in a PDM M-ary OCDM system is higher by a factor of more than 10^{28}, if two different lookup tables have been used for the two polarizations, and the eavesdropper does not know how the 8-bit

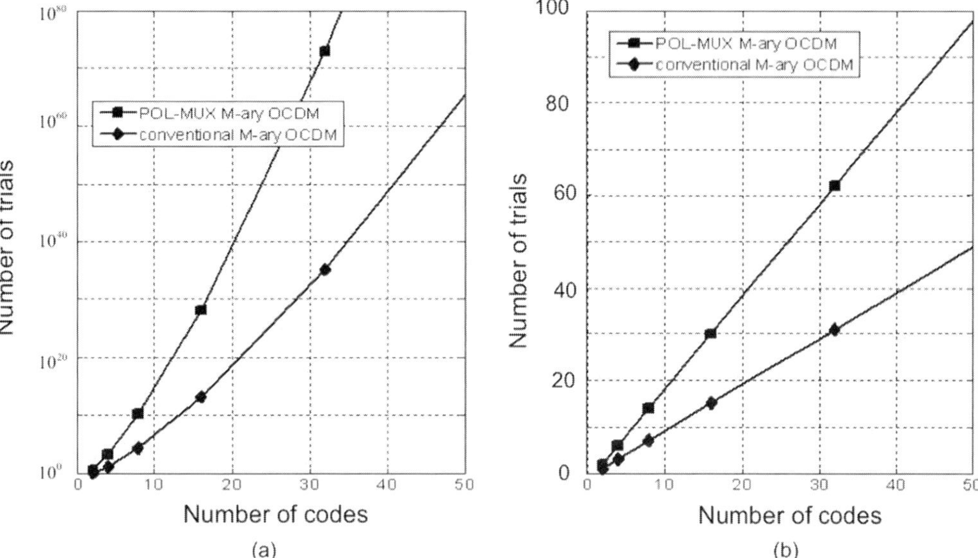

Figure 8.22 (a) Number of trials necessary to break confidentiality with a COA and (b) number of trials necessary to break confidentiality with a CPA [13]. © Reprinted by permission of the Optical Society of America.

sequence has been split in the LX and HX blocks. On the other hand, 10^{13} trials are needed to break the confidentiality of a conventional M-ary system that uses 16 optical codes.

The lower bound security parameter of modern cryptanalysis is the number of plaintexts that an eavesdropper needs to know in a CPA in order to break the system confidentiality. This attack assumes that the eavesdropper has the capability of choosing arbitrary plaintexts to be encrypted to obtain the corresponding cipher-texts, i.e., the optical codes. In a conventional M-ary system, a CPA could reveal the cryptographic secret key, i.e., the scheme that has been used to couple each sequence of m bits with one of the M OCs. We assume that the lookup table is completely arbitrary (i.e., no recursive scheme has been used for the secret key), so that the adversary has to be able to encrypt all the codewords, except one, i.e., $M - 1$ codewords, to intercept the data. As an example, considering $m = 8$ bits, the eavesdropper should encode all the sequences 00000000, 00000001, . . . , 11111111 minus one to find all the information, and this operation requires $M - 1 = 255$ trials. In a PDM M-ary OCDM system, the eavesdropper can easily reveal how the message is split into the HX and LX blocks, just by encoding a single message. For each polarization, the adversary has to find all the correspondences (minus one) between the sequences of $m/2$ bits and the $\sqrt{M}$ optical codes, making $\sqrt{M} - 1$ attempts. From Fig. 8.22(b), if we assume that the lookup tables of the two polarizations are independent, the total number of trials required to decrypt all the codewords in a PDM M-ary OCDM system is $2(\sqrt{M} - 1)$, and it is 30 in our case.

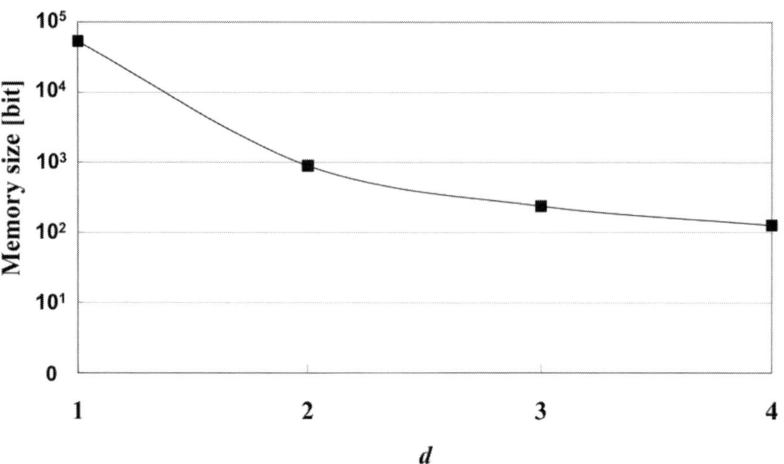

Figure 8.23 Memory size against parameter d [14]. © Reprinted by permission of IEEE.

8.6.3.2 Multi-dimensional codes

The multi-dimensional code described in Section 8.3.2 provides another approach to scale up M-ary OCDM with respect to the ary count. When n ($n \leq N$) optical short pulses are launched to different input ports of an N-port optical encoder, the total number of available n-dimensional codewords that can be generated is $2^N - 1$, from Eq. (8.5). Table 8.5 shows the code lookup table that transforms, for example, a sequence of $m = 12$ bits into $M = 4096$ ($= 2^{12}$) multi-dimensional optical codes [14]. The columns represent the input ports of an encoder, and the check-mark denotes the input port where a short optical pulse is launched. For the sake of simplicity, a 12-bit block is divided into three 4-bit blocks, and each 4-bit block is encoded. To assign all sixteen 4-bit blocks, five optical codes ($2^5 - 1 > 16$) are needed, and the combinations of optical codes are assigned sequentially to bit blocks.

The implementation of an electric code lookup table is based on a memory device, whose memory capacity S bits equals the number of its elements. The number of elements of the lookup table shown in Table 8.5 is given by

$$S = d \times 2^{m/d} \times \left(\frac{\log_2 M}{d} + 1 \right) \tag{8.8}$$

where d denotes the number of divisions of bit blocks. The numerical S is plotted as a function of d in Fig. 8.23 for $M = 4096$. It is evident that as d becomes larger, the required memory size decreases, and the electronic circuit will be simplified. The code lookup table in Table 8.5 corresponds to the case with $d = 3$, which requires a memory size $S = 240$ bits. The code lookup table can be implemented using a field programmable gate array (FPGA). In Fig. 8.24, for example, with $d = 3$, the configuration of an FPGA is shown using a Xilinx Inc. array, mode number XC4VLX25SF363, which has a response time of 10 ns, maximum interface frequency of 622.08 MHz, consisting of sets of

Table 8.5 Code lookup table for 4096-ary OCDMA for the case $d = 3$ [14], © Reprinted by permission of IEEE

LX	OC1	OC2	OC3	OC4	OC5	MX	OC1	OC2	OC3	OC4	OC5	HX	OC1	OC2	OC3	OC4	OC5	OC15
0000	✔					0000	✔					0000	✔					
0001		✔				0001		✔				0001		✔				
0010			✔			0010			✔			0010			✔			
0011				✔		0011				✔		0011				✔		
0100					✔	0100					✔	0100					✔	
0101	✔	✔				0101	✔	✔				0101	✔	✔				
0110	✔		✔			0110	✔		✔			0110	✔		✔			
0111	✔			✔		0111	✔			✔		0111	✔			✔		
1000	✔				✔	1000	✔				✔	1000	✔				✔	
1001		✔	✔			1001		✔	✔			1001		✔	✔			
1010		✔		✔		1010		✔		✔		1010		✔		✔		
1011		✔			✔	1011		✔			✔	1011		✔			✔	
1100			✔	✔		1100			✔	✔		1100			✔	✔		
1101			✔		✔	1101			✔		✔	1101			✔		✔	
1110				✔	✔	1110				✔	✔	1110				✔	✔	
1111	✔		✔		✔	1111	✔		✔		✔	1111	✔		✔		✔	

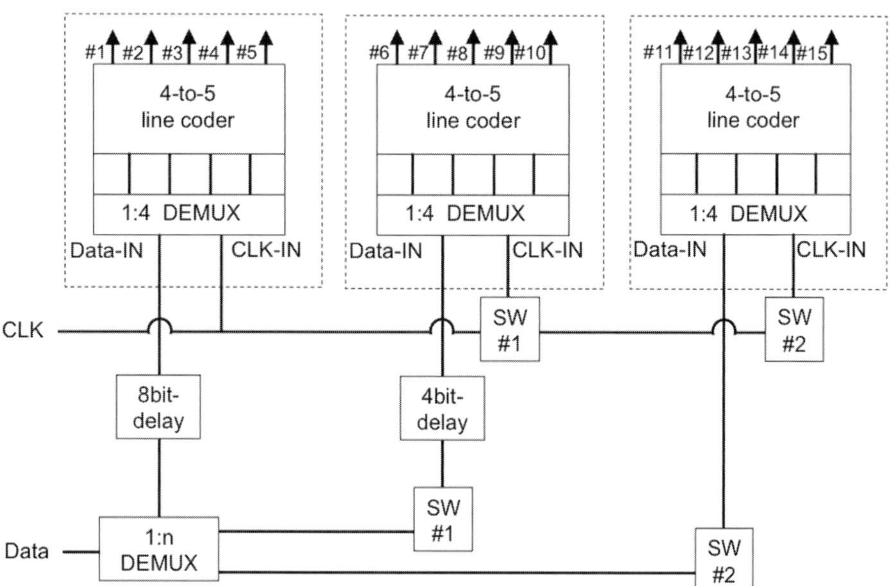

Figure 8.24 Detailed structures of FPGA (1:12 Demux and 12-to-15 line coder) [14]. © Reprinted by permission of IEEE.

4-to-5 line coders and demultiplexer (Demux). It consists of three sets of the 4-to-5 line coders and 1:n ($n = 1$, 2, 3) demultiplexer (Demux). This scheme provides scalability, because the ary count M can be changed from 16 to 4096, according to the switching status of the electrical switches shown in the inset of Fig. 8.24. For the case with $m = 12$ and $d = 3$, fifteen codes are required from Table 8.5. Therefore, switches 1 and 2 are both in the state ON, and the input 12 bits are segmented every 4 bits and fed to the 4-to-5 line coder. The output of the line coder is guided to a multiport encoder to assign a set of five optical codes. In Fig. 8.25 the architecture of a 4096-ary OCDM system using FPGA-based line coders is shown. The preliminary proof-of-concept experiment of 4096-OCDM was conducted at 2.488 Gb/s. A 50-port optical en/decoder was used.

We analyze the data confidentiality of M-ary OCDM using multi-dimensional optical codes. According to Kerckhoffs' principle, we assume that the eavesdropper knows everything about the OCDM encoding technique, in terms of the data rate and chip rate, code length, modulation formats, and wavelengths. In addition, if the message of m bits is split into d parts, the adversary knows the segmentation rule. The d groups of m/d bits are encoded separately onto $1 + (m/d)$ optical codes, using d independent lookup tables, so that the total number of possible choices is $d(2^{(m/d)}!) = d(\sqrt[d]{m!})$. Figure 8.26(a) shows the number of trials necessary to break the system confidentiality in a 4096-ary system, as a function of d. When the 12-bit message is split into $d = 3$ parts, the number of COA trials necessary is $3 \cdot (16!) = 7 \cdot 10^{13}$. The lower bound security parameter of modern cryptanalysis is the number of plaintexts that an eavesdropper needs to know in

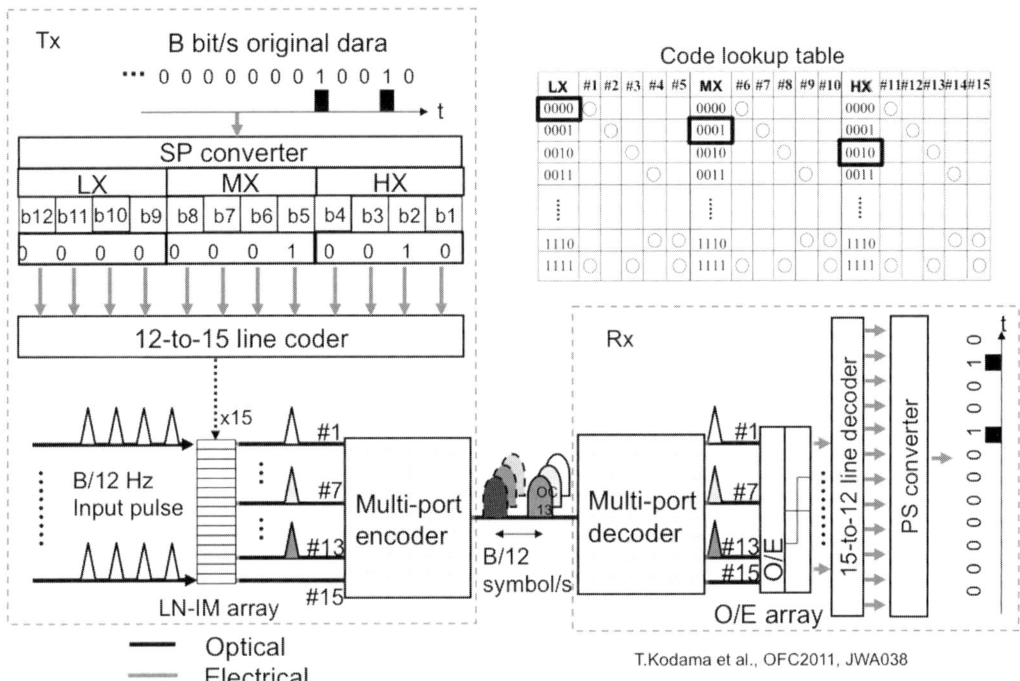

Figure 8.25 Architecture of the 4096-ary OCDM system [14]. © Reprinted by permission of IEEE.

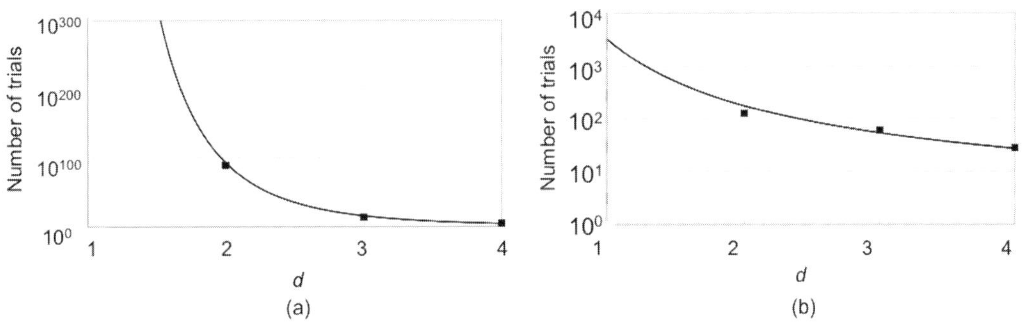

Figure 8.26 (a) Number of trials necessary to break confidentiality with a COA and (b) number of trials necessary to break confidentiality with a CPA [14]. © Reprinted by permission of IEEE.

a CPA, to break the system confidentiality. This attack assumes that the eavesdropper has the capability to choose arbitrary plaintexts, and he or she can encrypt them to obtain the corresponding cipher-texts, i.e., the optical codes. In an M-ary OCDM that uses multi-dimensional PSK codes, the eavesdropper can easily reveal how the message is split into blocks. For each block, the adversary has to find all the correspondences (minus one) between the sequences of bits m/d and the optical codes, making $\sqrt[d]{M} - 1$

attempts. If we assume that the lookup tables of the blocks are independent, the total number of trials required to decrypt all the codewords in a multi-dimensional M-ary OCDM system is $d(\sqrt[d]{M} - 1)$. Figure 8.26(b) shows the confidentiality against CPAs for a 4096-ary system as a function of d. For $d = 3$, the number of attempts required is 45.

9 Testbeds of OCDMA and hybrid systems

A proof-of-concept experiment of an OCDMA system will confirm the validity of its operation principle. The testbed experiments of OCDMA-PON systems have their own significance in demonstrating its practical feasibility. They show that it is possible to build operational systems based upon existing components and they bring up a number of issues that need to be solved before the systems can be commercialized. Without such experimental demonstrations, the systems will have difficulty gaining the confidence of the market.

In this chapter, experimental testbed demonstrations of OCDMA-PON and its hybrid systems are introduced. The network architectures of the testbeds are shown schematically in Figs. 9.1 and 9.2. In Fig. 9.1, OCDMA-PON using two types of optical en/decoder, the SSFBG and multiport en/decoders, is illustrated. Note that the multiport en/decoders are located at a remote node (RN), and this configuration can allow en/decoder-free ONUs. This is a motivation for heterogeneous en/decoders for realizing cost-effective OCDMA-PON. In Fig. 9.2 the architectures of hybrid OCDMA systems, WDM-OCDMA-PON and TDM-WDM-OCDMA-PON, are shown. Here, the optical code is designated as OC n ($n = 1, \ldots, N$), the wavelengths as λ_k ($k = 1, \ldots, K$), and the time slot as m ($m = 1, \ldots, M$). The total capacity is the same for WDM-OCDMA-PON and TDM-WDM-OCDMA-PON, and it is enhanced by a factor of K when OCDMA is overlayed onto WDM. However, the number of subscribers that can be accommodated by TDM-WDM-OCDMA-PON is increased by a factor of M compared with WDM-OCDMA-PON.

9.1 OCDMA testbed

A 12-subscriber 10 Gb/s asynchronous OCDMA experiment using multiport en/decoders was conducted [1]. The experimental setup is shown in Fig. 9.3. Up to sixteen 10 Gb/s signals can be combined in an asynchronous manner under practical conditions of balanced power, random delay, random bit phase, and random polarization state. In order to facilitate error correction, Optical Channel Transport Unit 2 (OTU2) frame at a bit rate of 10.709 Gb/s is used, which contains pseudo-random bit sequence payload data and forward error correction (FEC) parity. This is based on the Optical Transport Network (OTN), standardized by ITU-T G.709. OTN enables transparent transport of services over optical wavelengths in WDM systems. The hierarchical frame structure is

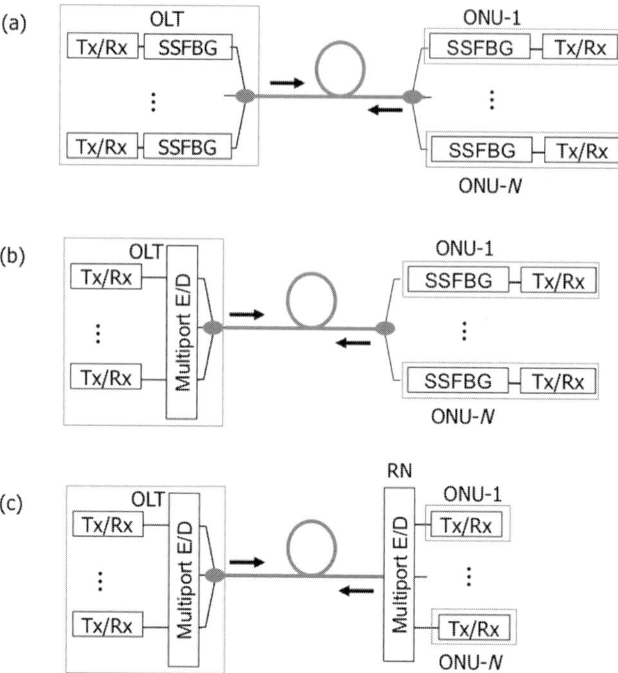

Figure 9.1 OCDMA-PON using two types of optical en/decoder, the SSFBG and multiport en/decoders. (a) SSFBG en/decoder at both OLT and ONUs, (b) heterogeneous combination of multiport en/decoder at the OLT and SSFBG en/decoder at each ONU, and (c) multiport en/decoders at both OLT and ONUs.

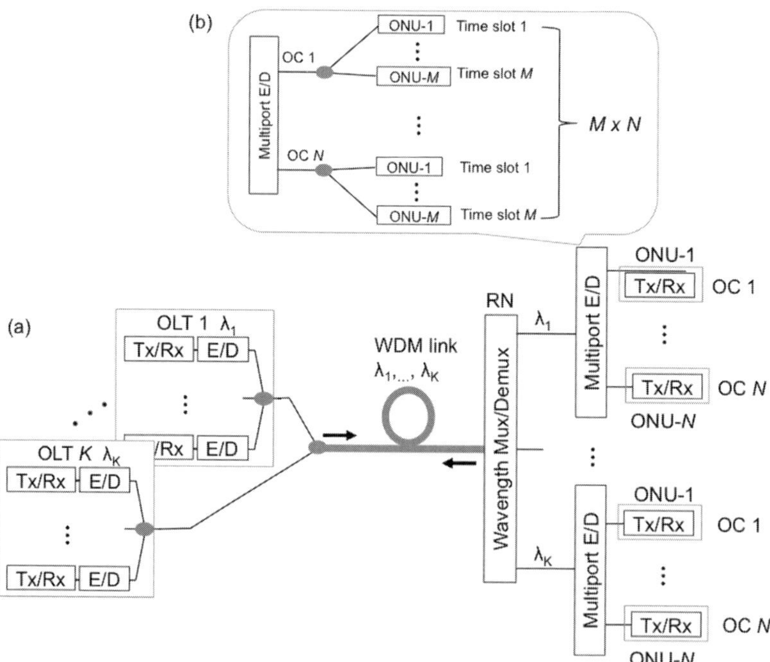

Figure 9.2 Architectures of hybrid OCDMA systems: (a) WDM-OCDMA-PON and (b) TDM-WDM-OCDMA-PON.

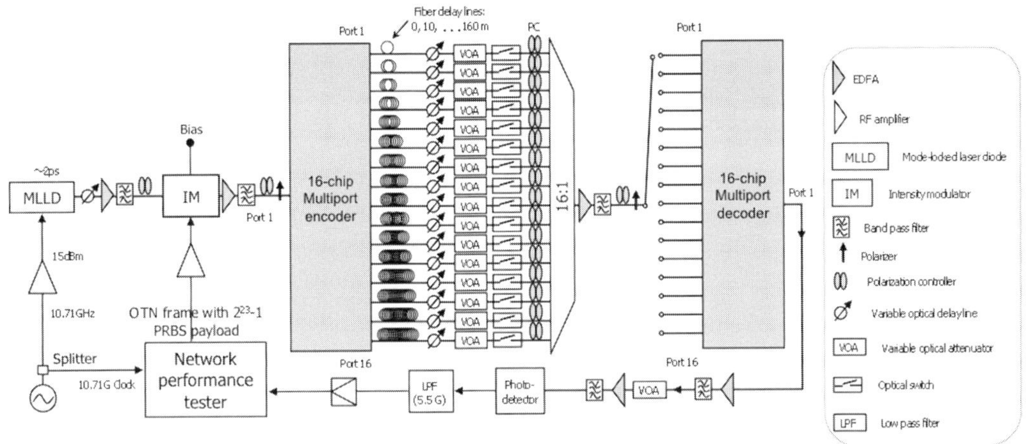

Figure 9.3 Experimental setup for a 10 Gb/s OCDMA system [1]. © Reprinted by permission of IEEE.

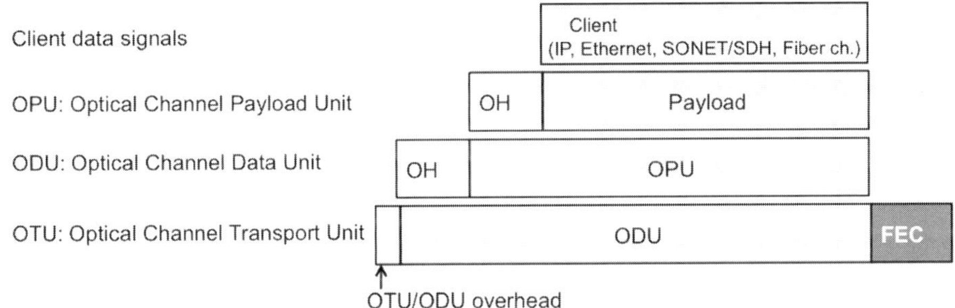

Figure 9.4 OTN hierarchy and its frames.

shown in Fig. 9.4. The transmission rates of OTN are summarized in Table 1.1. The FEC is mandatory in EPON and 10 G-EPON, and it uses Reed–Solomon (RS) (255, 223) with 14% overhead. In G-PON and 10G-PON, RS (255,239) with 7% overhead is optional as summarized in Table 2.1. The longer the code becomes, the larger the coding gain, but the penalty to be paid is the longer latency and high cost as well as large overhead. It is noteworthy that the coding gain for the stream data in the downlink is larger than for the burst data in the uplink. RS (255,239) has 3.4 dB net coding gain in Q-factor for a BER of 10^{-7}, and a BER of 10^{-15} is obtained with the input BER of 10^{-4} as shown in Fig. 9.5(a). RS (255,223) has a 3.6 dB net coding gain in Q-factor for a BER of 10^{-7} [2].

The multiport encoder generates 200 Gchip/s 16-chip 16-level PSK code sequences. The measured ACR shown in Fig. 7.34 is as high as 15–20 dB, and thus the MAI noise can be significantly suppressed. The measured temporal waveforms of encoded signal and auto-correlation signal are shown in Fig. 9.5(b). The measured spectra of encoded signal (upper row), decoded signals (middle row), and eye diagrams of the decoded

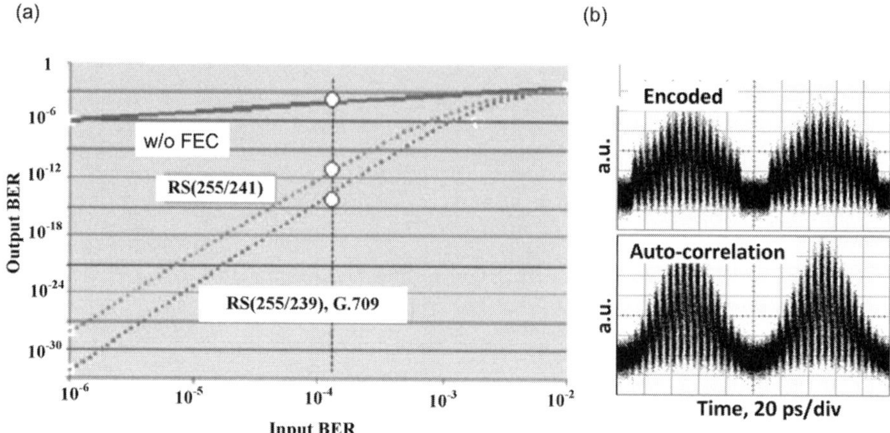

Figure 9.5 (a) Output BER versus input BER for RS (255,239) and RS (255,241) and (b) temporal waveforms of encoded and decoded signals [1]. © Reprinted by permission of IEEE.

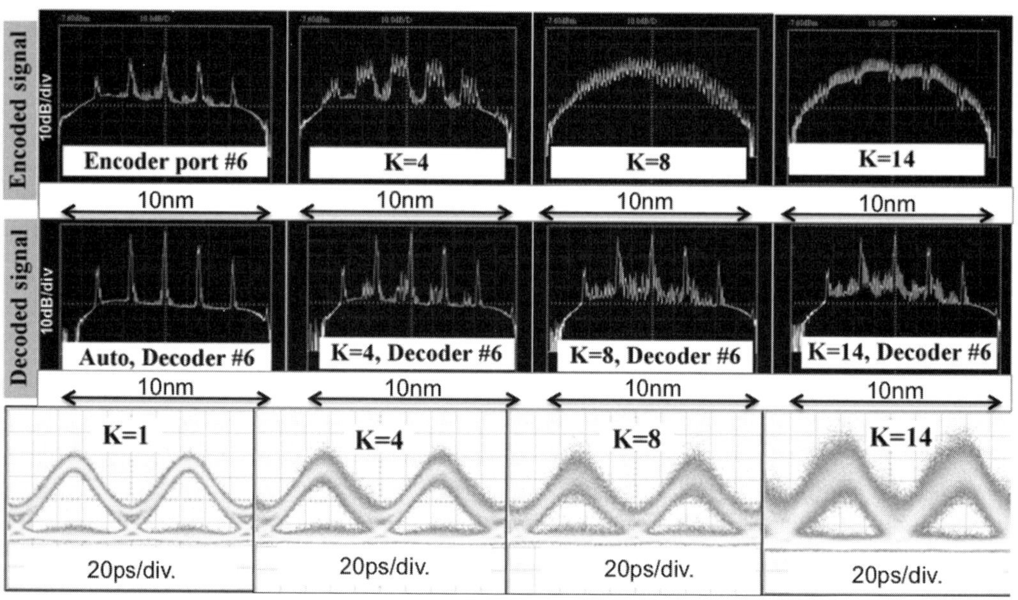

Figure 9.6 Measured spectra of encoded signals (upper row), decoded signals (middle row), and eye diagrams of the decoded signals (bottom) [1]. © Reprinted by permission of IEEE.

signals (bottom) for different numbers of active subscribers K are shown in Fig. 9.6. Finally, the BER performance for $K = 12$, for different subscribers (decoder Ports 2, 6, 10, 16), is plotted as a function of received power in Fig. 9.7. In all these cases, the BER achieves a value below 10^{-10} with the aid of FEC. This confirms that the OCDMA system can accommodate at most 12 subscribers.

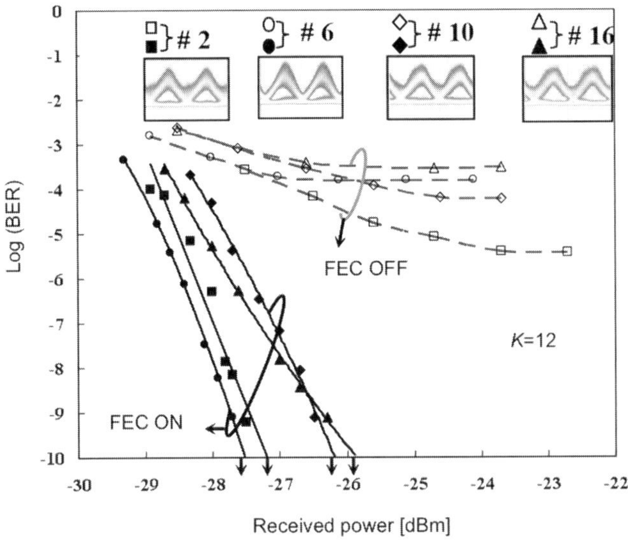

Figure 9.7 Measured bit error rate for number of active subscribers $K = 12$ [1]. © Reprinted by permission of IEEE.

9.2 Hybrid OCDMA systems

OCDMA can be integrated with other multiple access techniques such as TDMA, WDMA, and OFDMA described in Chapter 2. One of the motivations for the hybrid systems is to increase scalability of the PON architecture with respect to the number of subscribers. There are two scenarios: one is migration from an existing TDMA-based PON to a hybrid system, and the other is a "green field" deployment where there is no existing PON plant deployed in the field of interest. Hereafter, the main focus is on the migration from an existing TDMA-PON to a hybrid OCDMA system. In a TDM-PON system, the uplink bandwidth per subscriber decreases in proportion with the number of subscribers because they have to share the bandwidth. In contrast, a notable merit of OCDMA-based hybrid PON systems is that more subscribers can be accommodated without sacrificing the uplink bandwidth per subscriber. There are several options for the combination of OCDMA with other multiple access techniques. Here, TDM-OCDMA-PON, WDM-OCDMA-PON, and TDM-WDM-OCDMA-PON are introduced.

9.2.1 TDM-OCDMA-PON system

9.2.1.1 Architecture of TDM-OCDMA-PON

Figure 9.8 shows a scale-up scenario of a single 10 Gb/s-based TDM-PON by a factor of N over OCDMA [3]. If N conventional 10 Gb/s PON systems including M time-slotted ONUs are consolidated into a single PON as shown in Fig. 9.8(a), the

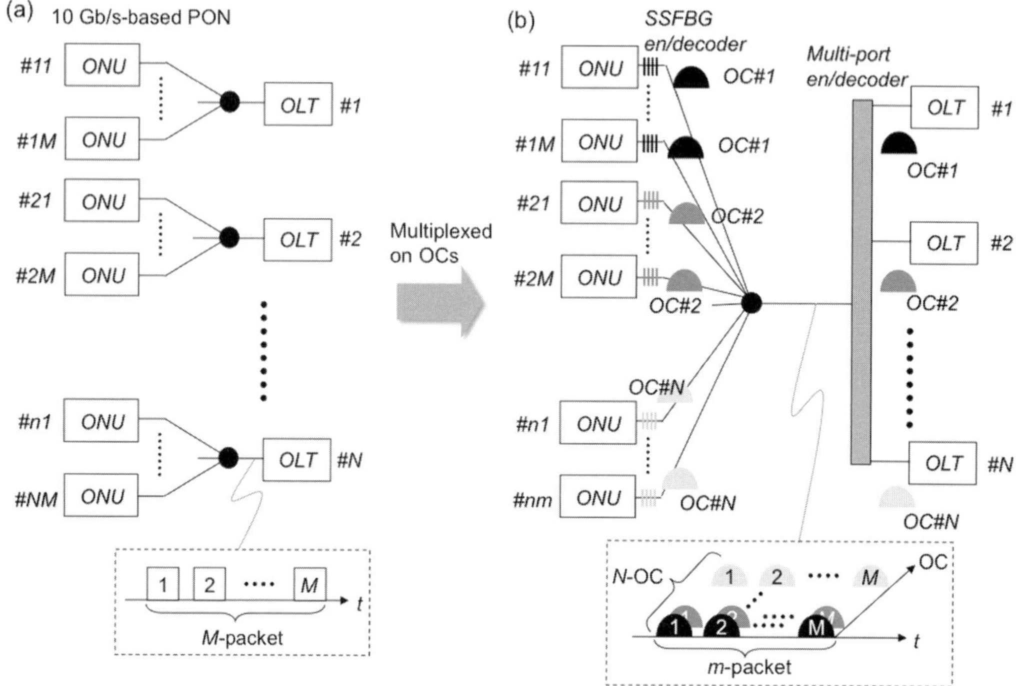

Figure 9.8 Architectures of scalable 10 Gb/s TDMA-PON system upgrades using OCDMA: (a) 10 Gb/s TDMA-PON and (b) TDMA-OCDMA-PON [3]. © Reprinted by permission of IEEE.

bandwidth per subscriber is reduced by a factor of N to $(M \times N)^{-1}$ due to the nature of TDMA. On the other hand, in the TDM-OCDMA-PON shown in Fig. 9.8(b), N conventional TDMA-PON systems are multiplexed on to N different optical codes, OC $1 \sim$ OC N. OC 1 is assigned to ONU $11 \sim 1M$, and OC N is assigned to ONU $N1 \sim NM$. This architecture adopts the heterogeneous combination of en/decoders described in Section 7.3.3. A single multiport en/decoder is used at the OLT. An SSFBG en/decoder is placed at each ONU. It is recalled that the TDM-OCDMA-PON can use the same wavelengths as the 10G-EPON downlink and uplink.

From the viewpoint of the upgrade cost of a 10 Gb/s-based PON system, additional components of a PON over OCDMA compared with conventional PON systems include the optical encoder/decoders at OLTs and ONUs, a pair of SSFBG en/decoders, which are less expensive than multiport en/decoders, placed at each ONU, and a multiport en/decoder placed at each OLT shared with all the ONUs. The SSFBG has the ability to process the ultra-long OC, polarization-independent operation, compact structure as well as low-cost capability for mass production. The system cost is discussed in Section 10.1.2. The correlation characteristics are different for SSSFBG en/decoders and multi-level PSK multiport en/decoders. Therefore, newly developed SSFBG en/decoders which generate multi-level PSK code, as described in Section 7.3.3, are used.

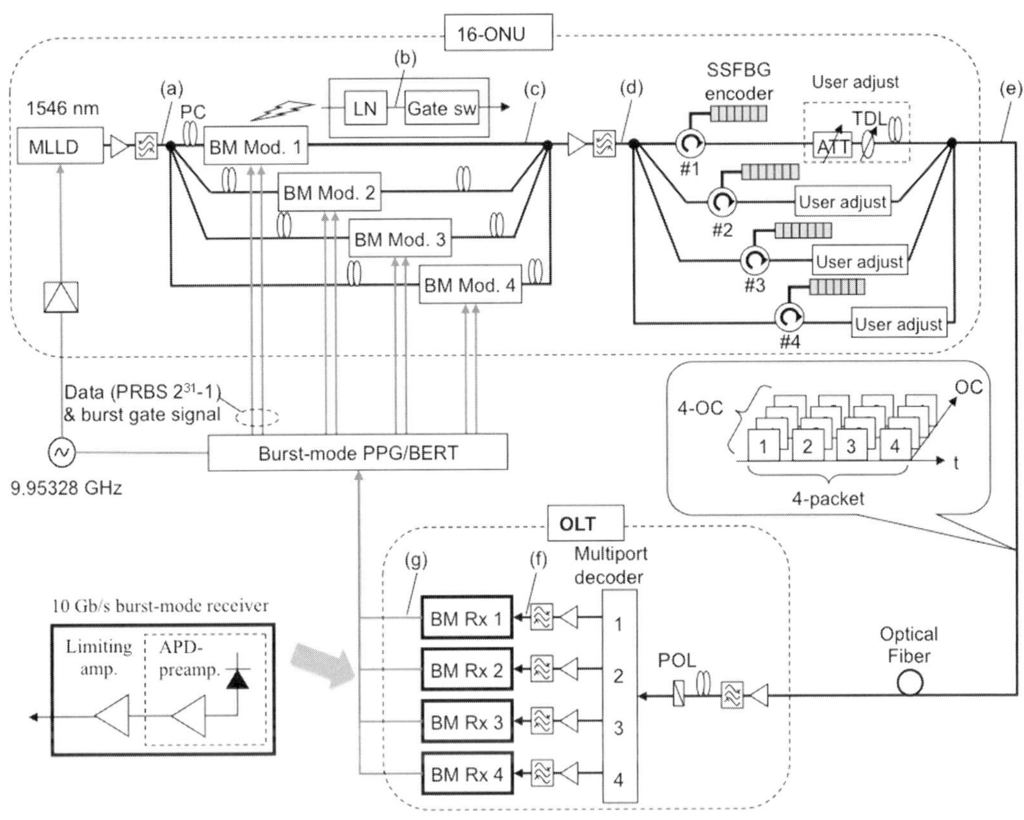

Figure 9.9 Experimental setup for 16-ONU 10 Gb/s PON over OCDMA uplink burst-mode transmission [3]. © Reprinted by permission of IEEE.

9.2.1.2 Experimental demonstration

In Fig. 9.9, the experimental setup for 16-ONU 10 Gb/s PON over OCDMA uplink burst-mode transmission is shown. The 4-packet, 10 Gb/s-based PON over 4-OC OCDMA is allotted an average uplink bandwidth of 2.5 Gb/s for each ONU, gaining four times as much bandwidth in comparison with a conventional 10 Gb/s TDMA-PON. The burst-mode transmission requires an optical burst-mode transmitter, burst-mode receiver, and clock and data recovery (CDR). A 2 ps pulse train is generated by a mode-locked laser diode (MLLD), shown in Fig. 9.10(a). The center wavelength is 1546 nm and the repetition rate is 9.95328 GHz. The output of the MLLD is modulated to four packets by the optical burst-mode modulators (BM-Mods), which consist of an LiNbO₃ (LN) intensity modulator and acousto-optic modulator (AOM) as a gate switch. The switching speed and the extinction ratio of the burst-mode gate switch are about 100 ns and over 40 dB, respectively. Therefore, the optical burst-mode modulators can realize fast burst turn-on/off time and sufficient power suppression during idle periods simultaneously. Figure 9.10(b) shows the LN intensity modulator output data with pseudo-random bit sequence (PRBS). Each packet length is 64 μs which includes 10 μs overhead as shown

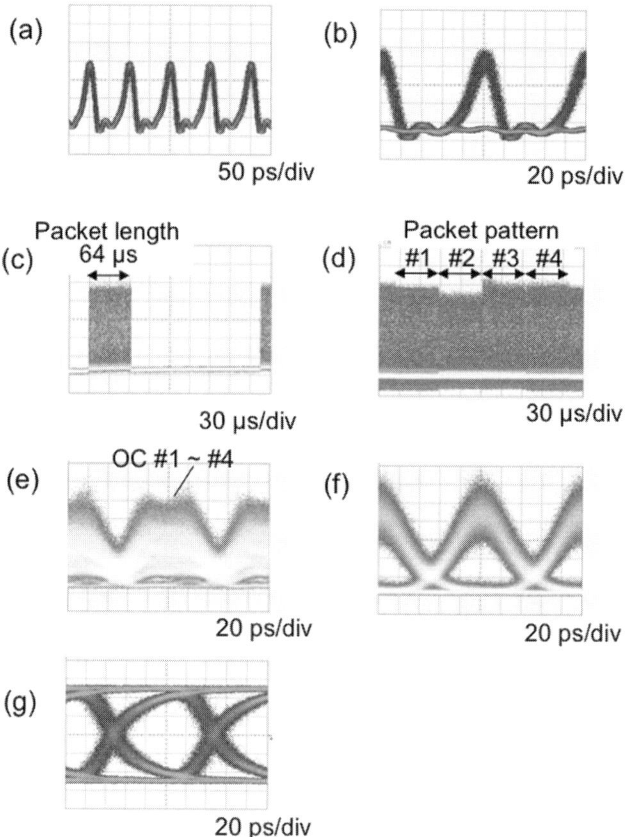

Figure 9.10 (a)–(g) Experimental results for a 16-ONU 10 Gb/s PON over OCDMA uplink [2]. © Reprinted by permission of IEEE.

in Fig. 9.10(c). Figure 9.10(d) shows the packet pattern for which the guard time is zero. This guard time is set to cope with the most severe condition for a fast response of the burst-mode receiver. More details of the burst-mode receiver are described in Section 5.6. These packets are encoded by four different 16-chip, 200 Gchip/s PSK codes using 16-level PSK SSFBG encoders. In the SSFBG encoder, the input optical pulse is time-spread into 16 chip pulses with 5 ps interval. Note that the Bragg wavelengths of four SSFBGs are set 12.5 GHz apart, equal to the center wavelengths of four input ports of the multiport decoder. The frequency interval is determined by the FSR of 200 GHz of the 16 ports of the multiport decoder which generates sixteen 16-chip, 200 Gchip/s, 16-level PSK codes. These four encoded signals are time-multiplexed into a TDM over OCDM signal. A tunable optical attenuator (ATT), tunable delay-line (TDL) and polarization controller (PC) are inserted in each path to investigate the system performance in the worst case scenario that the interference becomes most serious, as shown in Fig. 9.10(e).

At the OLT side, the received signal is decoded by the multiport decoder. Figure 9.10(f) shows the decoded signal of OC 1, showing the high-peak auto-correlation waveform with multiple access interference (MAI) noise skirt. Each decoded signal is processed

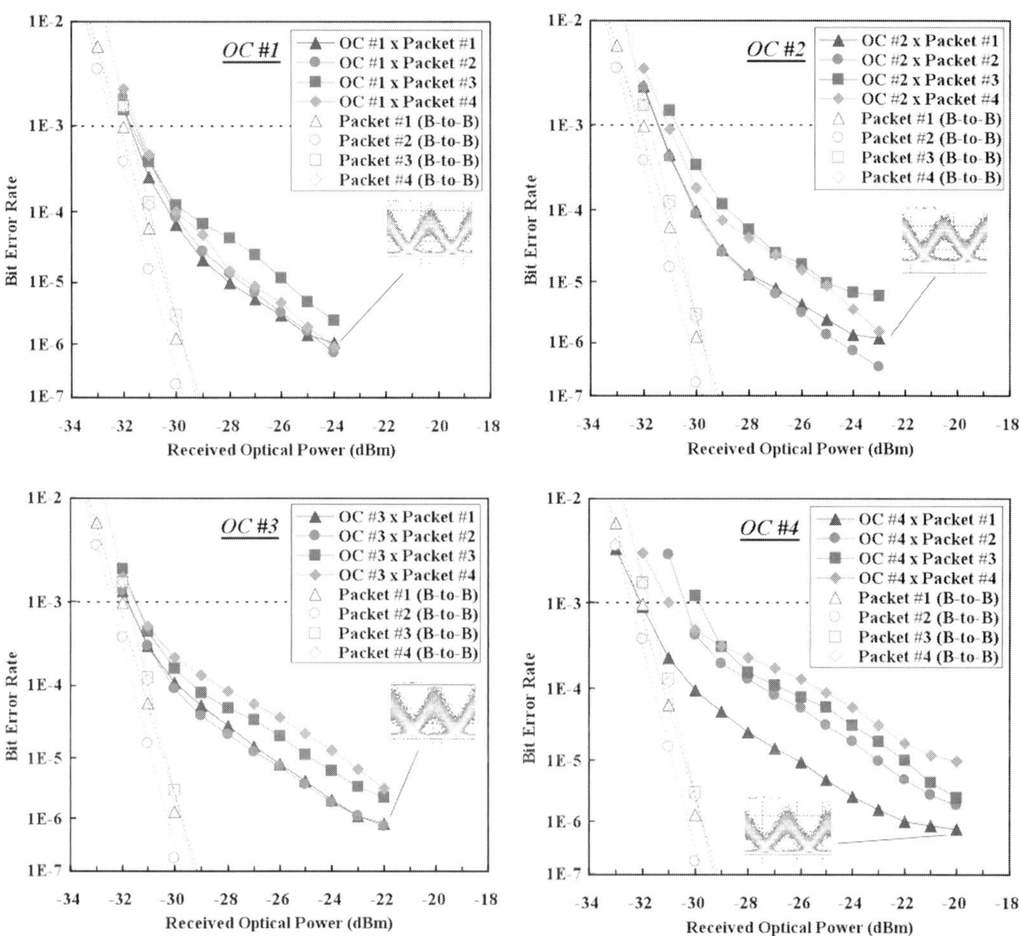

Figure 9.11 Measured BERs for all four codes and four packets [3]. © Reprinted by permission of IEEE.

at the 10 Gb/s burst-mode receiver. The 10 Gb/s burst-mode receiver consists of an avalanche photodiode (APD)-pre-amplifier module and a limiting amplifier, and it can provide a high-sensitivity burst-mode 2R function with optimal multiplication factor M = 8 of the APD. The bandwidth of the burst-mode receiver is set around 6.0 GHz for 10 Gb/s PON. A 10.3 Gb/s burst-mode 3R receiver which is fully compliant with IEEE 802.3av standard is introduced at the OLT for the uplink. This receiver can provide a burst-mode 3R function with 82.5 GS/s over-sampling. Figure 9.10(g) shows the good electrical eye opening from the decoded signal despite the MAI noise. In this experiment, EDFAs are inserted to compensate for the optical loss of components such as the SSFBG encoders and multiport decoders.

The measured bit error rates (BERs) of all 16 ONUs of the 4-OC, 4-packet system are shown in Fig. 9.11. The back-to-back system without optical en/decoding is also shown for reference. Error-free operation (BER $< 10^{-3}$) with FEC, a RS (255, 223), is obtained

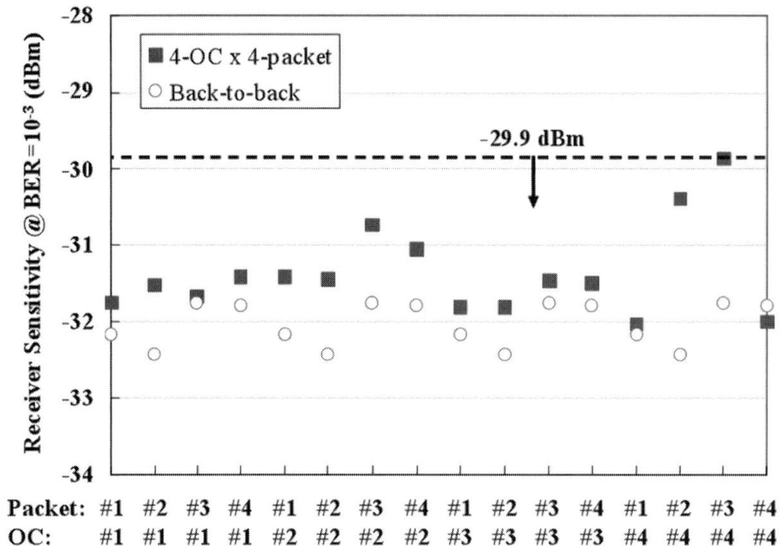

Figure 9.12 Measured receiver sensitivity for all data at BER=10^{-3} [3]. © Reprinted by permission of IEEE.

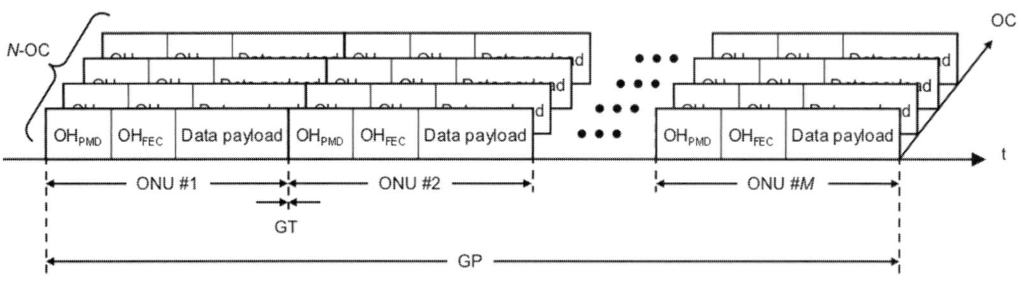

GT: Guard time
GP: Grant period

Figure 9.13 Uplink burst frame model [3]. © Reprinted by permission of IEEE.

for all four packets. In the 4-OC multiplexed case, a BER of less than 10^{-7} could not be obtained because of MAI noise. Good eye openings for all four decoded signals of OC 1–4 are obtained. In Fig. 9.12 the receiver sensitivity at a BER of 10^{-3} for all 16 ONU uplinks is shown. A receiver sensitivity of less than -29.9 dBm, the back-to-back sensitivity for 10^{-7}, is achieved by adapting the high-sensitivity burst-mode receiver. The power penalty between 4-OC 4-packet and back-to-back systems is less than 2.0 dB, which is caused by degradation of the optical signal-to-noise ratio due to ASE and MAI noise. The small deviation of receiver sensitivities is caused by the characteristic mismatch of optical components such as the burst-mode modulators and SSFBG encoders.

Let us consider the uplink performance, which can accommodate $N \times M$ subscribers by using M time slots and N optical codes. Figure 9.13 shows the uplink burst frame

Table 9.1 Parameters in the throughput calculation [3], © Reprinted by permission of IEEE

Quality	Value
Bit rate B	9.95328
PMD overhead OH_{PMD}	10
FEC overhead OH_{FEC}	12.9
Guard time GT	0
Grant period GP	1
Optical code N	4

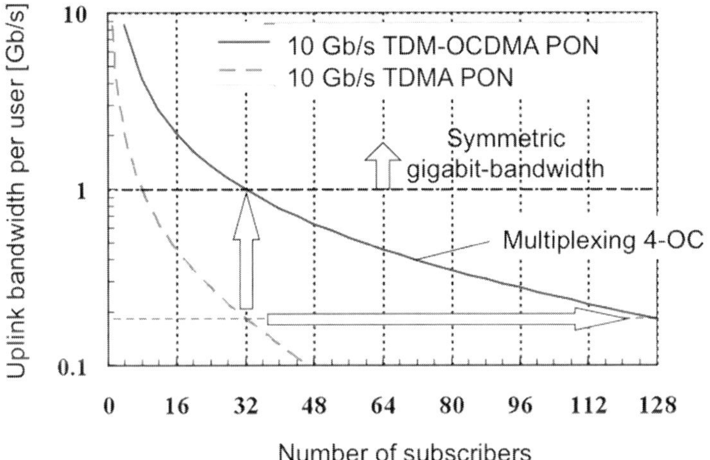

Figure 9.14 Calculated results of uplink bandwidth per user [3]. © Reprinted by permission of IEEE.

model. If the data payload is fully loaded, the uplink bandwidth per subscriber is given by

$$\mathrm{BW} = B \times \frac{T_{GP} - (M \times \mathrm{OH}_{PMD})}{T_{GP}} \times \frac{100 - \mathrm{OH}_{FEC}}{100} \times \frac{N}{M} \tag{9.1}$$

where B is the bit rate of the system, T_{GP} is the granted period which is equal to the optical burst frame length, M is the number of ONUs in a single 10 Gb/s-based PON system, OH_{PMD} is the physical medium dependent (PMD) overhead which consists of burst-mode turn-on/off and sync time, and OH_{FEC} is the FEC overhead ratio after FEC frame mapping. Table 9.1 shows the parameters of throughput calculation of the proposed system. OH_{FEC} and T_{GP} are assumed to be 12.9% and 1 ms, respectively. Other parameters are based on the experimental conditions. Figure 9.14 shows the calculated results of uplink bandwidth per subscriber. With $N = 1$ in Eq. (9.1), the bandwidth is equal to that of a conventional 10 Gb/s TDMA-PON system. The 10 Gb/s TDM-OCDMA-PON system can provide each subscriber with 1 Gb/s even when the system accommodates 32 subscribers. The scalability of TDM-OCDMA-PON is also

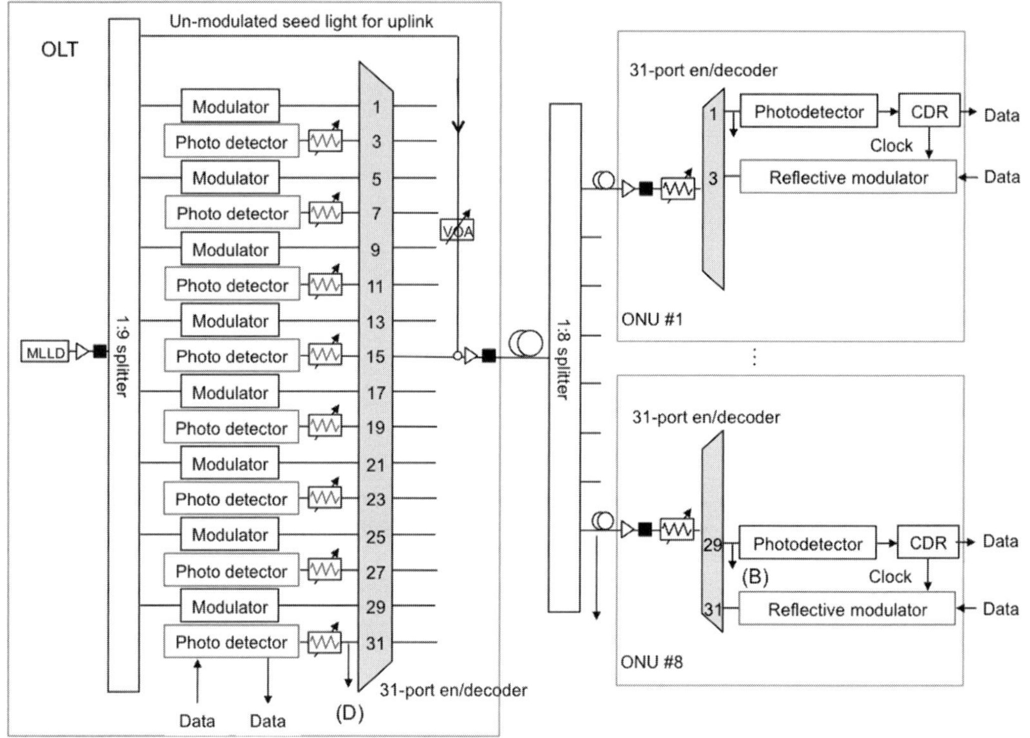

Figure 9.15 Architecture of colorless OCDMA-PON without light source at the ONUs [4].
© Reprinted by permission of ECOC2009.

demonstrated because it provides 128 subscribers with the same bit rate as TDMA-PON can offer to 32 subscribers.

9.2.1.3   Sourceless ONU without light source

To eliminate the need for an expensive light source at each ONU, a single MLLD at the OLT can seed light for the uplink to all the ONUs. In Fig. 9.15, a 10 Gb/s OCDMA-PON accommodates eight ONUs, and two optical codes are assigned to each ONU [4]. A 31-port en/decoder is placed at the OLT, and also at each ONU. Note that the multiport en/decoder at each ONU can be consolidated into one device and placed at a remote node, realizing en/decoder-free ONUs. An un-modulated seed pulse train from a MLLD of the same wavelength as the downlink signals is overlapped asynchronously to eight OCDMA downlink signals using a 3-dB coupler and delivered to the ONUs. The downlink signals, encoded with eight OCs 1, 5, 9, . . . , 29 from every four ports of the 31-port en/decoder at the OLT are decoded at each ONU, and the output auto-correlations from ports 1, 5, 9, . . . , 29 are received. The time-spread seed pulse, which is the encoded seed pulse, is also received with the desired downlink signal, but it has little impact on the quality of downlink signals. The seed pulse is decoded with OC 3 by the multiport en/decoder at ONU 1, yielding the cross-correlation waveform, and

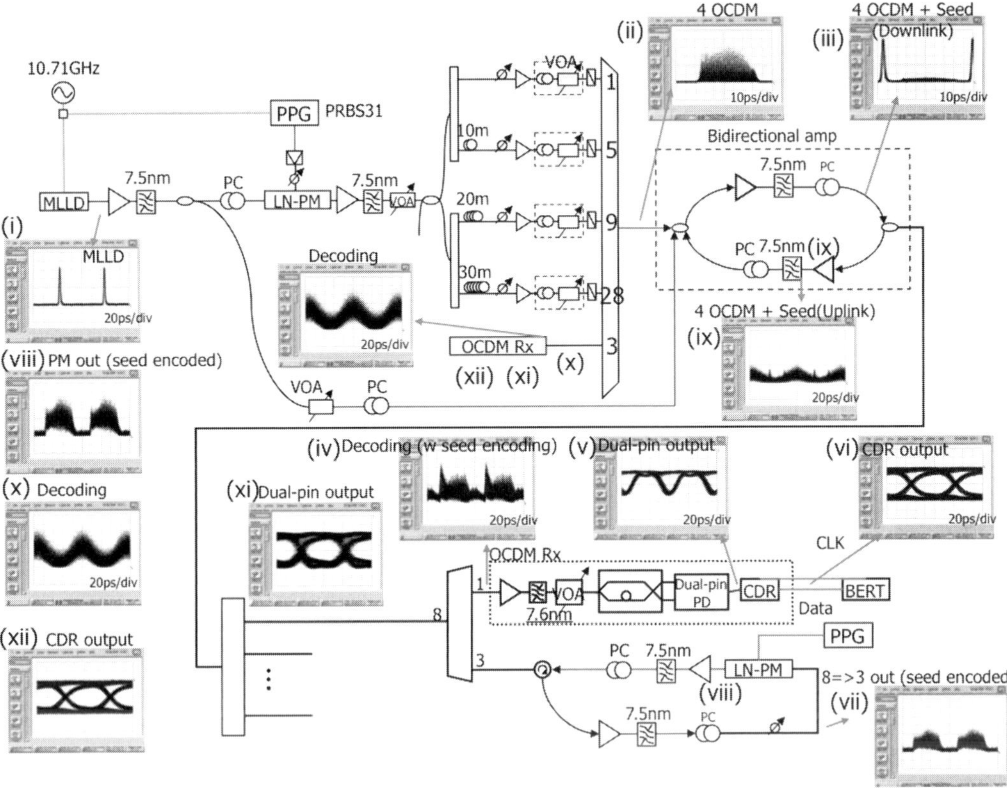

Figure 9.16 Experimental setup and the results for the uplink [4]. © Reprinted by permission of ECOC2009.

then modulated using the reflective-type modulator with each uplink data signal. Finally, eight uplink signals from ONUs are combined and transmitted to the OLT. At the OLT the uplink signals are decoded and the auto-correlation signals from the output ports OCs 3, 7, 11, ..., 31 are detected.

The experimental setup and results for a 10 Gb/s DPSK-OCDMA system with four downlink and an uplink subscribers is shown in Fig. 9.16. The OLT consists of an MLLD, an LiNbO$_3$ phase modulator (LN-PM), a single 31-port en/decoder, EDFA, and couplers. The MLLD generates 2 ps optical pulses, with 10.7 GHz repetition rate, at 1550 nm central wavelength (Fig. 9.16(i)). The pulse train is split into two by a 3-dB coupler. The seed pulse train is combined with four OCDMA downlink signals (Fig. 9.16(ii)) and transmitted to the ONU (Fig. 9.16(iii)). This signal is amplified by a bidirectional amplifier, which consists of a pair of EDFAs, polarization controllers (PCs), optical bandpass filters (OBPFs) and couplers. At the ONU, the downlink signal is sent into the input port of a 31-port en/decoder, and the DPSK signal is demodulated after decoding (Fig. 9.16(iv)). The decoded signal is DPSK detected by a fiber-based 1-bit delay-line and a dual-pin photodetector (PD) (Fig. 9.16(v)). For the BER measurements, the detected DPSK signal is recovered by the CDR circuit (Fig. 9.16(vi)) and forwarded to the bit error rate tester (BERT).

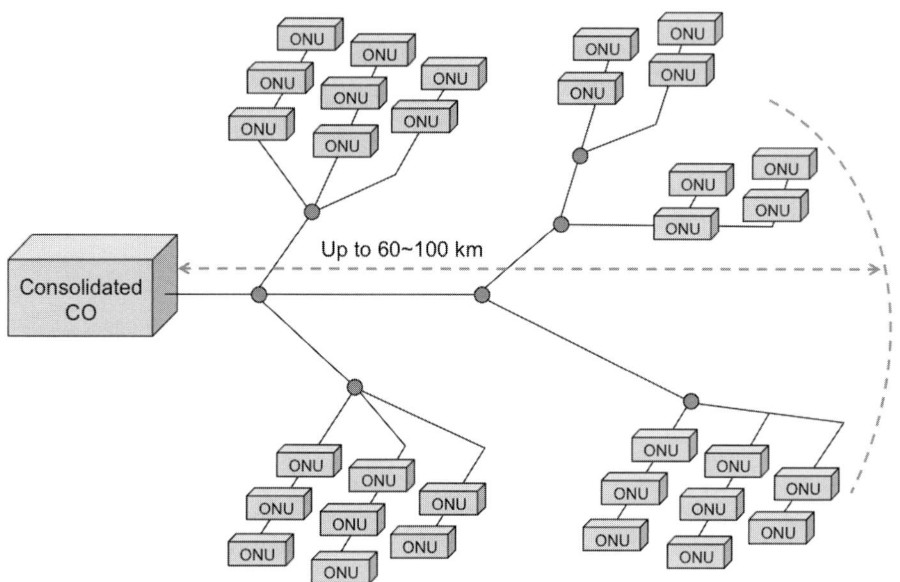

Figure 9.17 Long-reach PON architecture.

All the other unmatched en/decoder output ports simultaneously generate encoded signals (from the seed pulse train) overlapped with cross-correlation signals (from the 4-OCDMA downlink signals), that correspond to the MAI noise (Fig. 9.16(vii). The signal at output Port 3 is phase modulated with the uplink signal and loop-backed to the same 31-port en/decoder by a circulator (Fig. 9.16(viii)). The encoded waveform of the uplink signal (Fig. 9.16(ix)) is amplified by the bidirectional amplifier and sent into the input port of the 31-port en/decoder at the OLT. Finally, the decoded signal is detected at the matched port (Port 3) by the DPSK detector and CDR. Figures 9.16(x)–(xii) show the waveforms of the 31-port en/decoder, PD, and CDR outputs, respectively. Measured BERs of less than 10^{-9} are achieved for both downlinks and uplinks. This demonstrates the proper operation of the seeding light scheme for sourceless ONU.

9.2.1.4 Long-reach PON without dispersion compensation

An emerging issue of PON systems is scale-up with respect to geographic coverage, which has a significant impact on network operators with respect to holding down both CAPEX and OPEX. This is because a long-reach PON will allow several COs to be consolidated into one, as shown in Fig. 9.17, and hence the operation and network management will be simplified and less costly. This will eventually lead to the merger of metro and access networks. Therefore, PON systems without the long-reach capability will not be widely deployed in the future. The extension of the transmission distance between the OLT and ONUs from the conventional 20 km to 60 km has been standardized for XG-PON in ITU-T G.987.4. Furthermore, a 100 km reach is a desirable target. According to the calculation, extension of the transmission distance from 20 km to 100 km will enable the accommodation of 800 subscribers as many as 25 $(= (100/20)^2)$

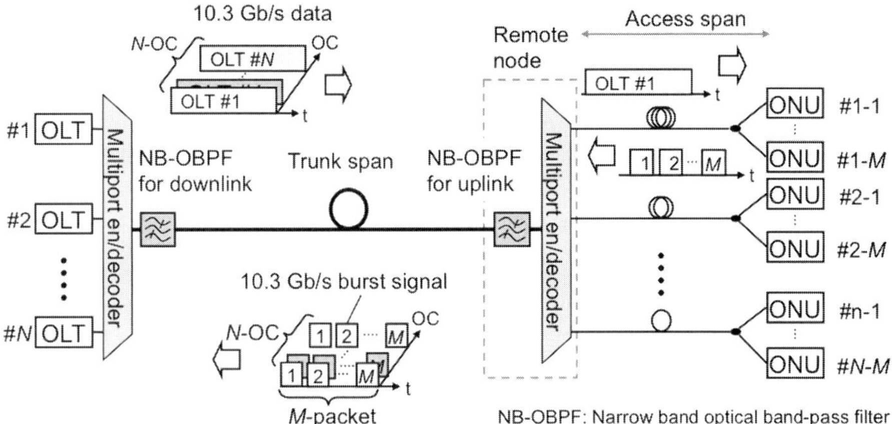

Figure 9.18 Long-reach 10G-TDM-OCDM-PON using only a pair of multiport en/decoders and NB-OBPF [5]. © IEEE by permission.

times. There are two main approaches to the aggregation of extended reach 10 Gb/s-class TDM-PON systems. One is the wavelength division multiplexing (WDM) aggregation approach with optical amplifiers. The other is the installation of 3R repeaters. However, it will be difficult to realize a long-reach 10 Gb/s-class TDM-PON by adapting WDM aggregation because both 10 Gb/s "colorless" ONUs and wideband optical amplifiers are required. The introduction of 3R repeaters will also be more costly due to the need to adapt optical transceivers at the remote node (RN).

To this end, a long-reach TDM-OCDMA-PON is a possible solution. A TDM-OCDM-PON with a single multiport en/decoder at the remote node (RN) without optical en/decoders at ONUs will be suitable not only for a long-reach system but also for low-cost ONUs. In Fig. 9.18 a long-reach 10 Gb/s TDM-OCDM-PON using only a pair of multiport en/decoders at the OLT and a remote node (RN) is shown [5]. Note that the multiport en/decoder can be located anywhere in the optical feeder cable between the OLT and the splitter. A narrow-band optical bandpass filter (NB-OBPF), which tailors the spectrum of the encoded signal, realizes long-reach transmission at 10 Gb/s without dispersion compensation. A long-reach, full duplex 4-packet × 4-OC transmission on a single wavelength over 65 km SMF without dispersion compensator has been demonstrated.

In the experimental setup shown in Fig. 9.19, a pair of 16-port en/decoders is used as optical en/decoders to generate 16-chip, 200 Gchip/s 16-chip 16-level PSK, which is the same as used in the experiment described in Section 9.2.1.2. The ports of the multiport en/decoder denoted U and D, respectively, are used for the uplink and downlink. The same wavelength is used on both uplink and downlink. In the downlink, the MLLD outputs a 2 ps pulse train, shown in Fig. 9.19(a), which is modulated into a $2^{31} - 1$ PRBS by the LN-IM. The modulated data are split into four branches and encoded by the multiport encoder in Fig. 9.19(b). In order to reduce the chromatic dispersion effect, the spectrum of each OC signal is tailored by filtering out the sideband components

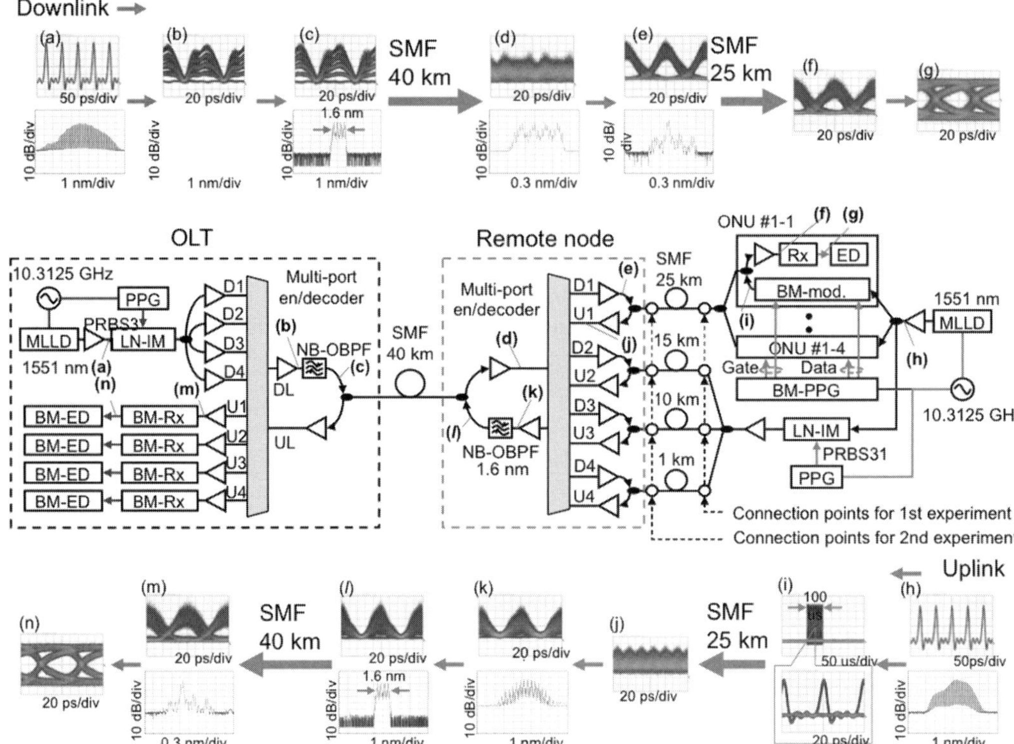

Figure 9.19 Experimental setup and results of long-reach 10G-TDM-OCDM-PON full-duplex transmission [5]. © Reprinted by permission of IEEE.

using the NB-OBPF of 1.6 nm bandwidth as shown in Fig. 9.19(c). Figure 9.19(d) shows a complete eye closure of four-OC data after SMF 40 km transmission due to chromatic dispersion. However, a good eye opening is obtained after decoding from the multiport decoder at the RN as shown in Fig. 9.19(e). At the ONU, clear eye opening even after the longest SMF 65 km transmission is achieved, as shown in Fig. 9.19(f).

The basic mechanism of this spectral tailoring scheme is to cut off the spectrum sidelobe of the code using NBPF, in order to reduce the dispersion effect. It is recalled that the center wavelength of the passband of each optical code is separated by 12.5 ($= 200/16$) GHz as seen in the power spectrum of the 16-port en/decoder in Fig. 7.27. The NB-OBPF, with pass bandwidth nearly equal to 200 GHz, is used to pass only one FSR component filtered without loss of the signal information. In the uplink, the 3 ps optical pulses generated by the MLLD shown in Fig. 9.19(h) are modulated into four optical packets by the burst-mode modulators. The packet length is 100 μs with an overhead length of 800 ns, which is compliant with IEEE802.3av standard shown in Fig. 9.19(i). These optical packets are combined, and each guard time between packets is set to less than 300 ns. Other uplink signals are modulated to the continuous data with a $2^{31} - 1$ PRBS for simplicity. However, this experimental setup demonstrates a proof-of-concept of this system without loss of generality. The pulse broadening waveform

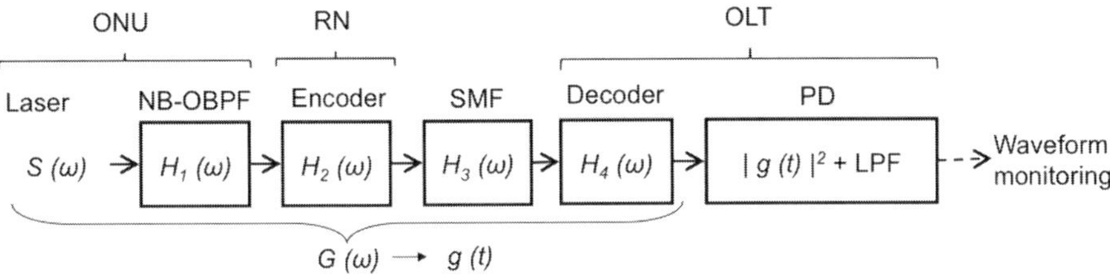

Figure 9.20 Transmission model expressed by cascading optical transfer functions [6]. © Reprinted by permission of the Optical Society of America.

after SMF 25 km transmission, shown in Fig. 9.19 (j), is due to chromatic dispersion. However, the envelope of the multiport encoder output is maintained like the downlink optical encoded waveform as shown in Fig. 9.19(k). Figure 9.19 (l) shows that the spectrum of four-OC data is limited to only four spectral components as the downlink. After the SMF 40 km transmission, four-OC data are decoded and recovered for BER measurement by the multiport decoder and burst-mode 3R receiver at the OLT as shown in Figs. 9.19(m) and (n), respectively.

The ultimate transmission distance of a spectrum-tailored optical encoded signal can be analyzed theoretically [6]. The transmission model is expressed by cascading optical transfer functions of the single-mode fiber, a couple of multiport encoders, NB-OBPF, followed by the photodetector and low pass filter (LPF) as shown in Fig. 9.20. The optical transfer function $G(\omega)$ before the photodetector is given by

$$G(\omega) = S(\omega)H_1(\omega)H_2(\omega)H_3(\omega)H_4(\omega). \tag{9.2}$$

The transfer functions of the multiport en/decoder $H_2(\omega)$ and $H_4(\omega)$ are given by Eq. (7.26). $H_3(\omega)$ is the transfer function of the single-mode fiber and is given in Section 3.7 by

$$H_3(\omega) = \exp\left(\frac{j}{2}\beta_2\omega^2 z\right) \tag{9.3}$$

where ω is the angular frequency and z is the transmission distance. Assume that a Gaussian chip pulse is the input $S(\omega)$, the auto-correlation waveform after decoding is obtained as

$$H_2(\omega) = H_4(\omega) = \sum_{l=0}^{N-1} \exp\left[-j\pi(2l - N + 1)\left(\frac{i + k + 1}{N} + \frac{\omega T_{chip}}{2\pi}\right)\right]$$

$$= \exp\left\{-\frac{j\omega(N-1)T_{chip}}{2}\right\} \frac{\sin\left\{\pi N\left(\frac{i+k+1}{N} + \frac{\omega T_{chip}}{2\pi}\right)\right\}}{\sin\left\{\pi\left(\frac{i+k+1}{N} + \frac{\omega T_{chip}}{2\pi}\right)\right\}} \tag{9.4}$$

where i denotes the encoder-input and the decoder-output port number, k the encoder-output and decoder-input port number, and N the total chip number. By substituting Eqs. (9.3) and (9.4) in Eq. (9.2), the auto-correlation after transmission can be obtained.

Table 9.2 Parameters used in the numerical simulations [6], © Reprinted by permission of the Optical Society of America

Laser	Pulse shape: Gaussian
	Center wavelength: 1550 nm
	Repetition rate: 10 GHz
	Pulse width: 2 ps
NB-OBPF	Transmission window: rectangular 3 dB bandwidth: 1.6 nm
Multiport en/decoder	16-chip, 200 Gchip/s
Single-mode fiber	Fiber length: 25~150 km GVD: 20 ps²/km
LPF	Transmission window: Gaussian −3 dB bandwidth: 8.5 GHz

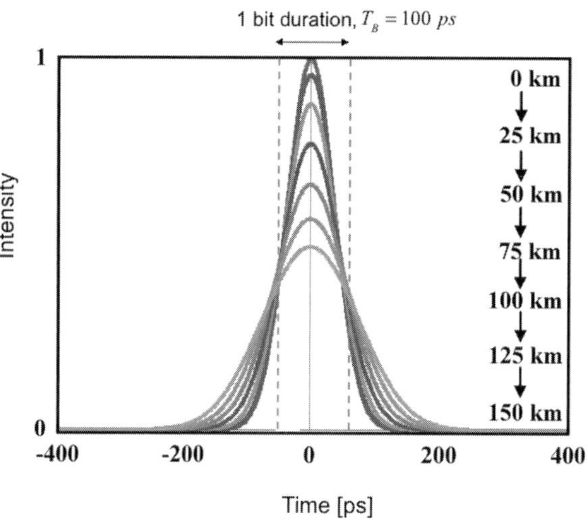

Figure 9.21 Calculated auto-correlation waveforms up to 150 km [6]. © Reprinted by permission of the Optical Society of America.

In Fig. 9.21, the numerically calculated auto-correlation waveforms after propagation up to 150 km are shown. Table 9.2 summarizes the parameters used in the numerical simulation. Figure 9.22 shows the evolution of the auto-correlation pulse train of a 7-bit data pattern along the fiber length for simplicity. The dispersion effects mainly appear in the adjacent bits. The longer the propagation becomes, the more the inter-symbol interference (ISI) becomes visible, as indicated by dashed circles. The results from the numerical simulations, for all distances of z ($z = 0$, 50, 100 and 150 km), confirm that discrimination between "0"-level and "1"-level signals is feasible. The margin is defined as the difference between the minimum power of the "1"-level signal $P_{L1\text{-}min}$ and the maximum power of the "0"-level signal $P_{L0\text{-}max}$ as shown in Fig. 9.23:

$$\text{Margin} = 10 \log_{10} \frac{P_{L1-\min}}{P_{L0-\max}}\ [\text{dB}]. \tag{9.5}$$

The margin is decreased slightly by extending the transmission reach. In the case of 100 km transmission, it assumes a value of 3.1 dB. This result confirms the decoding in the experiment with a good margin of thresholding.

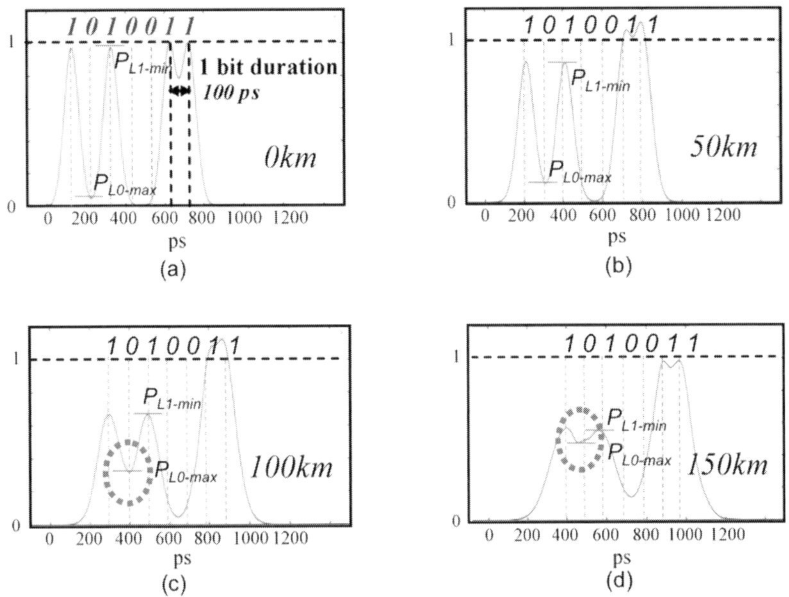

Figure 9.22 ISI effect of the auto-correlation train: (a) 0 km, (b) 50 km, (c) 100 km, (d) 150 km [6]. © Reprinted by permission of the Optical Society of America.

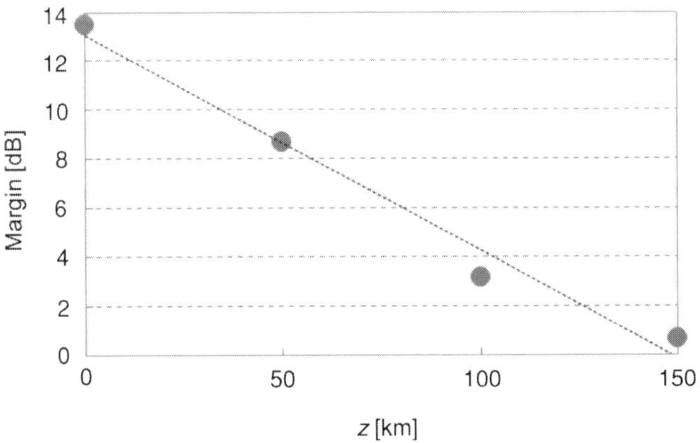

Figure 9.23 Margin against transmission distance [6]. © Reprinted by permission of the Optical Society of America.

9.2.2 WDM-OCDMA-PON system

A migration to WDM-OCDMA-PON from OCDMA-PON is a promising approach to scaling up a PON system further in the quest for more subscribers and more total system capacity. The architecture of WDM-OCDMA-PON is illustrated in Fig. 9.24. OCDMA-PON is scaled up by a factor of K by using K wavelengths [7]. The total number of subscribers that can be accommodated in the PON becomes $K \times N$. Since a WDM channel is shared with n users, the same optical code sequences OC $1 \sim$ OC N can be reused on K wavelength grids.

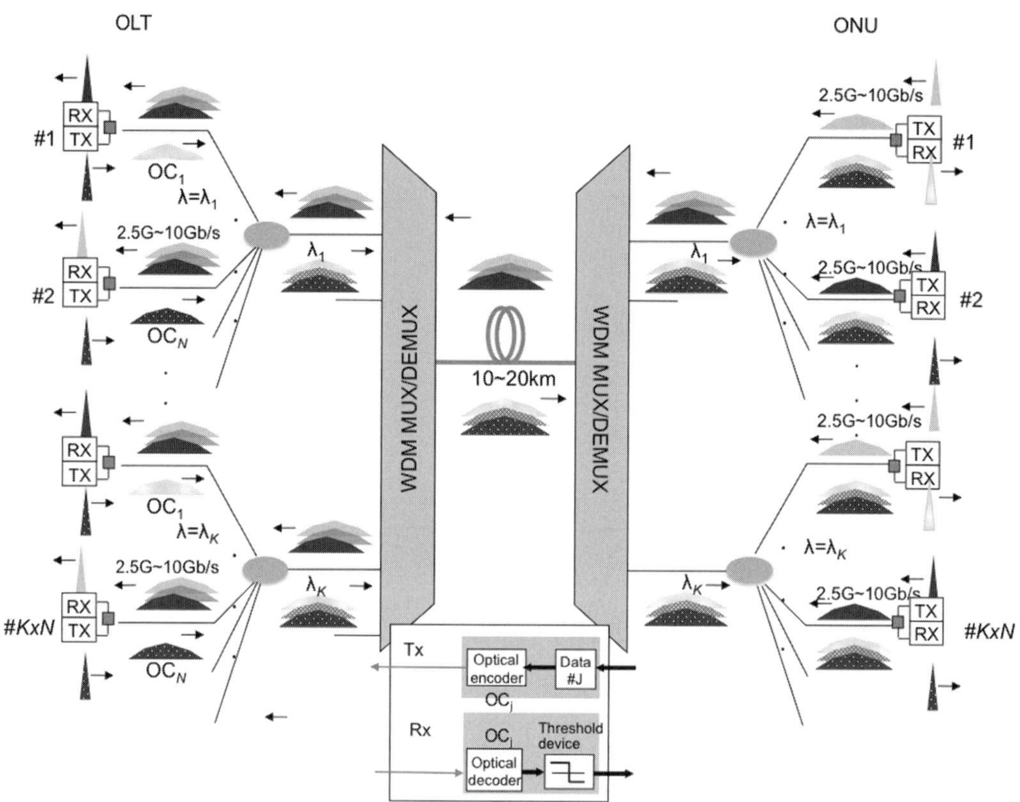

Figure 9.24 Architecture of WDM-OCDMA-PON [7]. © Reprinted by permission of IEEE.

There are two options for WDM: one is CWDM, and the other is either conventional WDM or DWDM. CWDM has been linked with PON by ITU-T G.694-2, in which the wavelength interval is set at 20 nm in the spectral range from 1270 nm to 1610 nm as shown in Fig. 9.25. This relatively wide wavelength interval is set so as to eliminate the temperature control required to stabilize the oscillation wavelength of the laser diode used as the light source in PON systems. However, a number of wavelengths of only 18 is obviously insufficient to accommodate even a moderate number of subscribers. Therefore, the conventional WDM grid will be preferable.

Let us consider the number of subscribers of OCDMA on WDM grids. The same code sequences can be reused on all the wavelength grids by taking advantage of the frequency periodicity of optical code sequences. As illustrated in Fig. 9.26, the optical code sequences OC 1~OC 16, generated by the multiport encoder, are allocated with the free spectral range (FSR) of 200 GHz, in which each code is separated with a frequency interval of 12.5 (=200/16) GHz. The interchannel crosstalk between the wavelength grids will impose the limitation on the spectral efficiency. As the neighboring grids become closer to each other, the crosstalk will degrade the signal-to-noise ratio. The goal will be to see how close the neighboring grid can become. In Fig. 9.27, the

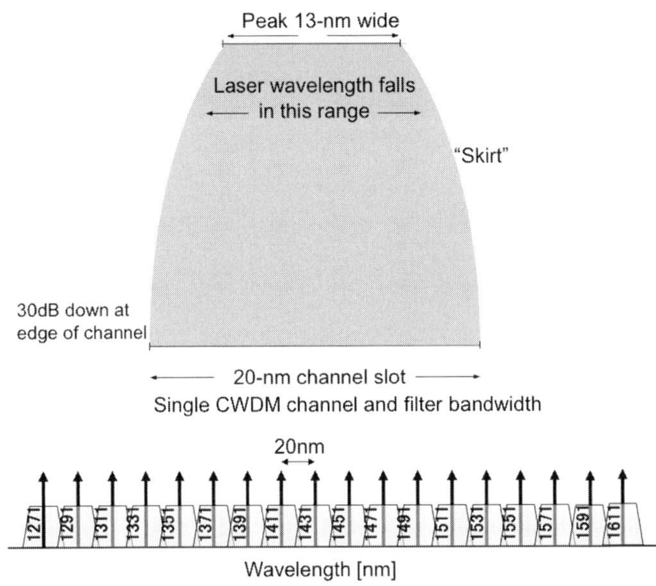

Figure 9.25 CWDM grid allocation over the spectral range from 1270 to 1630 nm [7]. © Reprinted by permission of IEEE.

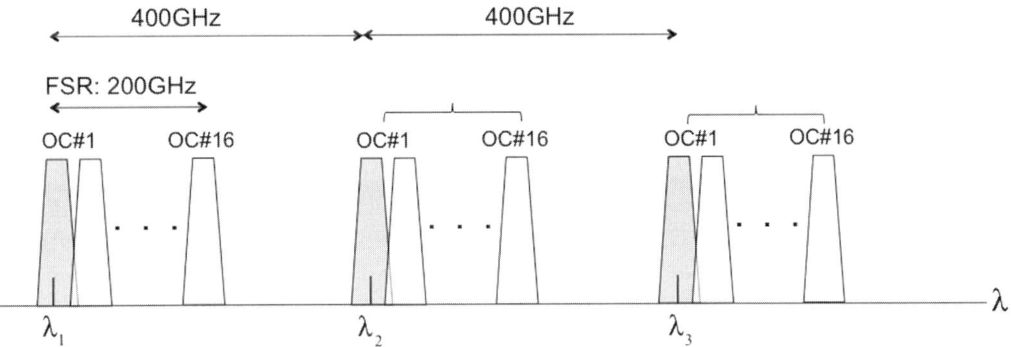

Figure 9.26 Wavelength allocation of optical codes generated by a multiport encoder.

numerical crosstalk powers $\langle P_x \rangle$ from the same optical code on different wavelengths are plotted as a function of WDM channel interval for optical codes generated by 127-chip, 160 Gchip/s and 511-chip, 640-Gchip/s SSFBG encoders. From the power spectrum of the optical code in Fig. 7.27, it can be seen that the neighboring optical codes overlap in their spectra. As the first notches of 127-chip and 511-chip SSFBGs are located 160 and 640 GHz apart from the central peaks, respectively, the crosstalk power reflects this spectral profile. $\langle C \rangle$ and $\langle W \rangle$, respectively, represent the average powers of maximum MAI noise among the optical codes and the auto-correlation wings on the same grid, normalized by their auto-correlation peaks. The criterion for WDM channel interval for tolerable interchannel crosstalk is set so that $\langle P_x \rangle$ is equal to or lower than

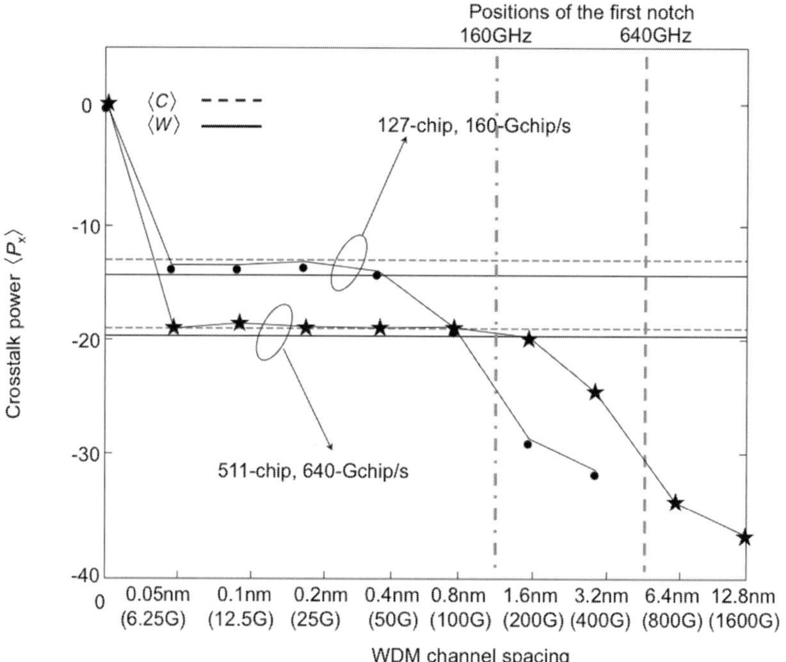

Figure 9.27 Average peak power of interchannel crosstalk versus WDM channel interval for 127-chip and 511-chip SSFBG en/decoders [7]. © Reprinted by permission of IEEE.

the values of both $\langle C \rangle$ and $\langle W \rangle$. For the 127-chip SSFBG, $\langle C \rangle$ is -14 dB, and $\langle W \rangle$ is -14.8 dB, while for the 511-chip SSFBG, $\langle C \rangle$ and $\langle W \rangle$ are reduced to -18.8 and -19.4 dB, respectively. For the 127-chip SSFBG, therefore, the channel spacing can be reduced down to 50 GHz, while, for the 511-chip SSFBG, it has to be 200 GHz or larger. In Fig. 9.28, the power penalty for BER $= 10^{-9}$ is plotted as a function of the number of active subscribers K for the 511-chip, 640-Gchip/s SSFBG with different WDM channel spacings. For $K = 16$, the power penalty is 2 dB with 200 GHz interval, but it can be made negligible for a 400 GHz interval. The spectral efficiency with 200 GHz interval is 0.8 ($=10 \times 16/200$) [b/s/Hz] at the bit rate of 10 Gb/s. It is also confirmed that, with the channel interval of 200 GHz, the interchannel crosstalk has a smaller impact on the BER than the intrachannel crosstalk, that is, the MAI noise.

A field trial of 3-wavelength $\times$ 4-optical code WDM-OCDMA-PON has been conducted in an optical testbed of the *Japan Gigabit Network II (JGNII)* [8]. JGNII is a nationwide open testbed network as shown in Fig. 9.29(a), which is operated by the National Institute of Communication and Information Technology (NICT), Japan, as an ultrahigh-speed testbed network for R&D collaboration between industry, academia, and government. The fiber used in this experiment is installed in the field between NICT laboratories in Koganei City and Otemachi of downtown Tokyo in a loop-back configuration of 100 km standard SMF. The dispersion of total 1720 ps/nm shown in Fig. 9.29(b) is compensated for by using a DSF. The configuration of the

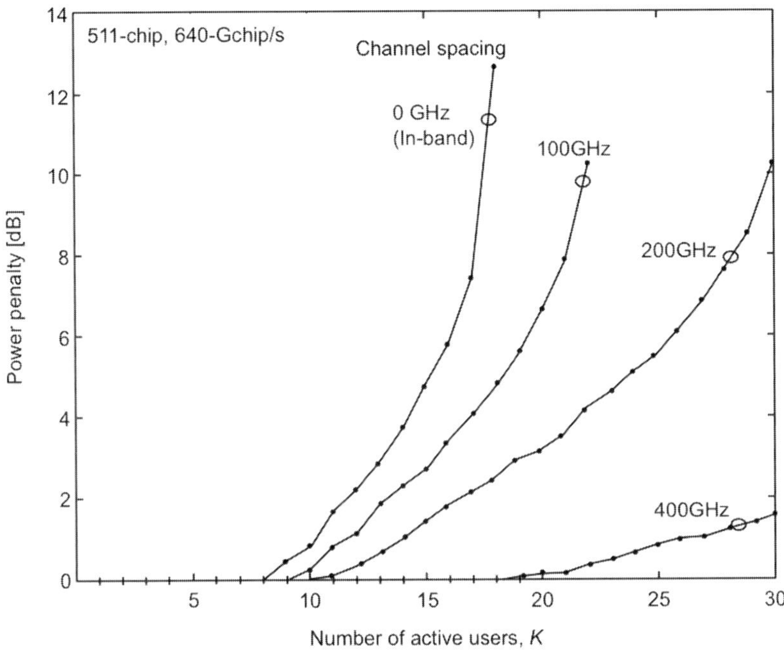

Figure 9.28 Power penalty versus number of active users K for a 511-chip SSFBG with different WDM channel intervals [7]. © Reprinted by permission of IEEE.

WDM-ODMA-PON experiment is shown in Fig. 9.30. Three wavelength grids, 1550.2 nm, 1553.4 nm, and 1556.6 nm, with an interval of 400 GHz (~3.2 nm) are set, and the same group of four optical code sequences is assigned on each wavelength grid using a 16-chip, 200 Gchip/s 16-port encoder. For decoding, a tunable optical tapped delay-line en/decoder with phase shifters is used. It has 16 taps and a phase shifter on each arm, as introduced in Section 7.2.2. Figure 9.30 shows the spectra of the three wavelengths, the temporal waveforms of 12-multiplexed signals including four codes from every four output port, Ports 4, 8, 12, and 16 of the 16-port encoder on three wavelengths, the decoded signals, and the eye diagrams after the balanced detector. Figure 9.31 shows the eye diagrams of the encoded (top row), decoded (middle row), and the electrical (bottom row) signals with three WDM and different numbers of active subscribers at each wavelength (K). As a consequence, all 12 channels, 3-wavelength × 4-codes achieve BER $< 10^{-9}$.

9.2.3 TDM-WDM-OCDMA-PON system

A migration from TDM-PON to WDM-TDM-PON is a practical scenario in the short term. A possible solution path to further scaling up of the WDM-TDM-PON system will be WDM-TDM-OCDMA-PON. The architecture of WDM-TDM-OCDMA-PON adopting a pair of multiport en/decoders is illustrated in Fig. 9.32. By introducing M

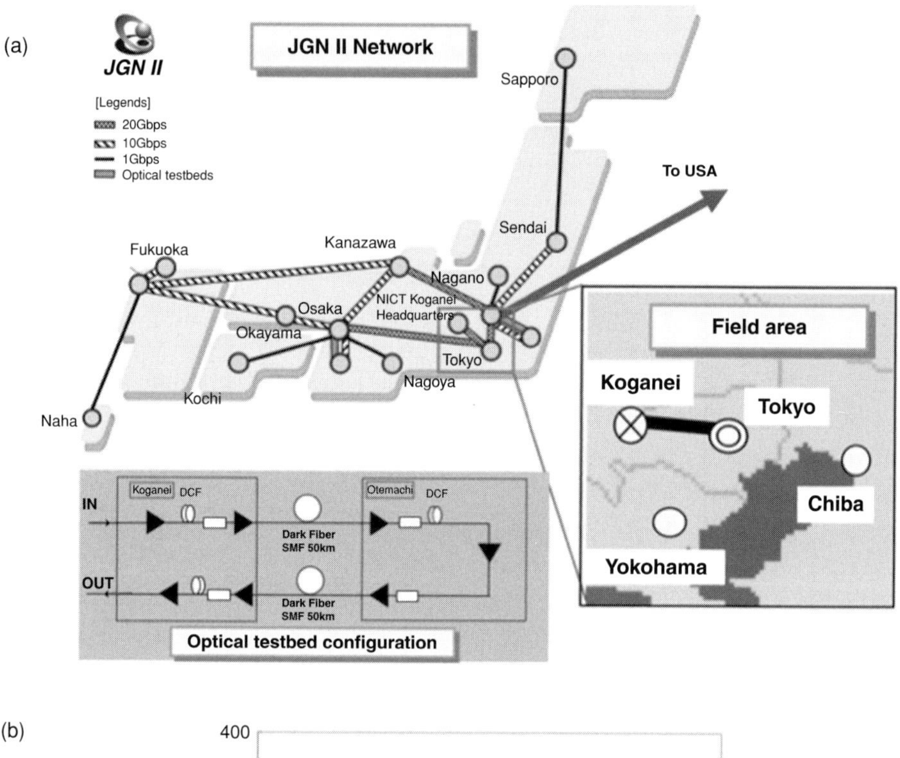

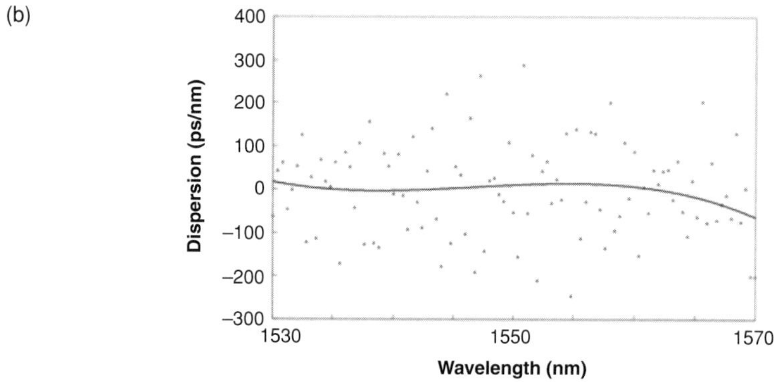

Figure 9.29 (a) WDM-OCDMA field testbed and (b) chromatic dispersion of installed fibers [8]. © Reprinted by permission of IEEE.

time slots over WDM-OCDMA-PON, the total number of ONUs becomes $M \times K \times N$, scaling up by a factor of M. For example, with 32 TDM packets ($M = 32$), 8 OCs ($N = 8$), the aggregate total capacity becomes $K \times 10$ Gb/s with $K \times 40$ Mb/s uplink for individual ONUs. The wavelength interval is set to be equal to the free spectral range (FSR) of the multiport E/D so that one can take advantage of the spectral periodicity of the multiport E/D. This is the same approach as used for WDM-OCDMA-PON described in Section 9.2.2. This allows OC reuse on all the wavelengths, thus requiring only N OCs.

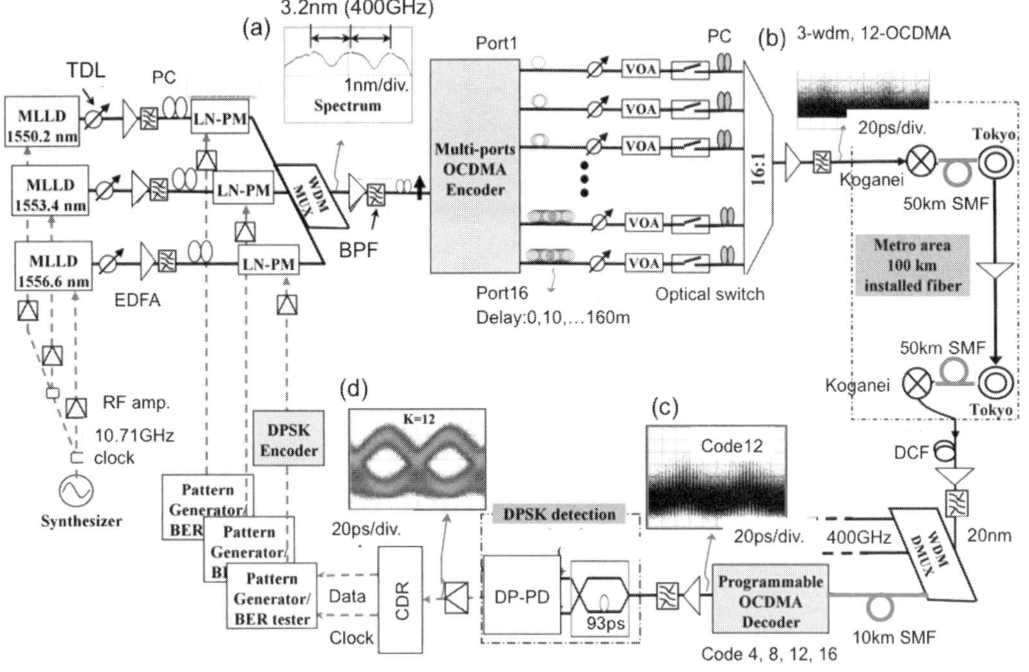

Figure 9.30 Experimental setup of 3-WDM-4-OCDMA-PON [8]. © IEEE by permission.

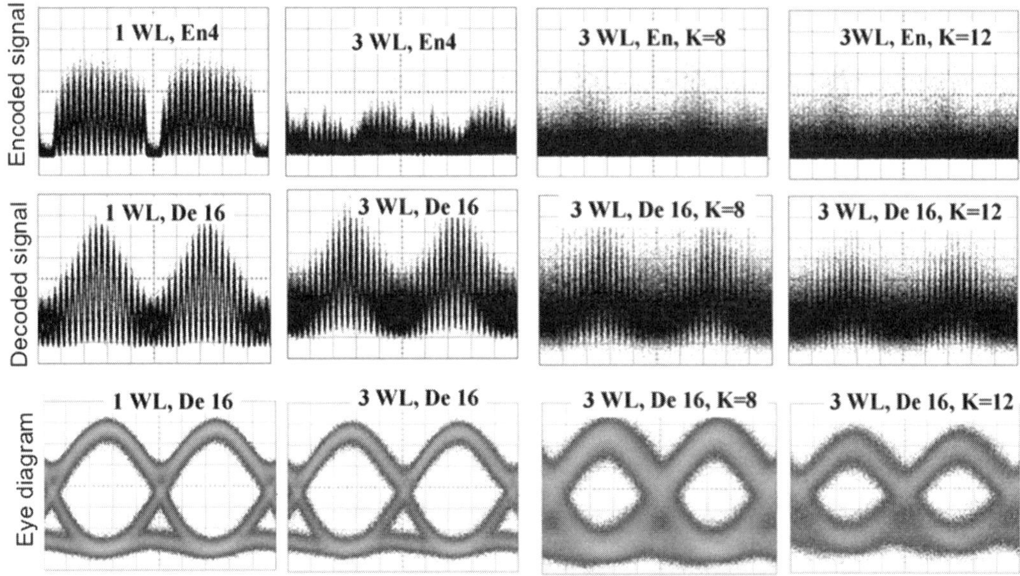

Figure 9.31 Experimental results: eye diagrams of (upper row) encoded, (middle row) decoded, and (lower row) electrical signals with 3 WDM and different numbers of active subscribers at each wavelength K [8]. © Reprinted by permission of IEEE.

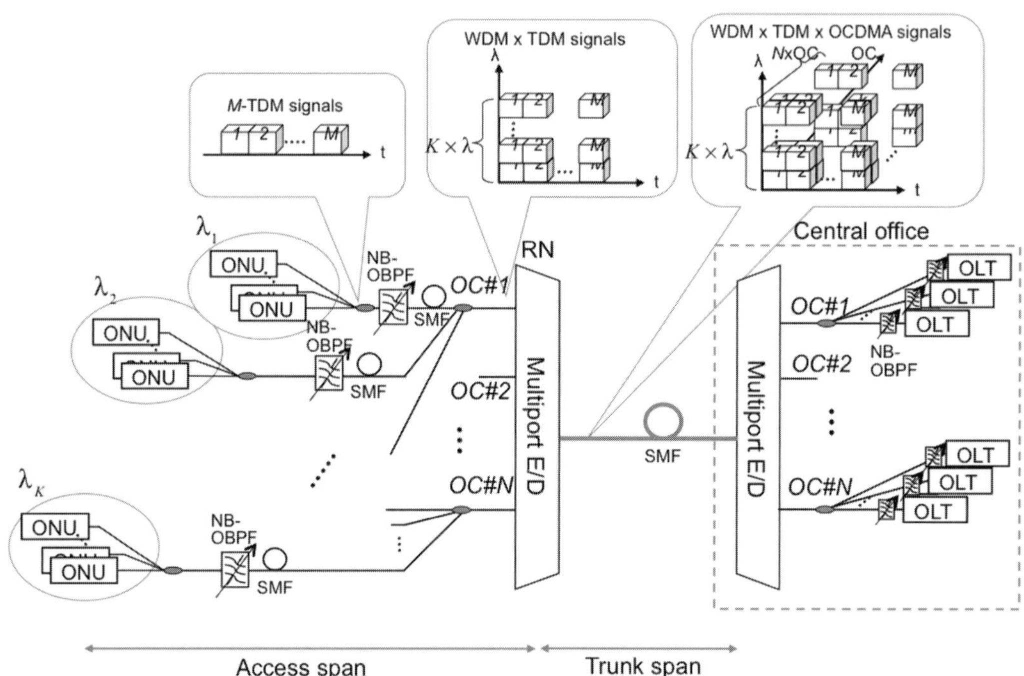

Figure 9.32 Architecture of WDM-TDM-OCDMA-PON [6]. © Reprinted by permission of the Optical Society of America.

The experiment was conducted for an uplink of 10 Gb/s, long-reach 2-WDM × 2-TDM × 4-OCDM-PON [6]. The experimental setup and results are shown in Fig. 9.33. A pair of multiport en/decoders is placed at the OLT and RN. Two MLLDs with a repetition rate of 10.3125 GHz at the central wavelengths of 1550 nm and 1552 nm, 250 GHz apart, are used. Two different lengths of trunk span SMF of 50 km and 75 km are tested, while the SMF between an RN and ONUs is 25 km long. The total chromatic dispersion of 100 km SMF at 1551 nm is 1935 ps/nm. 1.6 nm NB-OBPFs are placed at the output port of the en/decoder at the OLT to tailor the spectrum of the signal, thus enabling dispersion compensation free transmission. More details of the spectrum tailoring technique are described in Section 9.2.1.4. A 2 ps optical pulse train from the MLLD, shown in Fig. 9.33(a), is modulated into two optical packets with a $2^{31} - 1$ PRBS with no guard time, shown in Fig. 9.33(b). The packet length is 100 μs with overhead length of 800 ns which is compliant with IEEE802.3av standard. For simplicity, the other three signals, modulated by the LN-IM, are the bit stream data. The waveform and spectrum of pulse broadening after SMF 25 km transmission are shown in Fig. 9.33(c). Four data at wavelengths of 1550 nm and 1552 nm are encoded with four different optical codes at the RN and multiplexed as shown in Fig. 9.33(d). After traversing another 50 km or 75 km single-mode fiber the eye diagram at the OLT is completely closed in, Fig. 9.33(e). However, a clear eye opening is obtained after decoding and NB-OBPF as shown in Fig. 9.33(f). The data are recovered after the burst-mode reception

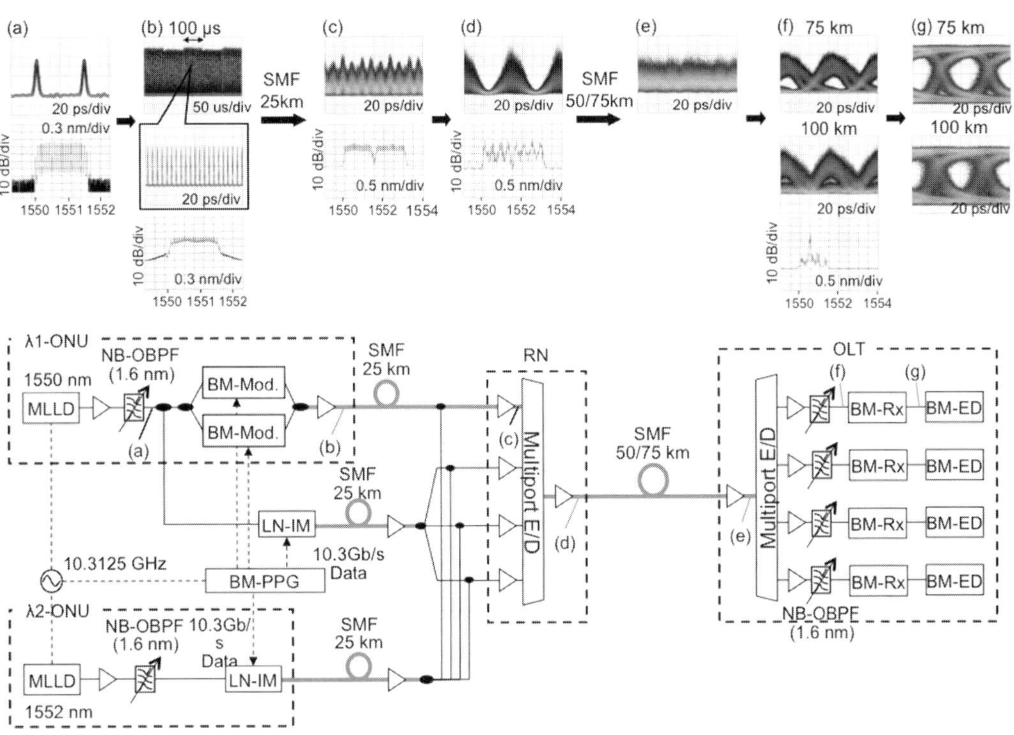

Figure 9.33 Experimental setup and results for 2-WDM × 2-TDM × 4-OCDM-PON [6].
© Reprinted by permission of the Optical Society of America.

in Fig. 9.33(g). The measured BERs after propagation of 75 km and 100 km are less than 10^{-3}, which are within the limit of error-free transmission with the aid of FEC of RS (255, 223).

9.3   40 Gb/s OCDMA systems

As NG-PON2 has a target bit rate of 40 Gb/s and beyond, 40 Gb/s OCDMA systems are under development. A main focus is the optical en/decoder, which has one-fourth shorter chip interval than that of a 10 Gb/s en/decoder. In this section, a 40 Gb/s OCDMA system testbed using two types of en/decoder, multiport and SSFBG en/decoders is introduced. One of the challenges for a 40 Gb/s bit rate is the design of the optical en/decoder. The chip interval has to be shrunk to squeeze as many chips into the one bit time duration of $T_B = 25$ ps as are squeezed into 100 ps at 10 Gb/s. Otherwise, the encoded signals in the neighboring time slots overlap each other, and this will cause inter-symbol interference (ISI). Figure 9.34, for example, compares two cases of 16-chip, 200 Gchip/s time-spread optical code used in 10 Gb/s and 40 Gb/s systems. The encoded signal extends over 80 ps, and thus there is no ISI at 10 Gb/s, but a significant overlap is seen at the bit rate of 40 Gb/s. To mitigate the ISI, therefore, the number of chips has to be limited to roughly

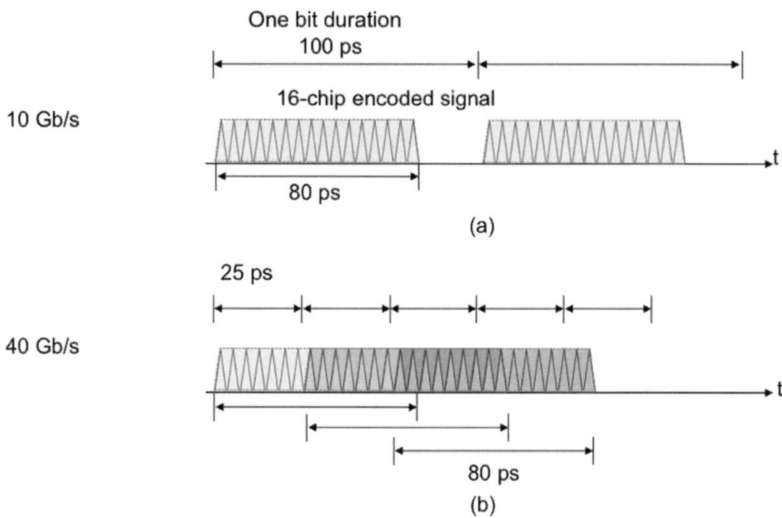

Figure 9.34 16-chip, 200 Gchip/s time-spread optical code in the cases of (a) 10 Gb/s and (b) 40Gb/s systems.

ten, as far as an optical pulse from a conventional MLLD having a pulsewidth of typically 2 ps is used as the light source. It is not a realistic option from a practical perspective to use a femtosecond pulse source in order to increase the chip count. Nevertheless, it is worth developing 40 Gb/s OCDMA because this enables the realization of various 40 Gb/s types of hybrid OCDMA systems.

9.3.1   Multiport encoder and decoder configuration

An 8-chip, 320 Gchip/s, 8-port en/decoder is developed, which generates 8-level PSK optical code. As shown in Fig. 9.35, the module is W100 $\times$ H12 $\times$ D50 mm^3 in size [9]. The FSR is 320 GHz, and as the chip interval is 3.1 ps, the encoded signal fits precisely in a 25 ps time slot of 40 Gb/s. The auto-correlation extends over 47 ps corresponding to a 15-chip time period, as shown in Fig. 9.35.

The experimental setup of a 40 Gb/s 4 $\times$ DPSK-OCDMA downlink is shown in Fig. 9.36. The output pulse from the 10 GHz MLLD in Fig. 9.36(a) is time-multiplexed to 39.81312 GHz by a 10G-to-40G optical time-division-multiplexer (OTDM-Mux) in Fig. 9.36(b) and then phase modulated by a PRBS $2^{31}-1$ data using an LiNbO$_3$ phase modulator (LN-PM). The signal is split into four branches, and four fully asynchronous encoded signals are emulated after passing the random delay. It is recalled that fully synchronous transmission will be the worst case scenario, which contrasts with the synchronous transmission of the best case as discussed in Section 6.4.4. The four signals are launched into the odd input ports of the 8-port encoder to generate four different encoded signals. The noise-like waveform and spectrum of the 4-user OCDMA-multiplexed signal in Fig. 9.36(c) is transmitted into the 50 km long dispersion-compensated single-mode fiber.

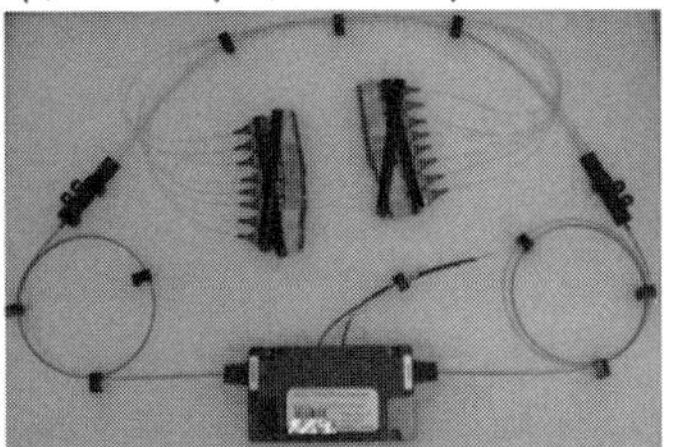

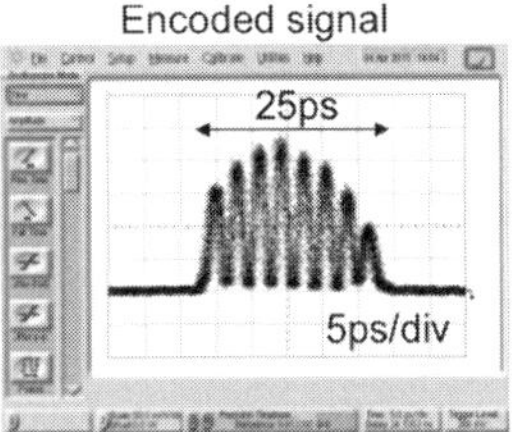

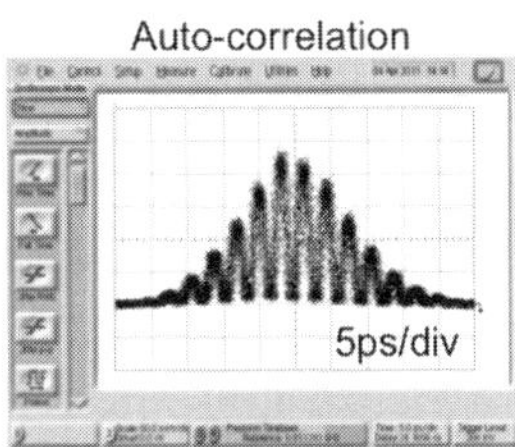

Figure 9.35 8-chip, 320 G chip/s, 8-port en/decoder. The module, FSR, and waveforms of the encoded signal and its auto-correlation [9]. © Reprinted by permission of the Optical Society of America.

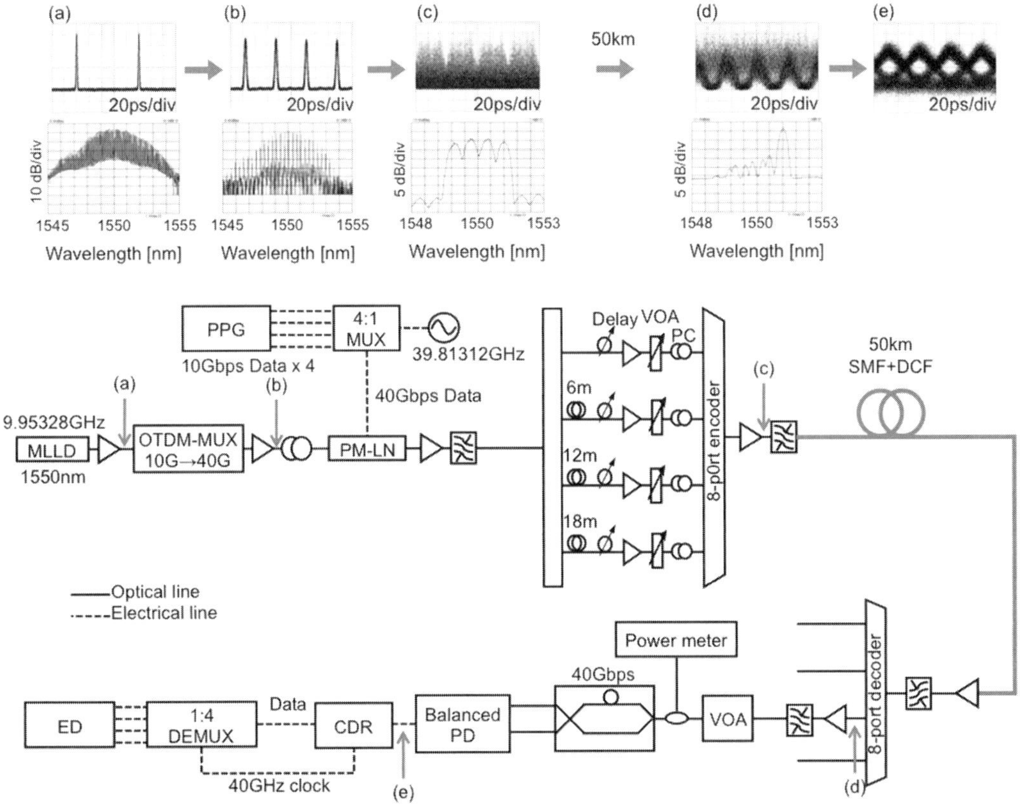

Figure 9.36 Experimental setup of 40 Gb/s 4 × DPSK-OCDMA downlink and the experimental results [9]. © Reprinted by permission of the Optical Society of America.

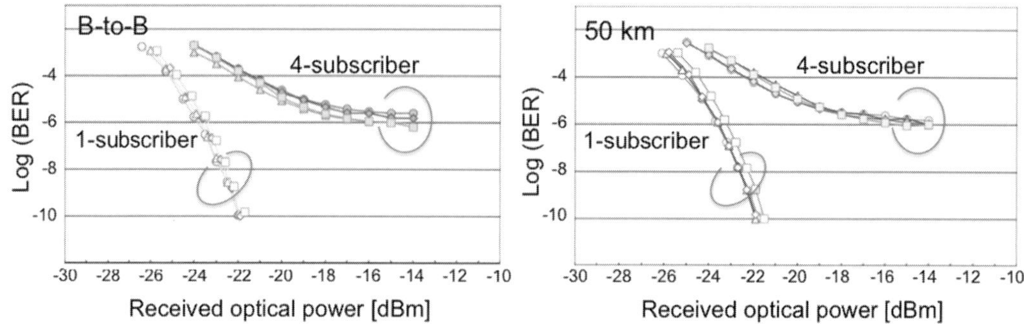

Figure 9.37 Measured BERs of all the users for the back-to-back (B-to-B) and 50 km transmission [9]. © Reprinted by permission of the Optical Society of America.

At the receiver, the signal is decoded by the 8-port decoder in Fig. 9.36(d) and detected using a fiber-based interferometer and a balanced photodiode (PD). The eye opening of the signal detected at the output of the balanced PD is shown in Fig. 9.36(e), followed by the 40G-to-10G electrical demultiplexer and BER measurement using a 40G error detector (ED). The measured BERs of all the users for the back-to-back (B-to-B) and 50 km transmission are shown in Fig. 9.37. After the 50 km transmission, a BER below 10^{-5} is achieved. This confirms that the system can realize error-free operation with the FEC. The power penalties between the single user and multiple user cases are caused by MAI noise. This experiment has been followed by the 40 Gb/s hybrid WDM-OCDMA-PON, demonstrating a total capacity of 2.56 Tb/s using 8-wavelength, four optical codes, and polarization division multiplexing [10].

9.3.2   SSFBG and multiport encoder and decoder configuration

A 40 Gb/s OCDMA-PON using a heterogeneous combination of SSFBG en/decoders at each ONU and multiport en/decoder at the OLT is described here. As the 8-port encoder generates 8-level PSK code, 320 Gchip/s 8-chip long, an SSFBG encoder which generates 8-level PSK has been developed for the first time. The correlation between heterogeneous multiport and SSFBG en/decoders is described in Section 7.3.3. The test SSFBG has eight grating sections, 0.325 mm long, and a phase difference of $2\pi \times 1/8$ is introduced between the adjacent grating sections. The total grating length is 2.6 mm, corresponding to a 25 ps time spread of the encoded signal as shown in Fig. 9.38 so there is no overlap between adjacent bits at 40 Gb/s [11]. The grating has an apodized profile, in which the modulation of the refractive index in each grating section is not uniform but has a pulse-like spatial profile whose peak is located at the center of each grating section. By comparing the encoded waveform and the auto-correlation waveform of the SSFBG with uniform grating (not shown) with that of an apodized SSFBG, a well-defined chip structure is observed, shown in Fig. 9.38(a). The spectrum of the apodized SSFBG is shown in Fig. 9.38(b). The peak reflectivity is located at a frequency interval of 320

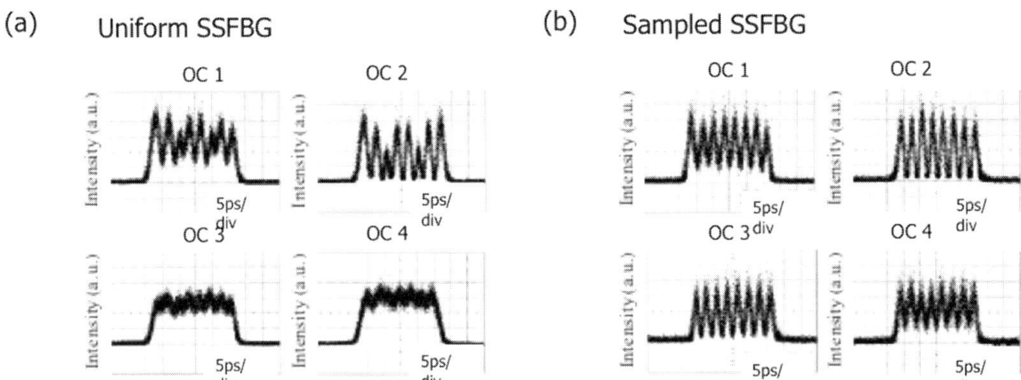

Figure 9.38 (a) Temporal waveforms of encoded signals and auto-correlation and cross-correlation and (b) spectrum of test 8-level PSK, 8-chip long, 320 Gchip/s SSFBG en/decoder [11]. © Reprinted by permission of SPIE.

GHz. Compared with the conventional SSFBG with uniform spatial modulation depth of grating which has a single maximum reflectivity peak, shown in Fig. 7.15, for example, a clear distinction is that three maximum peaks with frequency interval 320 GHz have nearly the same reflectivity. This unique characteristic of the apodized SSFBG will be suitable for the en/decoder in WDM-OCDMA-PON.

In Fig. 9.39 the experimental setup of 4×40 Gb/s full-duplex OCDMA-PON using the apodized 8-level PSK SSFBG en/decoders at each ONU and an 8-port en/decoder at the OLT is shown [12]. Since the setup of the downlink is similar to that shown in Fig. 9.36, except for the SSFBG en/decoder, detailed descriptions are referred to Fig. 9.36. In the uplink, the waveforms correspond to outputs of the MLLD in (i) and OTDM-Mux in (ii), modulated DPSK signal in (iii), four multiplexed encoded signals with random delay in (iv), decoded signal after 50 km transmission in dispersion-compensated SMF in (v), and the detected eye pattern with the DPSK photodetector in (vi).

In the downlink, the waveforms correspond to the outputs of the MLLD in (vii) and the OTDM-Muxed signal in (viii), the modulated DPSK signal in (ix), four multiplexed encoded signals with random delay in (x), the decoded signal after 50 km transmission in the dispersion-compensated single-mode fiber link in (xi), and the detected eye pattern with the DPSK photodetector in (xii). From the measured BERs of uplink and downlink at back-to-back and 50 km transmission shown in Fig. 9.40, error-free (BER $< 10^{-9}$) operation is obtained by all codes in all cases of single ONU. Under the condition of four unidirectional ONUs, BERs of less than 10^{-5} are attained for the uplink and downlink. In the case of full-duplex four users, BERs of less than 10^{-4} are achieved after 50 km fiber transmission, and this result indicates that the system can realize error-free operation when the 7%-overhead FEC, standardized by ITU-T G.975, is employed. The power penalty between the single-user and the multiple-user cases is caused by the MAI noise.

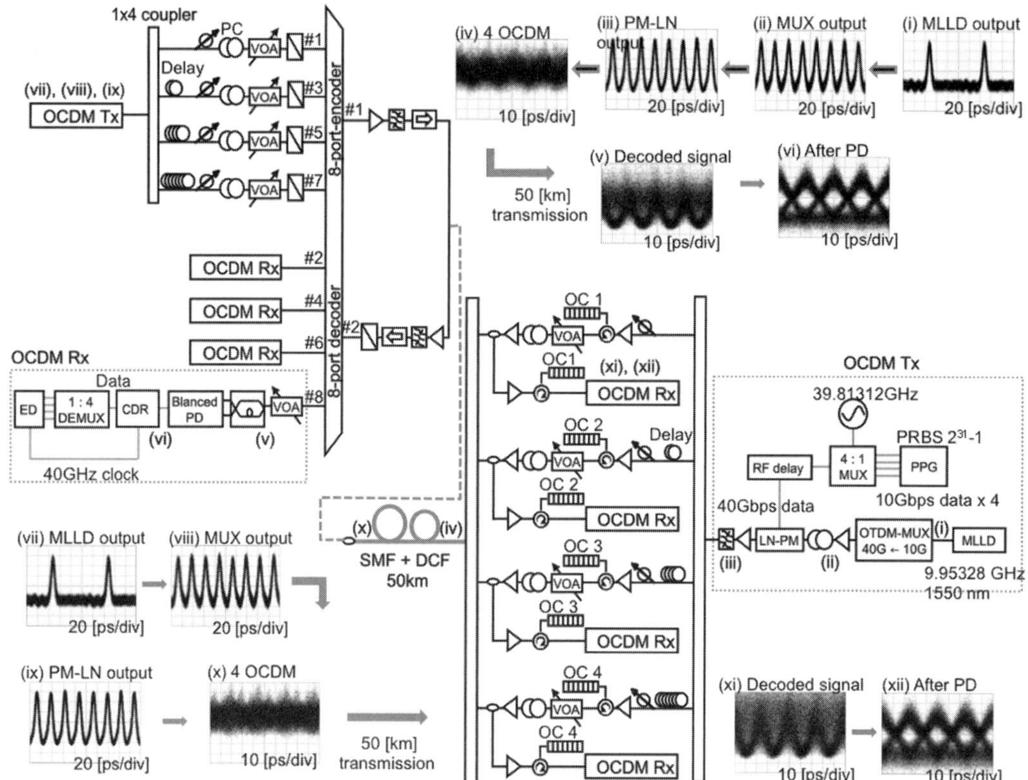

Figure 9.39 Experimental setup of 4 × 40 Gb/s full-duplex OCDMA-PON using the apodized 8-level PSK SSFBG en/decoders at each ONU and an 8-port en/decoder at the OLT [12]. © Reprinted by permission of the Optical Society of America.

9.4   Space OCDMA

Space OCDMA, an extension of OCDMA to a two-dimensional (2-D) space coding for image transmission and multiple access has been studied, which can exploit the inherent parallelism of optics [13]. The beauty of parallelism is that a light beam can carry the information of a 2-D array of pixels of a binary-digitized image, and hence the 2-D array of pixels of an image can be transmitted and processed simultaneously without parallel-to-serial conversion. The space-spread spectrum using 2-D signature patterns to encode each pixel of a binary image, instead of the temporal signature sequences used in conventional time-domain OCDMA, enables parallel image transmission and multiple access as shown in Fig. 9.41. This technique is distinctive from time-spread OCDMA because it spreads the spatial frequency spectrum rather than spreading either the time or frequency domain. One of the keys to space-OCDMA is the methodology in the construction of the 2-D optical orthogonal signature pattern (OOSP), and another key is the parallel image transmission using a multicore fiber, in which each core bears

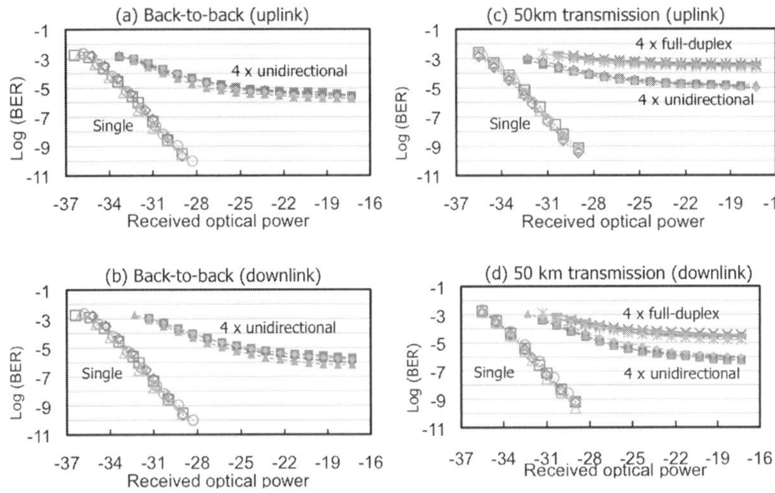

Figure 9.40 Measured BERs of back-to-back (a) uplink and downlink. Values after 50 km transmission (c) uplink and (d) downlink [12]. © Reprinted by permission of the Optical Society of America.

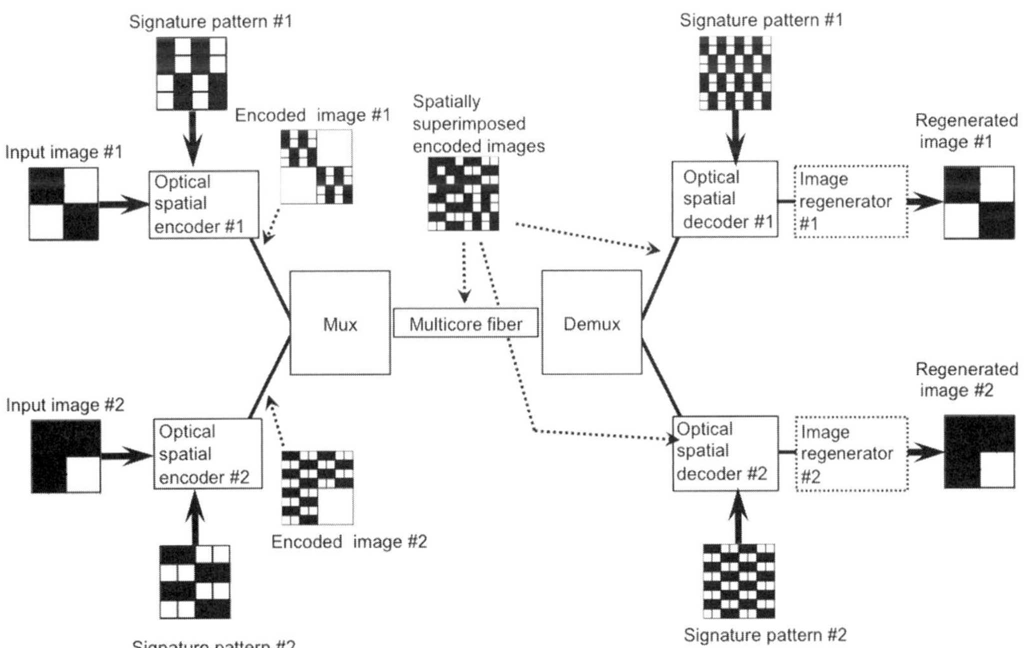

Figure 9.41 Schematic of a space OCDMA system based upon spatial spread spectrum. A 2-D signature pattern is used for encoding and decoding each pixel [13]. © Reprinted by permission of IEEE.

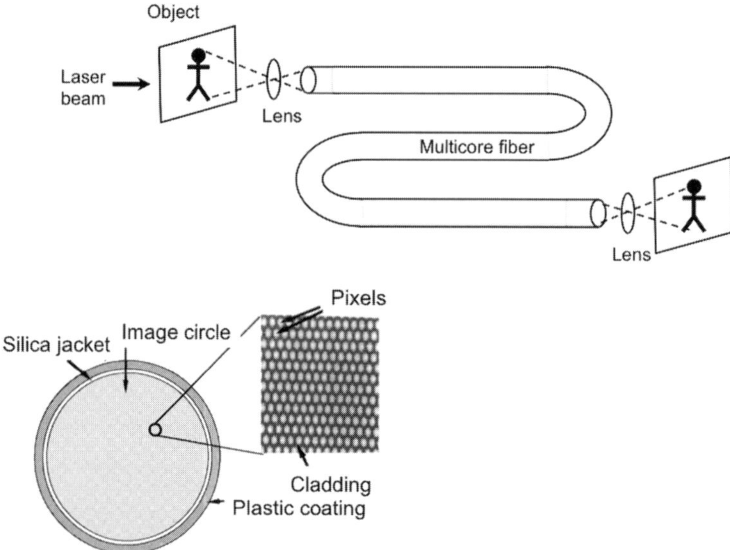

Figure 9.42 Multicore fiber for image transmission [13, 14]. © Reprinted by permission of IEEE.

one or several pixels of an image, shown in Fig. 9.42, and all the pixel data of an image are transmitted simultaneously in parallel.

9.4.1   Space encoding and decoding

A spatially discrete binary image is expressed in matrix form, in which the matrix element (i, j) corresponds to the logical binary value of the pixel located at position $[x_i, y_j]$. First, the encoding process is shown schematically in Fig. 9.43. Consider the 2×2 input images $A^{(1)}$ and $A^{(2)}$ are encoded into 8×8 encoded patterns with 4×4 signature pattern $E^{(1)}$ and $E^{(2)}$, respectively, and are transmitted after the pattern is multiplexed. The $N \times N$ input binary image A and the $M \times M$ signature pattern E with spreading factor M^2 are expressed as

$$A = \begin{bmatrix} a_{11} & \cdots & a_{1N} \\ \vdots & \cdots & \vdots \\ a_{N1} & \cdots & a_{NN} \end{bmatrix} \tag{9.6}$$

and

$$E = \begin{bmatrix} e_{11} & \cdots & e_{1M} \\ \vdots & \cdots & \vdots \\ e_{M1} & \cdots & e_{MM} \end{bmatrix} \tag{9.7}$$

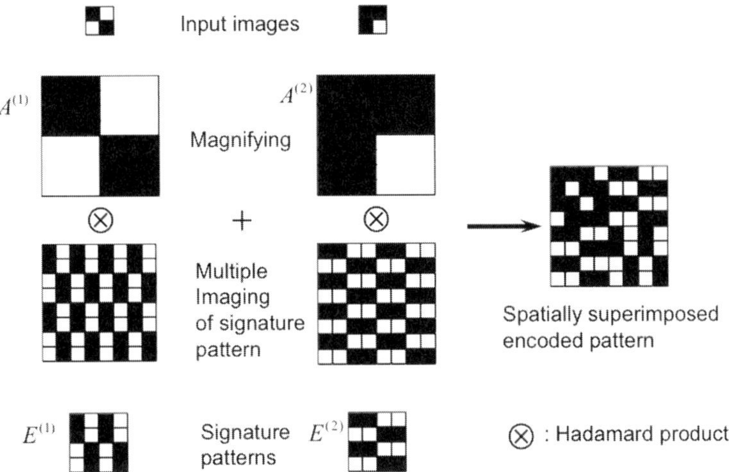

Figure 9.43 Optical encoding process. Operations of the Hadamard product followed by spatial superimposing of the encoded patterns [13]. © Reprinted by permission of IEEE.

where all the elements of A and E take logical binary values of "0"or "1". The magnified input image A_{mag} and the signature pattern E_{mlt} after multiple imaging are defined as follows

$$A_{mag} = \begin{bmatrix} A_{11} & \cdots & A_{1N} \\ \vdots & \cdots & \vdots \\ A_{N1} & \cdots & A_{NN} \end{bmatrix} \tag{9.8}$$

and

$$E_{mlt} = \begin{bmatrix} E & \cdots & E \\ \vdots & \cdots & \vdots \\ E & \cdots & E \end{bmatrix} \tag{9.9}$$

where A_{ii} $(i = 1, 2, \ldots, N)$ is the $M \times M$ matrix in which all the elements are a_{ii}. It is noteworthy that these operations of magnification and multiple imaging can be performed in parallel using conventional optics such as lenses. When each pixel of the input image A is encoded by the signature pattern E, after being encoded the image is formed by the mutual product of the respective matrix elements, and is expressed using the so-called Hadamard product as shown in Eq. (9.10) where $\otimes$ denotes the Hadamard

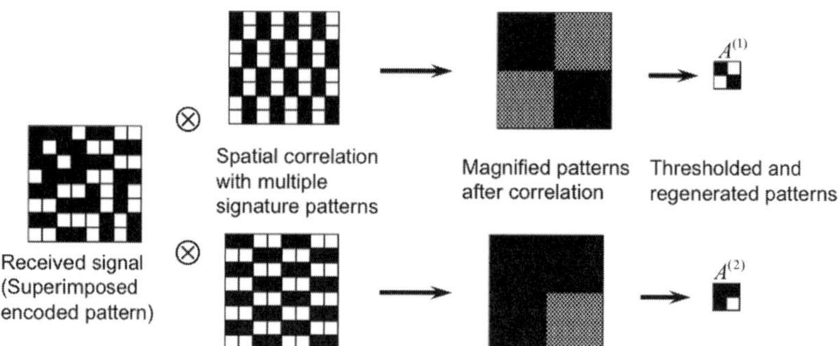

Figure 9.44 Optical decoding process: correlation followed by thresholding [13]. © Reprinted by permission of IEEE.

product:

$$
\tilde{A} = \sum_q A_{mag}^{(q)} \otimes E_{mlt}^{(q)} =
\begin{bmatrix}
\sum_q A_{11}^{(q)} \otimes E^{(q)} & \cdots & \sum_q A_{1N}^{(q)} \otimes E^{(q)} \\
\vdots & \cdots & \vdots \\
\sum_q A_{N1}^{(q)} \otimes E^{(q)} & \cdots & \sum_q A_{NN}^{(q)} \otimes E^{(q)}
\end{bmatrix}
$$

$$
=
\begin{bmatrix}
\sum_q a_{11}^{(q)} E^{(q)} & \cdots & \sum_q a_{1N}^{(q)} E^{(q)} \\
\vdots & \cdots & \vdots \\
\sum_q a_{N1}^{(q)} E^{(q)} & \cdots & \sum_q a_{NN}^{(q)} E^{(q)}
\end{bmatrix}.
\tag{9.10}
$$

The summation with respect to (q) is carried out over all input images which are to be transmitted concurrently. The summation corresponds to multiplexing in which the encoded patterns are spatially superimposed.

The decoding process is shown schematically in Fig. 9.44. The receiver correlates its own signature pattern with the received pattern, and thresholds the correlation output to regenerate the original image. The correlation is performed in such a manner that the signature pattern is multiplied with the incoming pattern, and the product is integrated for the bit duration T_B. This process can be regarded as pattern recognition. When a receiver q which possesses the signature pattern $E^{(q)}$ receives the encoded signal $\tilde{A}'$, it extracts the input image $A^{(q)}$ from $\tilde{A}'$. Correlating the received pattern $\tilde{A}'$ with a specific signature pattern, the output is given by Eq. (9.11) where $\sum_{M \times M}$ is the summation over M^2 matrix elements, and w_{q0} is the weight (the number of binary "1"s) of the signature pattern $E^{(q_0)}$. The weight is defined in Eq. (9.12). If there is no space shift between the incoming and signature patterns, the first and second terms in Eq. (9.11) represent the auto-correlation and cross-correlation of the signature pattern $E^{(q_0)}$, respectively. By thresholding the correlation output, the resultant output can be reduced to only the auto-correlation term

$$\sum_{M \times M} \tilde{A}' \otimes E_{mux}^{(q)}$$

$$= \begin{bmatrix} \sum_{M \times M} \sum_q [A_{11}^{(q)} \otimes E^{(q)}] \otimes E^{(q_0)} & \cdots & \sum_{M \times M} \sum_q [A_{1N}^{(q)} \otimes E^{(q)}] \otimes E^{(q_0)} \\ \vdots & \cdots & \vdots \\ \sum_{M \times M} \sum_q [A_{N1}^{(q)} \otimes E^{(q)}] \otimes E^{(q_0)} & \cdots & \sum_{M \times M} \sum_q [A_{NN}^{(q)} \otimes E^{(q)}] \otimes E^{(q_0)} x \end{bmatrix}$$

$$= w_{q0} \begin{bmatrix} a_{11}^{(q_0)} & \cdots & a_{1N}^{(q_0)} \\ \vdots & \cdots & \vdots \\ a_{N1}^{(q_0)} & \cdots & a_{NN}^{(q_0)} \end{bmatrix}$$

$$+ \begin{bmatrix} \sum_{M \times M} \sum_{q \neq q_0} [A_{11}^{(q)} \otimes E^{(q)}] \otimes E^{(q_0)} & \cdots & \sum_{M \times M} \sum_{q \neq q_0} [A_{1N}^{(q)} \otimes E^{(q)}] \otimes E^{(q_0)} \\ \vdots & \cdots & \vdots \\ \sum_{M \times M} \sum_{q \neq q_0} [A_{N1}^{(q)} \otimes E^{(q)}] \otimes E^{(q_0)} & \cdots & \sum_{M \times M} \sum_{q \neq q_0} [A_{NN}^{(q)} \otimes E^{(q)}] \otimes E^{(q_0)} \end{bmatrix} \tag{9.11}$$

Finally, based upon Eq. (9.11), the receiver declares each pixel value of the incoming pattern as "0" or "1". Note that the thresholding level must be chosen in between w_{q0} and the total energy of the interference from the other $(K-1)$ users.

The rules for constructing 2-D optical orthogonal signature patterns, or OOSPs having a large cardinality, are as follows.

(i) Signature patterns must be distinguishable from any space-shifted versions of themselves in the 2-D plane.

(ii) Any two different signature patterns in a set must be distinguishable from each other, even with the existence of any vertical and/or horizontal space shifts in the 2-D plane.

Here, we rule out the possibility of encoded pattern rotation after transmission. The rules require that the auto-correlation must be much higher than the correlation sidelobes, and any peaks of the cross-correlation function must be much lower than the peak of the auto-correlation function. These requirements are expressed in binary discrete correlation forms as

$$\sum_{i=1}^{M} \sum_{j=1}^{M} e_{i,j}^{(q)} e_{i+k,j+l}^{(q)} \begin{cases} = w_q & \text{for } k = l = 0 \\ \leq \delta_A & \text{for } 1 \leq k, l \leq M - 1 \end{cases} \tag{9.12}$$

and

$$\sum_{i=1}^{M} \sum_{j=1}^{M} e_{i,j}^{(q)} e_{i+k,j+l}^{(q')} \leq \delta_c \quad \text{for } q \neq q', \ 0 \leq k, l \leq M - 1 \tag{9.13}$$

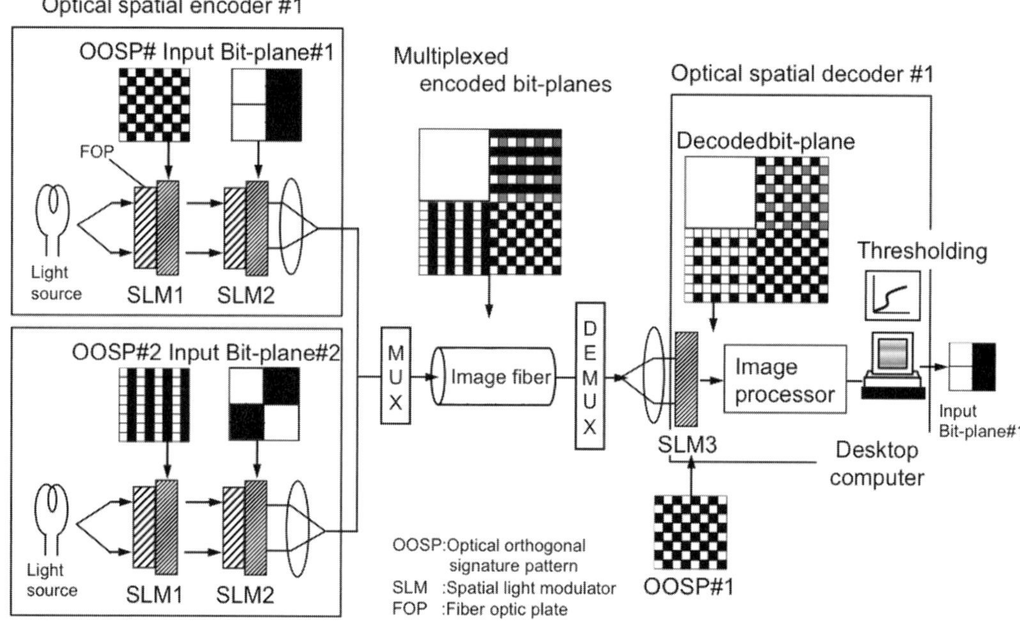

Figure 9.45 Optical implementation of a space-CDMA network [14]. © Reprinted by permission of IEEE.

where e_{ij} is the matrix element of E of Eq. (9.7). The methodology is detailed in the literature [13].

9.4.2   Experimental demonstration

Figure 9.45 shows a schematic of the optical implementation of the space OCDMA-PON [14, 15]. The transmitter consists of an optical spatial encoder, and the receiver consists of an optical spatial decoder and a thresholding device. A multicore fiber is used as the transmission medium having 30,000 cores embedded in a common cladding. The specifications of the fiber are summarized in Table 9.3. The encoder and the decoder are implemented with liquid crystal display (LCD) spatial light modulators (SLMs). The light source is a xenon lamp. In the encoder an input bit plane and an OOSP are addressed electrically on SLM1 and SLM2, respectively. Reading out SLM1 and SLM2 consecutively with the light beam allows encoding for all bits to be done optically in parallel. The encoded patterns are superimposed by means of spatial multiplexing, and the multiplexed bit plane is launched into an image fiber and transmitted in parallel. At the receiver side, the multiplexed bit plane is broadcast to every decoder by means of spatial demultiplexing. The decoder performs, in parallel, the correlations of all bits of the received bit plane and its own OOSP addressed on SLM3. After performing the 2-D surface integral and thresholding operations, the intended input bit plane is regenerated.

Table 9.3 Specifications of the test multicore fiber [15], © Reprinted by permission of the Optical Society of America

Parameter	Specification
Length	$10\sim100$ m
Number of pixels	30,000
Fiber diameter	850 μm
Picture diameter	790 μm
Minimum bending radius	80 mm
Numerical aperture	0.4
Spacing between pixels	4.4 μm
Core diameter	2.6 μm
Core material	GeO_2–SiO_2
Cladding material	F–SiO_2

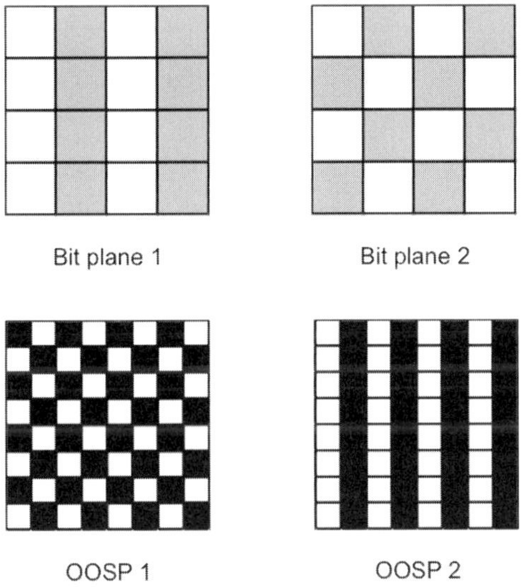

Bit plane 1   Bit plane 2

OOSP 1   OOSP 2

Figure 9.46 Input bit planes 1 and 2 and OOSP 1 and 2 [14]. © Reprinted by permission of IEEE.

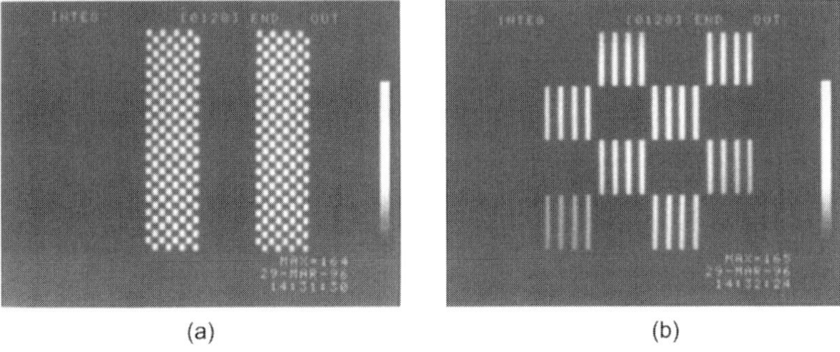

(a)   (b)

Figure 9.47 Encoded bit planes: (a) bit plane 1 with OOPS 1 and (b) bit plane 2 with OOPS 2 [14]. © Reprinted by permission of IEEE.

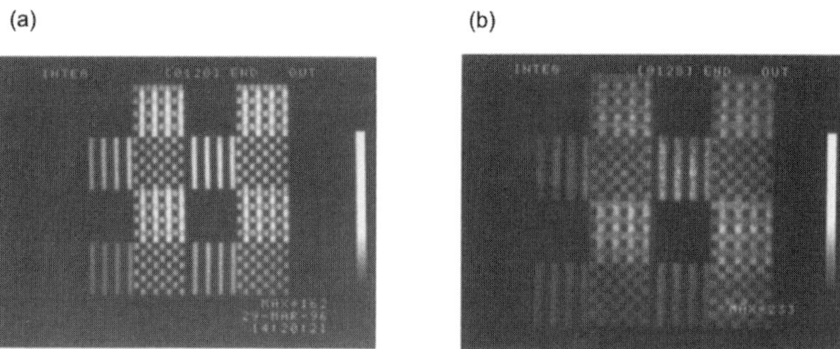

Figure 9.48 Multiplexed encoded bit planes: (a) before and (b) after propagation through an image fiber [14]. © Reprinted by permission of IEEE.

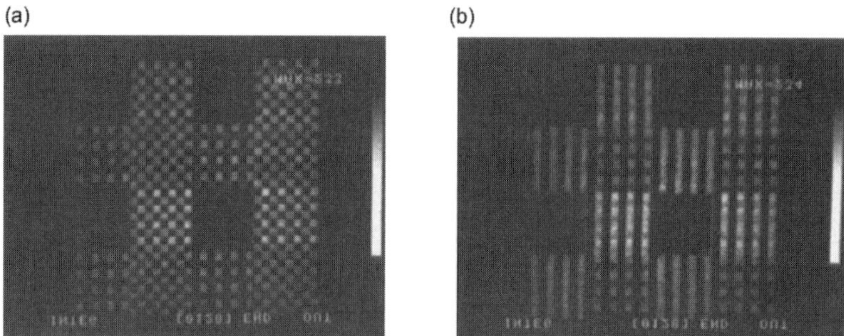

Figure 9.49 Images after the correlation with (a) OOSP 1 and (b) OOSP 2 before thresholding [14]. © Reprinted by permission of IEEE.

Two input 4×4 bit planes and two 8×8 OOSPs used in the experiment are shown in Fig. 9.46 [14]. Pixels of input bit planes 1 and 2 are encoded by different OOSPs 1 and 2, respectively, as shown in Fig. 9.47. The two 32×32 encoded planes are multiplexed and transmitted as shown in Fig. 9.48(a) and (b), respectively. By comparing the distributions before propagation having well-defined pixel value either zero, unity, or two it can be seen that discrimination of the three-value after propagation becomes poor in (b). Obviously, this will cause some error in the regeneration of the input bit planes. Each receiver can recover the intended bit plane by 2-D optical correlation of the received signal with its OOSP counterpart, followed by the thresholding process to cut off interference noise. The input bit planes 1 and 2 decoded with OOSPs 1 and 2 before thresholding are shown in Fig. 9.49. The values of bit planes 1 and 2 which are obtained by surface integral of the pixel values without and with normalization are shown in Figs. 9.50(a) and (b), respectively. Without the normalization, some of the pixel values, shown in bold in both Fig. 9.50(a) and (b), are apparently faulty. The threshold value cannot be determined without freedom of error. To improve the discrimination of the pixel value, a compensation is made for non-uniformity of the brightness. From the pixel values,

(a)

Without normalization

6.70	11.58	8.79	**10.00**
2.57	23.51	3.12	20.99
7.21	14.26	**11.15**	14.95
2.25	13.37	2.47	14.83

With normalization

12.43	17.60	12.62	17.38
4.02	25.78	3.05	26.44
12.22	17.28	11.81	19.38
5.81	25.89	4.24	28.36

(b)

Without normalization

2.36	14.37	8.79	**10.00**
11.06	**10.30**	17.70	9.81
2.84	25.80	3.27	22.16
10.13	8.53	13.12	7.27

With normalization

6.02	27.70	4.22	24.58
18.44	12.47	18.45	13.08
4.38	28.10	3.22	28.29
18.83	12.63	19.17	12.47

Figure 9.50 Pixel values of bit planes 1(a) and 2(b) without and with normalization [14]. © Reprinted by permission of IEEE.

normalized by taking into account the non-uniformity of the pixel brightness, all the pixel values "1" are corrected to be larger than the values "0."

When a larger matrix size of 2-D arrays of vertical-cavity-surface-emitting laser diodes (VCSELs) at the wavelength of 850 nm and PD arrays become available for the transmitter and receiver, a simpler and higher-speed system can be implemented [16]. Figure 9.51 shows the experimental setup using an 8×8 VCSEL array and 8×8 PD array. As each VCSEL acts as a pixel, 2×2 bit planes are encoded with 4×4 OOSPs and transmitted through a 1 m long image fiber. With this VCSEL array up to 1 Gb/s transmission is feasible. At the receiver the decoder correlates, in parallel, all bits of the received signals and addressed OOSPs. After the correlation and thresholding operation in the decoder, the intended 2×2 bit planes 1 and 2 are retrieved. The 4×4 mutually orthogonal OOSPs 1 and 2 shown in Fig. 9.52(a) are used. Two bit planes encoded with OOPSs and the multiplexed bit plane after propagation through the image fibers are also shown in Fig. 9.52(b) and (c), respectively. It is seen that there is some crosstalk between the pixels. The eye diagram of a channel or pixel at 64 Mb/s of $2^{11} - 1$ PRBS at the decoder is shown in Fig. 9.53(a). The measured BERs of eight channels are plotted as a function of the lateral misalignment of the PD array in Fig. 9.53(b). The misalignment tolerance for a BER of 10^{-9} is 25 μm, which is large enough for the practical fabrication of interconnection modules. Larger tolerances will be possible at the expense of coupling efficiency by defocusing the optical spots on the PD array. In this experiment, the total throughput was 512 Mb/s (64 Mb/s per channel × 8 channels). The transmission speed is limited only by the processing speed of the electronic circuits of the encoders and decoder. However, the bit rate can be increased further by employing high-speed logic circuits, such as emitter-coupled logic. For multiple access there is a tradeoff between the number of subscribers and the spatial resolution of the image to be transmitted. This requires larger cardinality of the signature pattern. As the size of

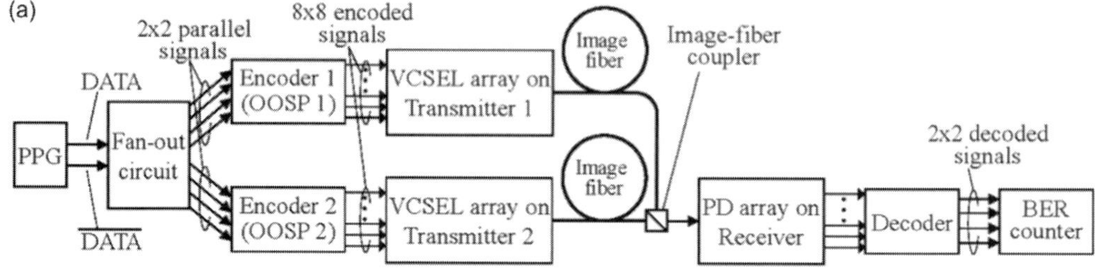

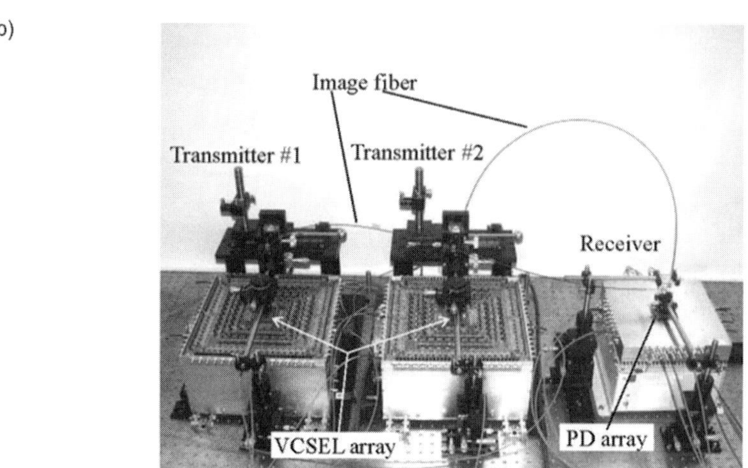

Figure 9.51 (a) Experimental setup of a space-OCDMA system using 2D-VCSEL and PD arrays. (b) Photograph of the experimental setup [16]. © Reprinted by permission of the Optical Society of America.

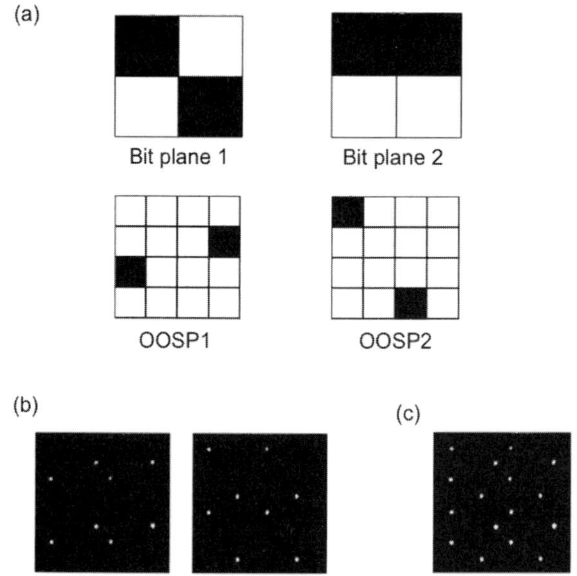

Figure 9.52 (a) 4×4 OOSPs 1 and 2, (b) 2 × 2 optical signals encoded by OOSP 1 and OOSP 2 after propagation through 1 m long image fibers and (c) encoded and multiplexed optical signals [16]. © Reprinted by permission of the Optical Society of America.

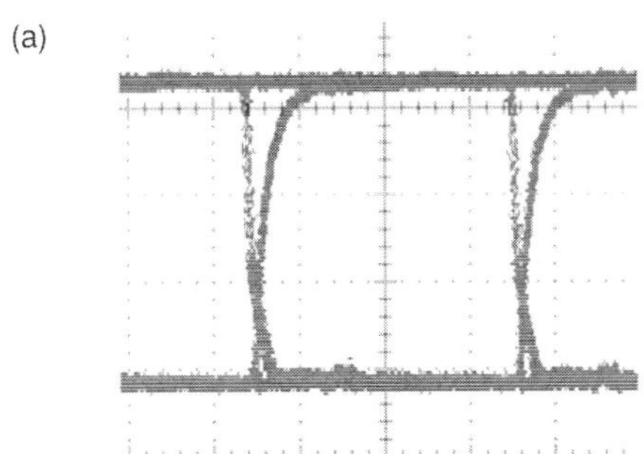

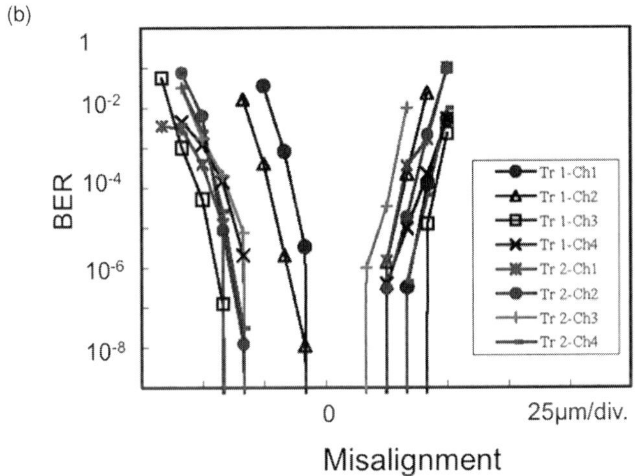

Figure 9.53 (a) Observed eye diagram of channel 1 of transmitter 1 and (b) measured BER versus lateral misalignment of the PD array [16]. © Reprinted by permission of the Optical Society of America.

the OOSP is increased, the cardinality of the signature pattern increases, resulting in more subscribers. Within the limited number of available pixels, however, the spatial resolution of the image system decreases.

9.4.3 Multicore erbium-doped image fiber amplifier

Since the operation wavelength of 2-D VCSEL is in the 1550 nm band, then an EDFA can be used as the optical amplifier as the 1R repeater for 2-D image transmission. A schematic of a generic 2-D parallel optical data link adopting multicore optical amplifiers is illustrated in Fig. 9.54. For this purpose a multicore erbium-doped image fiber amplifier (EDIFA) has been developed [17]. Two pumping schemes for EDIFAs are illustrated in Fig. 9.55. In Fig. 9.55(a) all the EDIFA cores are pumped with the pump beam through a large-core fiber that has the same diameter as the EDIFA. Therefore, not only discrete optical spots, but also spatially continuous images are amplified. The output endface of the fiber is imaged on the input endface of the EDIFA through

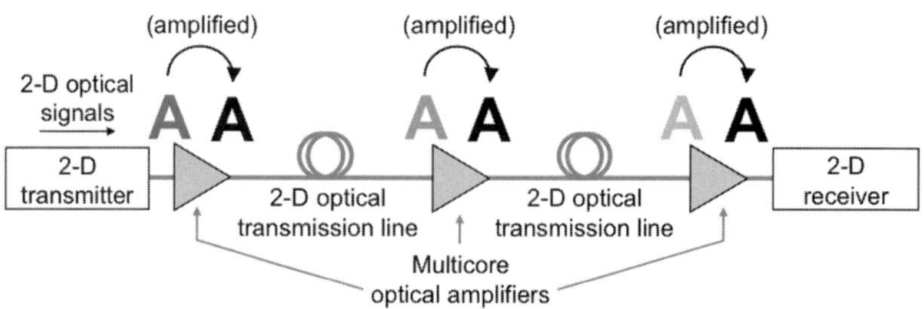

Figure 9.54 Generic 2-D parallel optical data link with multicore optical amplifiers [17]. © Reprinted by permission of IEEE.

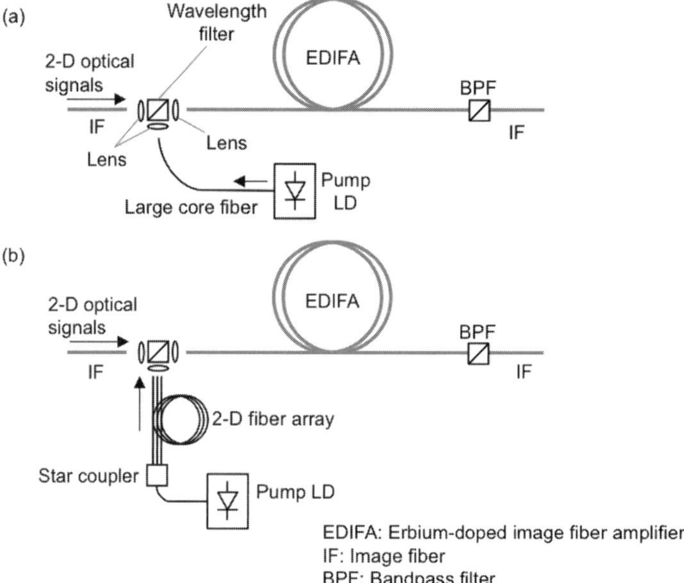

Figure 9.55 Pumping methods for EDIFAs in forward pumping. (a) All cores are pumped. (b) The pump power is concentrated on signal beams [17]. © Reprinted by permission of IEEE.

a wavelength filter used to multiplex the signal and pump beams. However, a large number of cores requires an extremely high pump power. When the 2-D optical signal consists of discrete optical spots, the pumping method in Fig. 9.55(b) will be useful, where the pump beams are launched from a 2-D fiber array. This pumping scheme is more energy efficient because the pump power is concentrated only on the optical signal beams.

Test EDIFAs having the parameters listed in Table 9.4 have been prepared. The doping concentrations were approximately 500 p/m for the 0.8 m fiber and 1000 p/m for the 3.0 m fiber. Aluminum is co-doped in the core preform with 0.1% concentration. The erbium dopant in the core changes the viscosity of the preform, resulting in bubbles in the EDIFA during the drawing process. To avoid this problem, the viscosity of the cladding is balanced by tailoring the concentration of fluorine and matched to that of the

Table 9.4 Parameters of a test erbium-doped image fiber amplifier (EDIFA) [17], © Reprinted by permission of IEEE

Parameter	Specification
Number of pixels	3000
Fiber diameter	630 μm
Picture diameter	590 μm
Core diameter	4.5 μm
Space between pixels	9.0 μm
NA	0.25

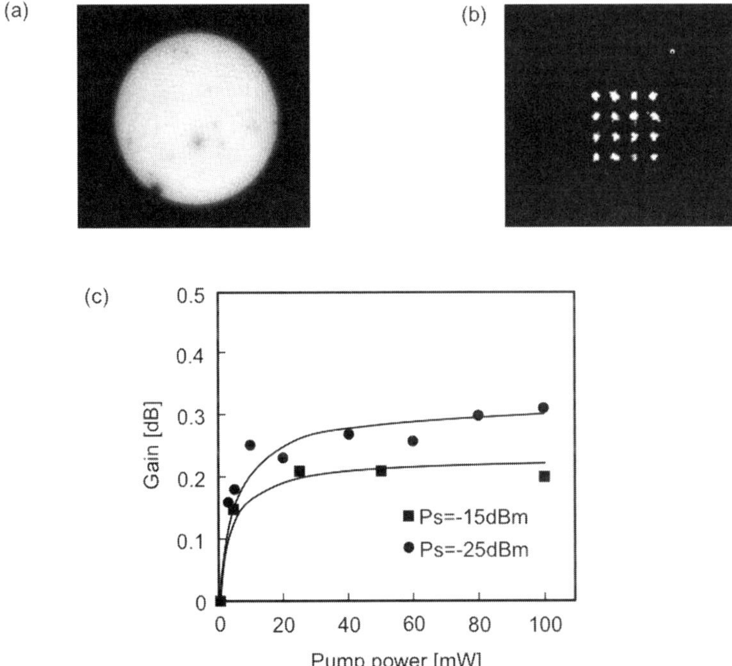

Figure 9.56 (a) Cross-section of 0.8 m long EDIFA. (b) Output signals after propagation through the 0.8 m EDIFA. (c) 3 m long EDIFA signal gain as a function of input pump power [17]. © Reprinted by permission of IEEE.

cores. Figure 9.56(a) and (b) show a cross-sectional view of the EDIFA and the optical signals after amplification. Some dark spots still remain due to the bubbles. To measure the gain characteristics of the test EDIFAs, signal (1550 nm) and pump (1480 nm) beams are launched from a multimode fiber butt-jointed on the input endface of the EDIFA. The measured gain of 3.0 m long EDIFA along with the theoretical prediction, denoted by the solid curves, is shown in Fig. 9.56(c). Here, the gain is defined as the ratio of the output-signal power with pumping to the output-signal power without pumping as

$$G = \frac{P_s(\text{with pumping})}{P_s(\text{without pumping})} \tag{9.14}$$

where P_s is the signal power. The gain is marginal and saturates at less than 1.2 dB. The pumping efficiency is low because the EDIFA cores are multi-moded at 1550 nm and 1480 nm. To achieve higher gain, the dopant concentrations used to control the viscosity need to be optimized to eliminate bubbles and to pull longer EDIFA. Furthermore, the core parameters need to be designed so that the EDIFA has single-mode cores at the wavelength of 1480 nm to increase the pumping efficiency.

10 Practical aspects

10.1 Capacity, cost and power consumption of hybrid PONs [1]

Comparison of the capacity, total system cost and the power consumption between several candidate systems for the NG-PON, including high-speed TDM-PON, WDM-PON, and TDM-WDM-PON or TWDM-PON systems as well as the OCDMA-PON system and its hybrid PON systems, is of practical interest. We will revisit the system architectures of these candidate PONs described in Chapters 2, 6, and 9, including TDM-PON, WDM-PON, TDM-WDM-PON, and OCDMA-PON, TDM-OCDMA-PON, WDM-OCDMA-PON, and TDM-WDM-OCDMA-PON. For the convenience of evaluation, TDM-PON, WDM-PON, and TDM-WDM-PON are modified slightly and reproduced in Figs. 10.1(a), (b), and (c), respectively. OCDMA-PON and its hybrids such as TDM-OCDMA-PON, and WDM-OCDMA-PON are reproduced in Figs. 10.2(a), (b), and (c), respectively. In this evaluation of the capacity, cost, and power consumption, in Figs. 10.1(b) and (c) it is assumed that a tunable light source and an optical tunable filter are equipped at each ONU for the upstream and downstream, respectively. To make it cost-competitive, the architectures of OCDMA and its hybrids are modified slightly to be sourceless at an ONU, and the short pulse train from the mode-locked laser diode (MLLD) can be looped back from the OLT to ONUs for cost reduction as well as ease of maintenance of the light source. OCDMA-PON having two MLLDs, one for the downlink and the other for the uplink in the loopback configuration, is illustrated in Fig. 10.2. It is recalled that

- N is the number of optical codes,
- M is the number of time-slots,
- K is the number of wavelengths.

10.1.1 Total capacity

To increase the total capacity, polarization division multiplexing (PDM) is a possible option in PONs in the future because a commercial digital coherent PDM-QPSK 100 Gb/s long-haul transmission system has been deployed. However, PDM will be difficult to adopt in OCDMA as long as a polarization-dependent type of en/decoder, such as a multiport en/decoder fabricated on a silica substrate by PLC technology, is used. In

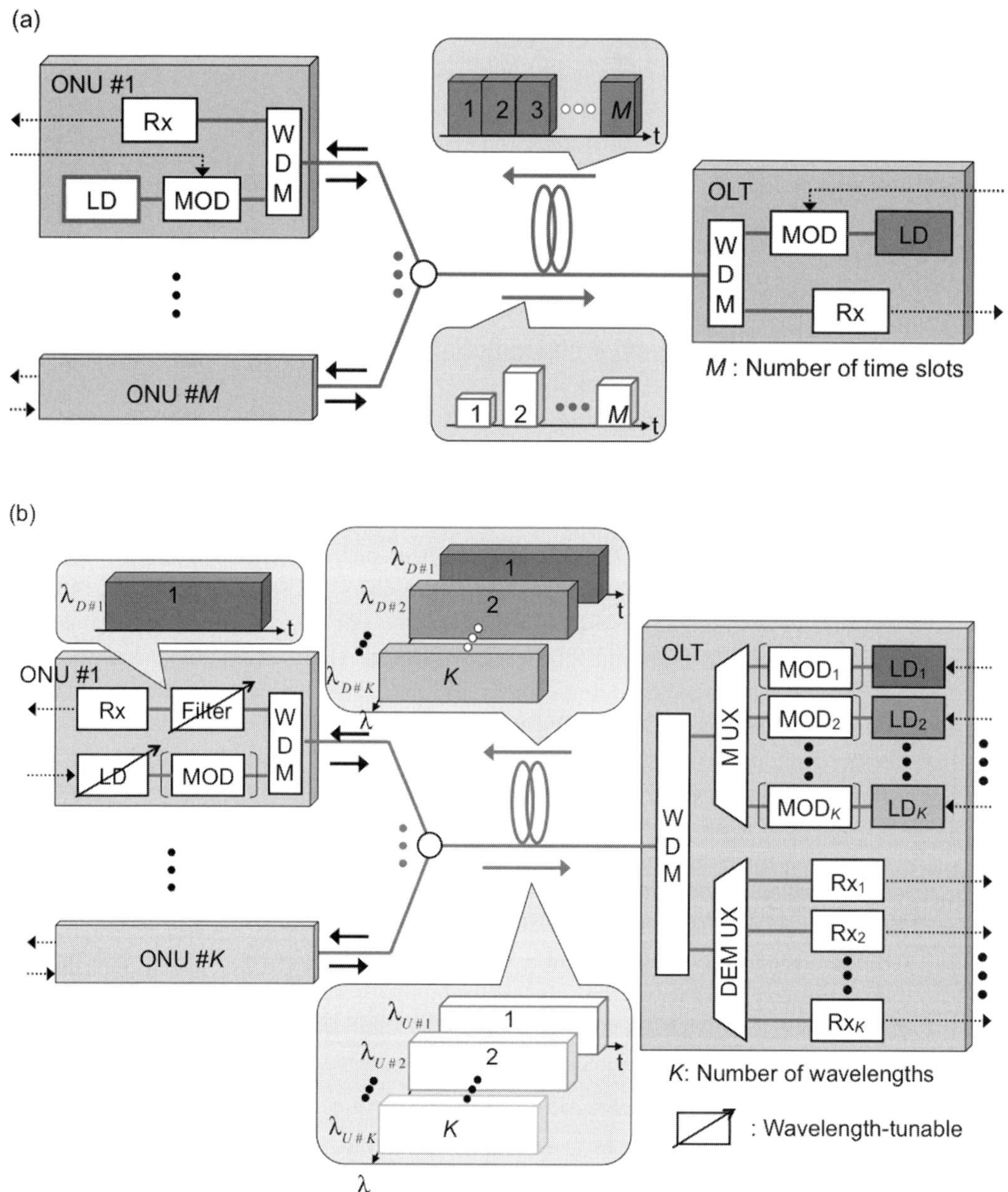

Figure 10.1 System configurations of PONs: (a) high-speed TDM-PON, (b) WDM-PON, and (c) hybrid TDM-WDM-PON [1]. © Reprinted by courtesy of S. Kaneko, N. Miki, H. Kimura, and N. Wada.

the evaluation of the total capacity, the following three assumptions are introduced, and the important constraints are summarized in Table 10.1.

Assumption 1 A symbol rate per wavelength of up to 100 GSymbol/s is assumed for TDMA and WDMA by considering the practically feasible frequency response of

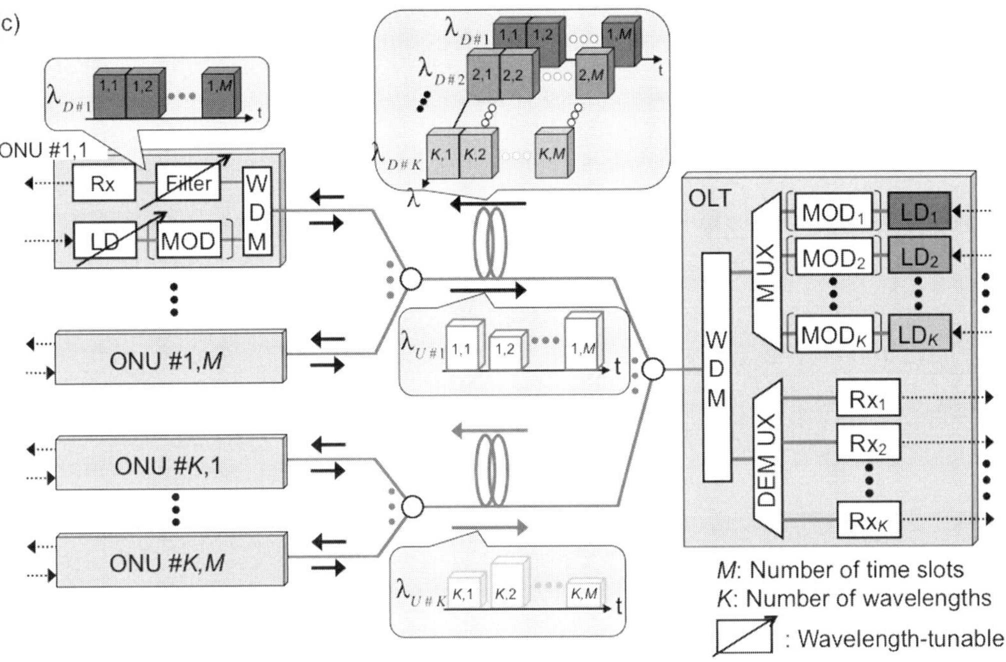

Figure 10.1 (*cont.*)

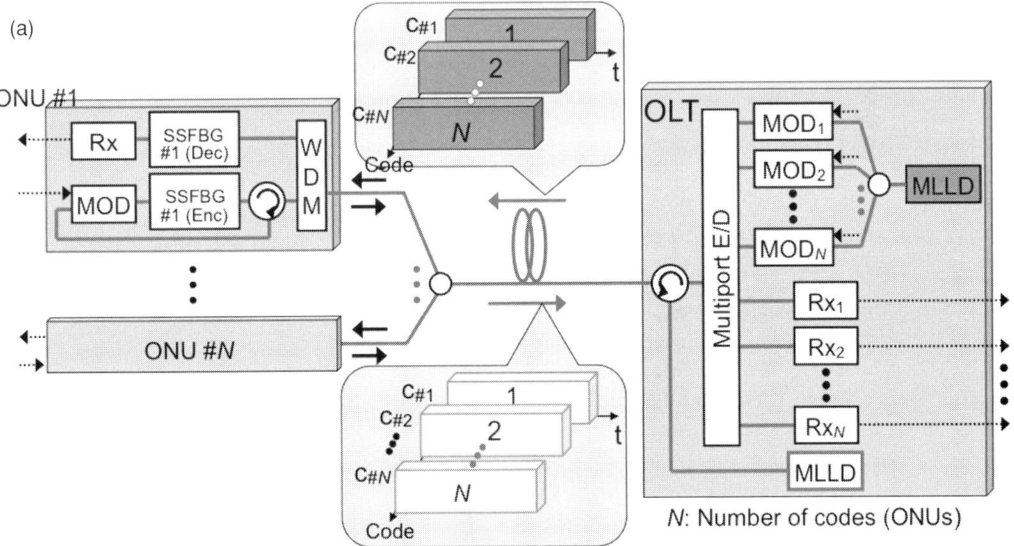

Figure 10.2 System configurations of PONs: (a) OCDMA-PON, (b) TDM-OCDMA-PON, and (c) WDM-OCDMA-PON [1]. © Reprinted by courtesy of S. Kaneko, N. Miki, H. Kimura, and N. Wada.

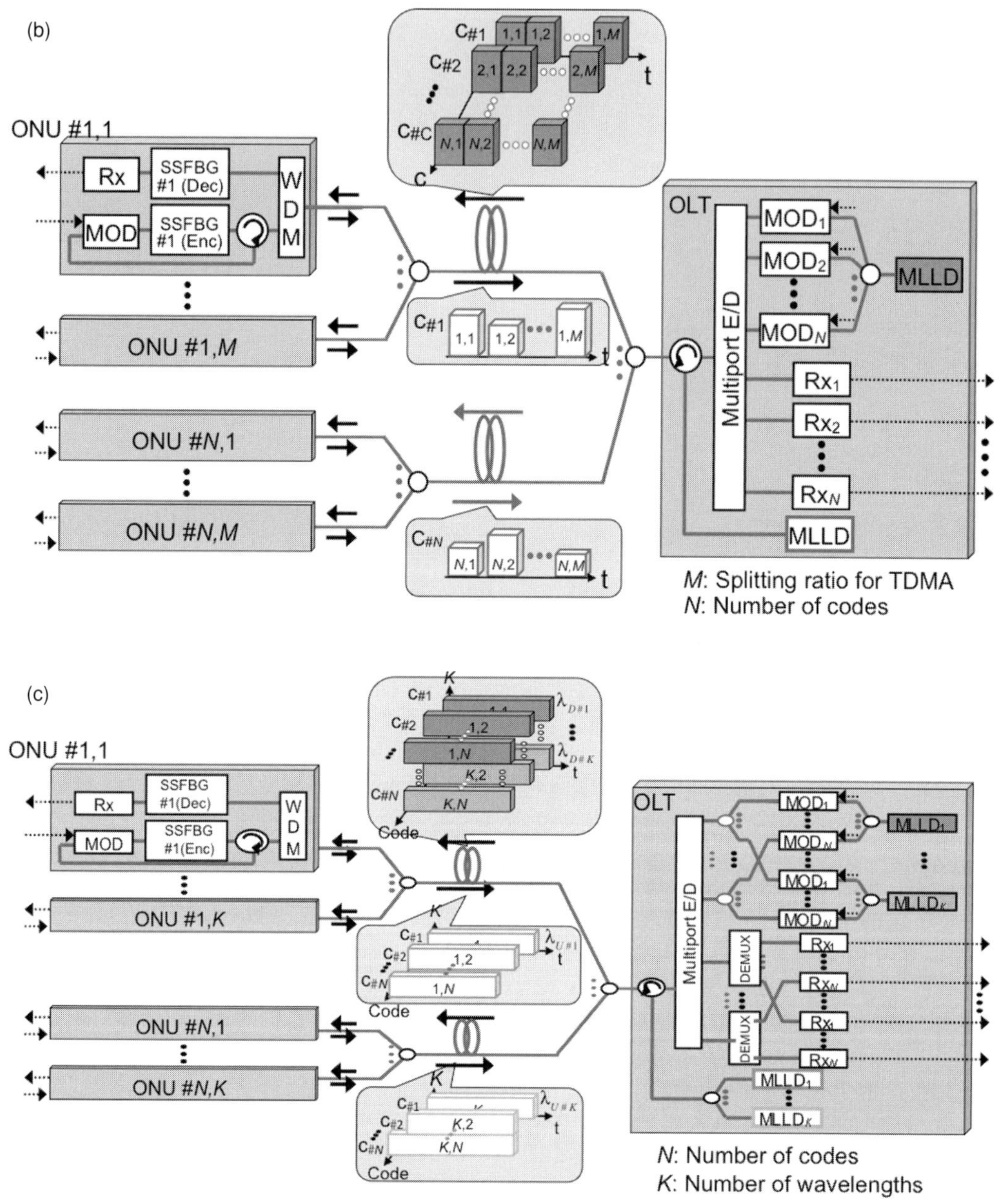

Figure 10.2 (*cont.*)

peripheral electrical circuits. A rate of up to 50 GSymbol/s is assumed for OCDMA, which is limited by the chip rate of the time-spread code. This is justified by the fact that 40 GSymbol/s OCDMA has been realized [2]. Note that in the current digital coherent long-haul transmission systems at bit rates of 100 Gb/s and beyond, the symbol rate is limited by the speed of the analog-to-digital converter (ADC) and digital-to-analog

Table 10.1 Assumptions of system specifications [1], © Reprinted by courtesy of S. Kaneko, S. Kataoka, N. Miki, H. Kimura, and N. Wada

Assumption	Parameter		Maximum value	Limiting factors
1	Symbol rate	TDMA and WDMA	100 GSymbol/s	Frequency response of electrical circuits
		OCDMA	50 GSymbol/s	Chip rate of time-spread
2	Wavelength range for uplink and downlink		30 nm (15 nm each)	Gain bandwidth of optical amplifier
3	Spectral efficiency		0.5 Symbol/s/Hz	Spectral overlap between neighboring channels

converter (DAC), but TDMA, WDMA, and OCDMA are free from this limitation because the digital coherent technique is not adopted.

Assumption 2 Since the wavelengths of all the uplink and downlink signals in a PON must be within the gain bandwidth of a single optical amplifier, the maximum value of the wavelength range is set as 30 nm in the C-band, assigning 15 nm each to the uplink and downlink.

Assumption 3 The maximum spectral efficiency is set to be 0.5 Symbol/s/Hz so that there is no spectral overlap between neighboring channels.

In Fig. 10.3 the evaluated total capacities of various PONs are summarized. The bit rate per ONU is indicated below the diagram. The number of ONUs is fixed at 32. The symbol rate of TDM-PON and WDM-PON is restricted by Assumption 1. The maximum total capacity of 100 Gb/s in the case of on-off-keying (OOK) is quadrupled to 400 Gb/s when PDM and differential quaternary phase-shift-keying (PDM-DQPSK) is employed. In TDM access the total capacity is determined by the line rate on a single wavelength of the feeder cable. As a consequence, the total capacity becomes 100 Gb/s and 400 Gb/s for OOK and PDM-DQPSK, respectively. Meanwhile, the limitation on the available spectral range has to be considered for WDM-PON. Since the wavelength range assigned to the uplink and downlink is up to 15 nm under Assumption 2, 937.5 GSymbol/s ($= 125 \times 15 \times 0.5$) is achieved based on Assumption 3. Therefore, the maximum total capacity reaches 3.75 Tb/s using PDM-DQPSK. In TDM-WDM-PONs, there are options depending on the combination of the numbers of wavelengths K and time slots M. For a number of wavelengths K up to 8, the maximum total capacity is represented as $K \times$ 100 GSymbol/s. But for $K = 15$ the total capacity is restricted to 937.5 GSymbol/s under Assumptions 2 and 3. The modulation formats are summarized in Table 10.2.

In OCDM-PONs, we assume the maximum number of available code sequences is 32. This assumption might be justified by the fact that up to 25 codes have been successfully tested [3]. Two modulation formats DBPSK and DQPSK are considered. Since the symbol rate is up to 50 GSymbol/s, the maximum total capacity of DBSPK is 1.6 Tb/s ($= 50$ Gb/s $\times$ 32 ONUs). By using DQPSK instead of DBPSK, the total capacity is doubled to 3.2 Tb/s. In the hybrid TDM-OCDMA, the number of codes, N, takes the values of 2, 4, 8, and 16. Due to the limit of the maximum symbol rate of 50 GSymbol/s,

Table 10.2 Modulation formats [1], © Reprinted by courtesy of S. Kaneko, S. Kataoka, N. Miki, H. Kimura, and N. Wada

	Total capacity				
	40 Gb/s	80 Gb/s	320 Gb/s	1.28 Tb/s	3.2 Tb/s
High-speed TDM	OOK	OOK	PDM-DQPSK		
WDM	OOK	OOK	OOK	PDM-DQPSK	PDM-DQPSK
TDM- WDM	OOK	OOK	OOK	PDM-DQPSK	PDM-DQPSK
OCDMA	DPSK	DBPSK	DBPSK	DBPSK	DQPSK
TDM- OCDMA	DPSK	DBPSK	DBPSK	DQPSK	
WDM- OCDMA	DPSK	DBPSK	DBPSK	DBPSK	DQPSK

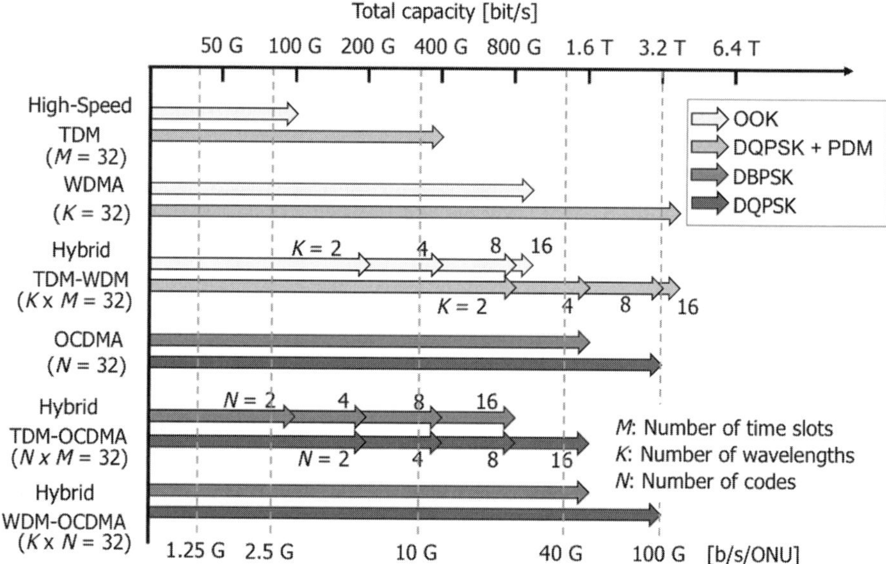

Figure 10.3 Evaluated total capacity of various PONs [1]. © Reprinted by courtesy of S. Kaneko, N. Miki, H. Kimura, and N. Wada.

the total capacity is represented as $N \times 50$ GSymbol/s. Therefore, the maximum total capacity becomes 1.6 Tb/s with DQPSK for $N = 16$. Note that OCDMA and WDM access can provide 100 Gb/s per ONU for both uplink and downlink but TDM access will not be able to provide such a high bit rate symmetry. From Fig. 10.3 OCDMA achieves a capacity comparable to those of WDM access and TDM-WDM.

10.1.2 Cost

The costs of OLTs and ONUs include the costs of optical and electronic components as well as the cost of assembly. Since the dominant cost factor is that of the optical components, their total cost is compared at total capacities of 40 Gb/s, 80 Gb/s, 320 Gb/s, 1.28 Tb/s, and 3.2 Tb/s. The following two prerequisites are taken into account.

Table 10.3 Specifications of laser diodes and photodetectors [1],© Reprinted by courtesy of S. Kaneko, S. Kataoka, N. Miki, H. Kimura, and N. Wada

Output power	DM*-DFB LD		+5 dB m	
	EA*-DFB LD		+5 dB m	
	cw DFB LD		+10 dB m	
Receiver sensitivity for		1.25 Gb/s	2.5 Gb/s	10 Gb/s
BER $< 10^{-12}$	PIN PD	−29 dBm	−26 dBm	−20 dBm
	APD	−37 dBm	−34 dBm	−28 dBm

* DM direct modulation, EA electroabsorption modulator.

- The transmitter configuration with the minimum total cost of optical components is selected among candidate configurations.
- A power budget over 29 dB is guaranteed between the OLT and the ONUs. Since the maximum optical link loss for the nominal class of NG-PON1 is 29 dB, a power budget exceeding 29 dB would be required for NG-PON2 and beyond.

Of the two modulation formats OOK and PSK in Table 10.2, the OOK transceiver is simpler in configuration and is used up to the bit rate of 10 Gb/s. The combination of light sources and photodetectors is summarized in Table 10.3, by considering the output power and the receiver sensitivity. The transceiver configurations are summarized in Table 10.4. From Table 10.4 it is seen that a direct modulation distributed feedback laser diode (DM-DFB LD) is used up to 10 Gb/s, while external modulation is employed beyond 10 Gb/s. An electro-absorption modulator (EAM) integrated with DFB-LD (EA-DFB) and LiNbO$_3$ (LN) modulator with cw DFB-LD (CW DFB + LN) are used between 10 and 40 Gb/s and over 40 Gb/s, respectively. There are two options for the receiver: an avalanche photodiode (APD) and pin-photodiode with an optical pre-amplifier (OA + PIN-PD). When the required power budget of 29 dB can be obtained with an APD, an APD is selected as the receiver. Otherwise, an OA + PIN-PD is used. For example, WDM-PON with total capacity 40 Gb/s and 80 Gb/s uses the DM-DFB LD and APD, since the line rates of each wavelength are 1.25 Gb/s and 2.5 Gb/s, respectively. The power budgets of 1.25 Gb/s and 2.5 Gb/s for the case with the DM-DFB LD and APD is 42 dB and 39 dB, respectively, according to Table 10.3.

In Table 10.5 the unit prices of the main optical components used in the evaluation are summarized. The values were obtained based upon the market prices of 10 Gb/s components, and it is assumed that the unit prices increase/decrease linearly with operating frequency. Note that the effect of mass production is not considered. Figure 10.4 summarizes the evaluation of the total cost. If the mass production effect is taken into consideration, the unit prices will drop by 10–30% as the volume of components doubles. Therefore, it is anticipated that the total cost will decrease by approximately 45–85%. The outcome from Fig. 10.4 is summed up as follows.

- The cost of high-speed TDM, WDM, and hybrid TDM-WDM-PONs increases rapidly beyond the total capacity of 320 Gb/s, since the costly PDM-DQPSK modulator is employed, instead of the OOK modulator.

Table 10.4 Transceiver configuration (modified) [1], © Reprinted by courtesy of S. Kaneko, S. Kataoka, N. Miki, H. Kimura, and N. Wada

	Total capacity				
	40 Gb/s	80 Gb/s	320 Gb/s	1.28 Tb/s	3.2 Tb/s
High-speed TDM	EA-DFB/ OA+PIN-PD	cw DFB+LN/ OA+PIN-PD	cw DFB +DQPSK mode/ OA+DQPSK-Rx		
WDM	DM-DFB/ APD	DM-DFB/ APD	DM-DFB/ OA+PIN-PD	cw DFB +DQPSK mode/ OA+DQPSK-Rx	cw DFB +DQPSK mode/ OA+DQPSK-Rx
TDM-WDM	DM-DFB/ OA+PIN-PD	DM-DFB/ OA+PIN-PD	EA-DFB/ OA+PIN-PD	cw DFB +DQPSK mode/ OA+DQPSK-Rx	cw DFB +DQPSK mode/ OA+DQPSK-Rx
OCDMA	MLLD +DPSK mode/ OA+DPSK-Rx	MLLD +DPSK mode/ OA+DPSK-Rx	MLLD +DPSK mode/ OA+DPSK-Rx	MLLD +DPSK mode/ OA+DPSK-Rx	MLLD +DPSK mode/ OA+DPSK-Rx
TDM-OCDMA	MLLD +DPSK mode/ OA+DPSK-Rx	MLLD +DPSK mode/ OA+DBPSK-Rx	MLLD +DPSK mode/ OA+DBPSK-Rx	MLLD +DPSK mode/ OA+DBPSK-Rx	
WDM-OCDMA	MLLD +DBPSK mode/ OA+DBPSK-Rx	MLLD +DBPSK mode/ OA+DBPSK-Rx	MLLD +DBPSK mode/ OA+DBPSK-Rx	MLLD +DBPSK mode/ OA+DBPSK-Rx	MLLD +DPSK mode/ OA+DBPSK-Rx

Table 10.5 Unit prices of optical components [1], © Reprinted by courtesy of S. Kaneko, S. Kataoka, N. Miki, H. Kimura, and N. Wada
Laser diode ($\times$100 US\$), photodetector ($\times$100 US\$), other components ($\times$100 US\$)

DM-DFB	1.25 G	9.1	PIN-PD	10 G	15	Optical amplifier			26
	2.5 G	10.6		40 G	52.5	WDM Mux/Demux			20
	10 G	20		80 G	102.5	Optical tunable filter			20
EA-DFB	1.25 G	4.1	APD	1.25 G	4.1	LN mode	OOK	10 G	25
	2.5 G	5.6		2.5 G	5.6			80 G	112.5
	10 G	15		10 G	15		DPSK	10 G	25
cw DFB		10						40 G	62.5
MLLD	10 G	200					DQPSK	20 G	62.5
	40 G	237.5						40 G	75
								50 G	81.3
								100 G	112.5
								160 G	150

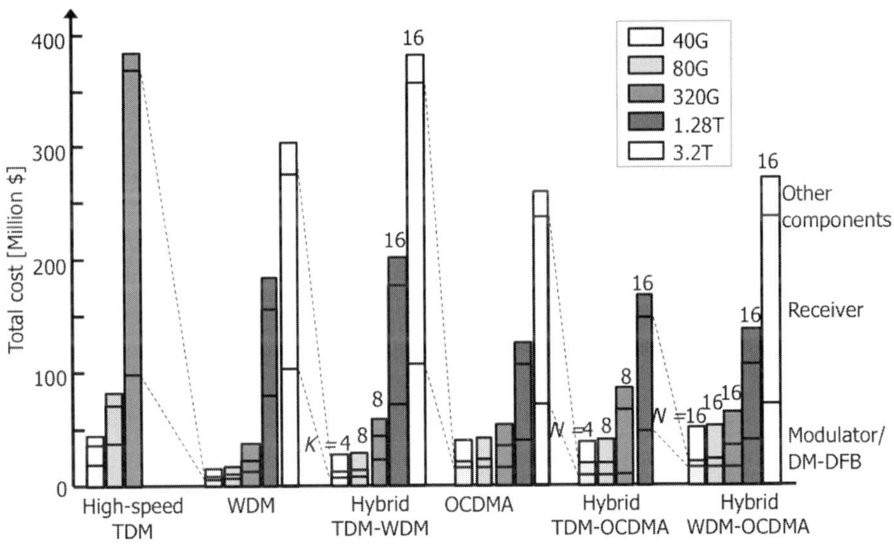

Figure 10.4 Evaluated total cost of various PONs [1]. © Reprinted by courtesy of S. Kaneko, N. Miki, H. Kimura, and N. Wada.

- TDM-PON is much more costly, compared to the other systems, because of the higher line rate.
- The cost of OCDMA systems is higher than that of WDM-PON up to the total capacity of 320 Gb/s, mainly due to the DBPSK transceiver, while OOK based on direct modulation can be employed in WDM-PONs. However, the trend reverses for the total capacity of 1.28 Tb/s and beyond because the DQPSK transceiver becomes necessary in WDM-PONs. It is noteworthy that the trend reverses at 1.28 Tb/s or more where the DQPSK modulator and receiver become necessary in WDM-PONs, and hence OCDMA systems become more cost effective.

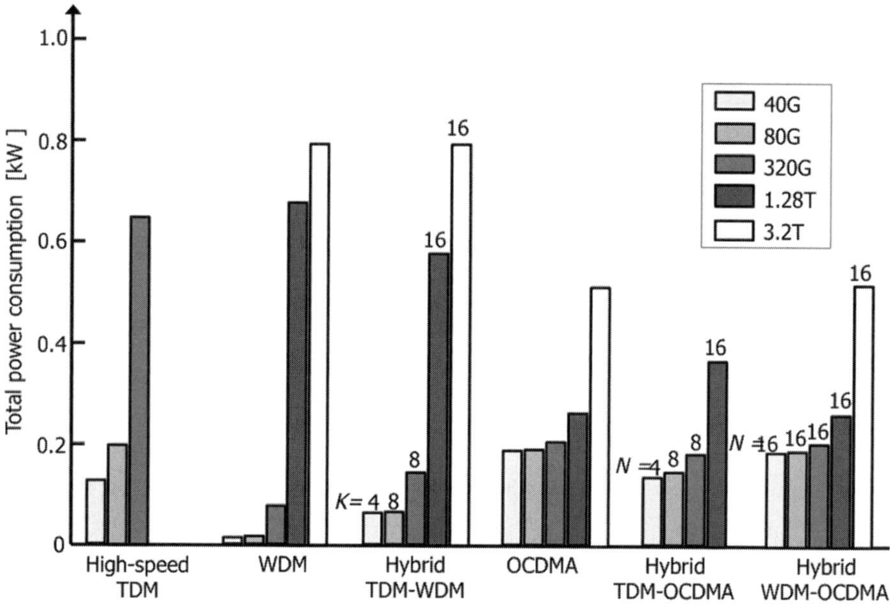

Figure 10.5 Evaluated total power consumption of various PONs [1]. © Reprinted by courtesy of S. Kaneko, N. Miki, H. Kimura, and N. Wada.

10.1.3   Power consumption

The total power consumptions of various PONs were also evaluated, using state-of-the-art device technologies [4]. The evaluation results for various types of PONs for the total capacity of 40 Gb/s, 80 Gb/s, 320 Gb/s, 1.28 Tb/s and 3.2 Tb/s are shown in Fig. 10.5. The sum of power dissipated on the physical layer at the OLT and 32 ONUs is represented on the vertical axis. The modulation format and transceiver configuration are the same as those used in the cost analysis. The outcome from Fig. 10.5 is summed up as follows.

- The power consumption increases rapidly for the total capacity of 320 Gb/s for high-speed TDMA, and for 1.28 Tb/s for WDM and TDM-WDM. This is mainly because PDM-DQPSK is employed.
- The power consumption in a high-speed TDMA-PON is higher than that in a WDM-PON in the examined region ($\geq$40 Gb/s) for the following reasons. External modulation where a high level of power is required by the driver circuit is employed even at 40 and 80 Gb/s in TDMA-PONs, while direct modulation continues to be used up to the total capacity of 320 Gb/s in WDM-PONs. Optical pre-amplifiers are also needed in high-speed TDM-PONs at 40 and 80 Gb/s. Needless to say, serializers (SERs) and deserializers (DESs) consume a significant amount of power due to the higher line rates.
- OCDMA systems require a relatively high level of power at the total capacity of up to 320 Gb/s, mainly due to the modulation drivers required for DBPSK and optical amplifiers to compensate for the insertion losses of the optical en/decoders. However,

OCDMA has a clear edge over the other options for a total capacity beyond 1.28 Tb/s. This is because the modulation format and transceiver configuration remain unchanged regardless of any increase in capacity. The main factor causing an increase in power consumption is the SERs and DESs. As a result, much less power is needed for the total capacity of 1.28 Tb/s or more for OCDMA compared to WDM-PONs.

10.2 Applications of optical codes in optical networks

An optical code serves as an identifier or label of an optical path, flow or packet in optical networks. It can be generated optically and identified without optical-to-electrical conversion and can provide new capabilities which the electronic counterpart cannot offer in routing of optical paths and optical packet switching. A unique feature is that the optical code label (OCL) processing is not based upon logic operation but is based upon an analog operation, and it uses only passive devices such as optical en/decoders. The OCL is recognized on the fly, and hence ultrafast processing can be performed in the time of flight of the optical code in the device. Owing to its inherent confidentiality, the OCL can enhance network security in the optical layer. The OCL can potentially enable very flexible and secure networking without restrictions imposed by wavelength channel and time slot assignments in conventional WDM and TDM networks, respectively.

10.2.1 Optical packet switching

The power consumption of a high-end electronic router, for example, Cisco CRS-3, is 2.7 nW/bit [5]. The routers are seen as major contributors to the high power consumption of ICT. Since IP traffic is increasing by CAGR of $30\sim40\%$, the power consumption of routers will eventually become a major factor in reducing the carbon footprint, as is evident from the power consumption of ICT in 2020 shown in Fig. 1.27. Besides the issue of energy efficiency, an increasing bit rate per subscriber in access networks will make the traffic profile more bursty, requiring new flexible techniques for MANs. Optical packet switching could present a potential solution for reducing the power consumption, footprint, and latency, while providing sub-wavelength granularity and transparency of the ideal characteristics of packet switched networks. A possible scenario for a solution would be to replace electronic routers with packet optical add/drop multiplexers (POADM) in MANs as shown in Fig. 10.6.

The architecture of an $N \times M$ optical packet router (OPR) shown in Fig. 10.7 consists of K optical packet router units, wavelength demultiplexers (Demuxs) at the input ports, and multiplexers (Muxs) at the output ports [6]. Here, K is the number of wavelengths of the WDM link. In Fig. 10.8 the architecture of an optical packet router unit is shown. There are three key building blocks: switch, OCL processor and swapper, and buffer. There are a wide variety of optical switches available in terms of switching speed and a large to small port count. For the buffer, optical fiber delay-line (FDL) can be used, although optical random access memory (RAM) might be available in the long term [7].

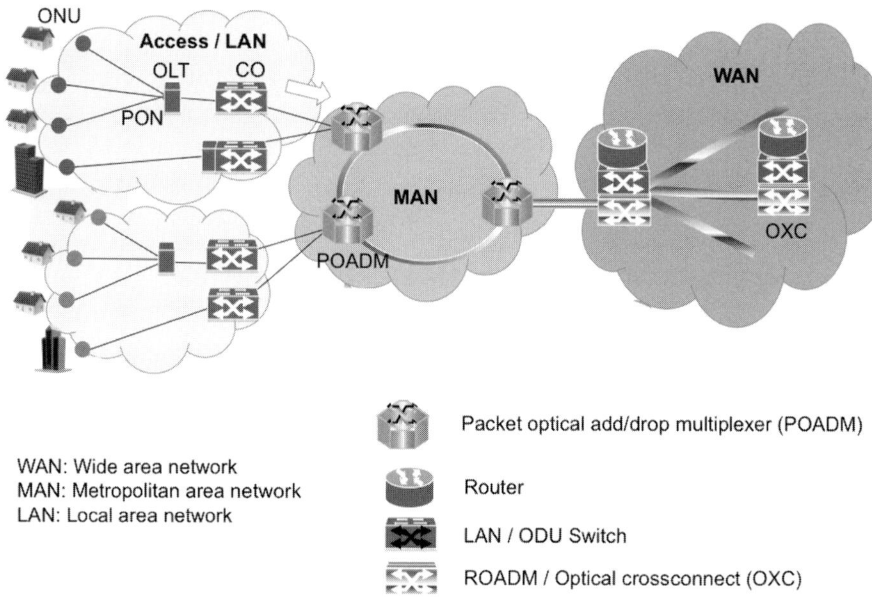

Figure 10.6 Optical packet routers introduced in MANs.

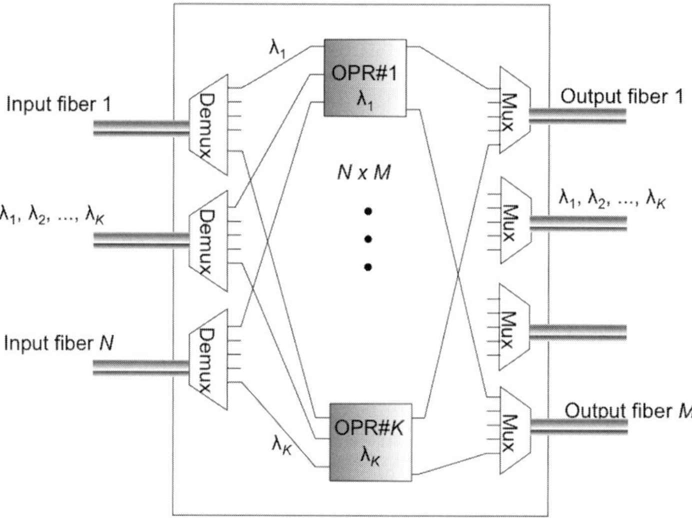

Figure 10.7 Generic architecture of an optical packet router for a WDM network [6]. © IEEE by permission.

Alternatively, a hybrid optoelectronic approach uses CMOS buffers [8]. For the OCL processor, several optical and optoelectronic techniques have been proposed recently, using spread spectrum [9], subcarrier pilot tone [10, 11], and cross gain modulation of a semiconductor amplifier with fiber Bragg grating-based correlator [12]. Here, the focus is on all-optical header processing using OCL. The OCL can carry the information of

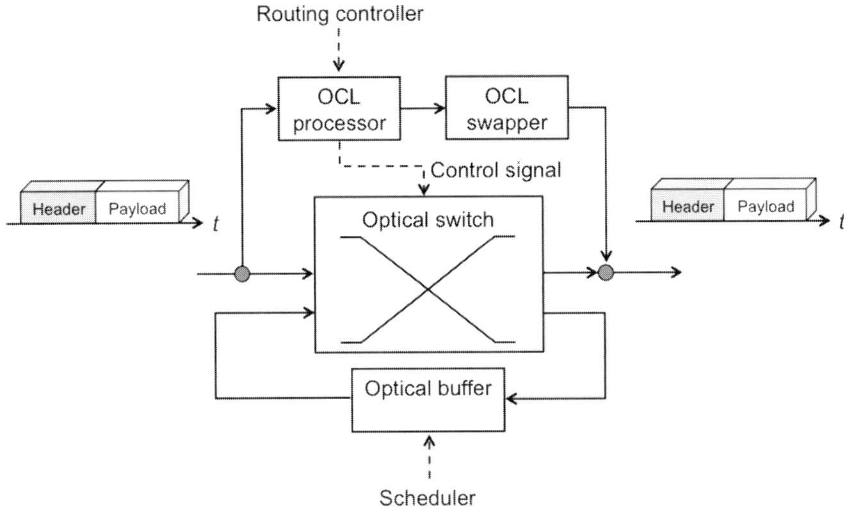

Figure 10.8 Architecture of an optical packet router unit.

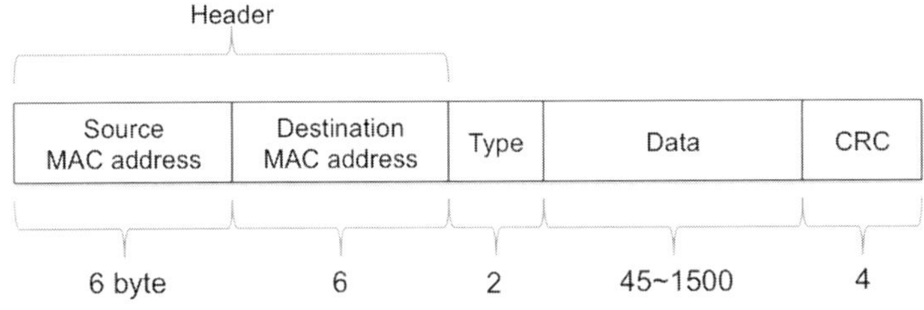

Figure 10.9 Frame of Ethernet ver2.

the header [13]. For example, the header is 20 bytes long, including 4-byte long source and destination addresses in IPv4, shown in Fig. 1.22. The OCL can also be introduced in the layer 2 (L2) switch, for example, the frame of Ethernet (ver2) shown in Fig. 10.9 includes in total an 18-byte identifier.

Optical correlation between the optical codes, described in Eq. (6.18), is a key to OCL processing. It is based upon matched filtering in the time domain and serves to recognize a temporal waveform of the received signal. In contrast to the decoding in OCDMA, the correlation output is free from noise due to the MAI and interference beating, except for additive noise. This is because the input OCL is a single entity without interfering codes. Therefore, the signal detected from the correlator is only the auto-correlation function, and the signal-to-noise ratio is much better than in the case of OCDMA. Optical correlation between the input OCL and a number of OCL entries in the lookup table can be performed in a parallel manner as shown in Fig. 10.10 [6]. This requires the optical decoder to stack as many OCL entries as possible. To mitigate this problem, a multiport decoder, as described in Section 7.3.2, can be adopted since an

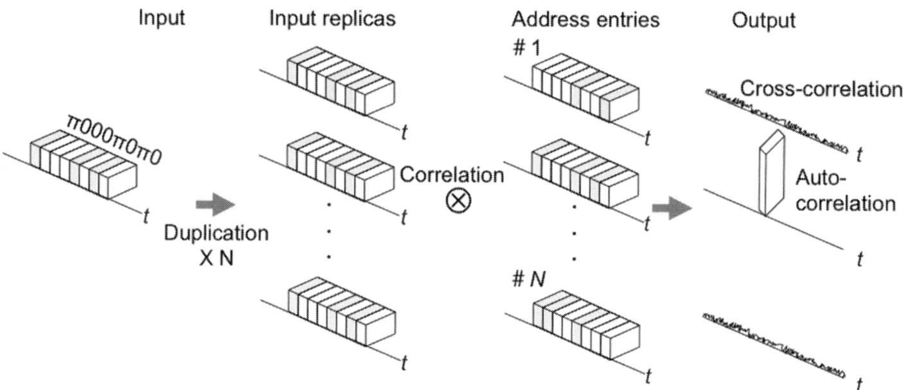

Figure 10.10 Parallel OLC processing [6]. © Reprinted by permission of IEEE.

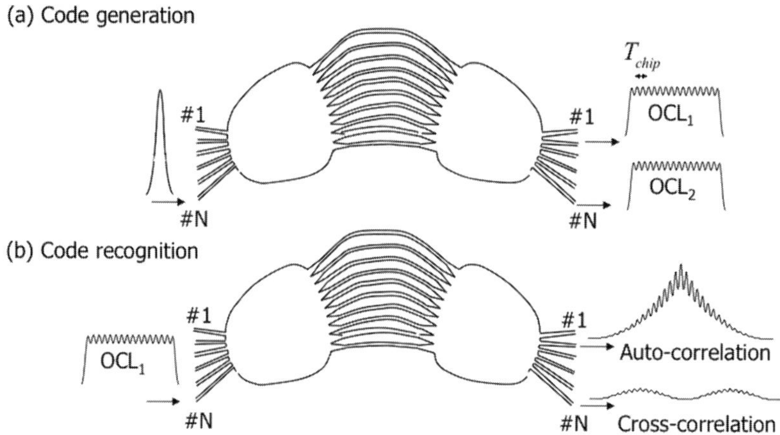

Figure 10.11 Simultaneous processing of OLCs with a multiport en/decoder [14]. © Reprinted by permission of IEEE.

N-port decoder can simultaneously process N OCLs in parallel within the time of flight with a single decoder as shown in Fig. 10.11.

Figure 10.12 shows the experimental setup for optical packet switching, consisting of the transmitter and optical switches along with the OCL processor using a 16 multiport en/decoder [14]. Three different 16-chip OCLs 1, 6, and 11, bearing 4-bit header information, are generated by the multiport encoder and prepended to three 10 Gb/s packet payloads, each of them composed of a 1016-bit long pseudo-random bit sequence (PRBS) as shown in the inset. Figure 10.13 shows a 10 GHz pulse train in (a) generated by a mode-locked laser diode (MLLD), which is split and launched into two LiNbO$_3$ intensity modulators (LN-IMs). The repetition rate of the pulse train on the upper arm is reduced to the packet rate in (b). The lower LN-IM generates the payload data of the time duration of 101.6 ns with 32-bit (3.2 ns) guard time between payload data, labeled (f)–(h). Three 16-chip long OCLs are shown in (c)–(e) (the chip waveforms are not clearly seen). The optical packet switch, indicated by the dotted line in Fig. 10.12, consists of three optical gate switches (LN-SWs). The OCL processor determines which

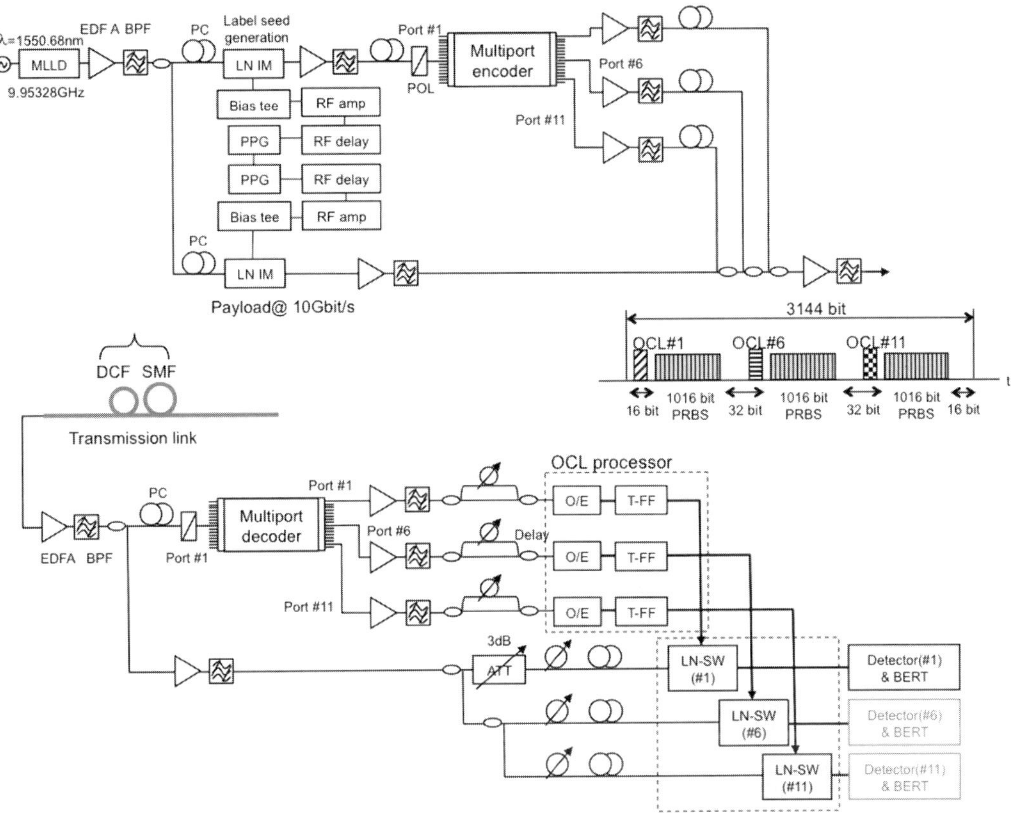

Figure 10.12 Optical packet switch consisting of three optical gate switches (LN-SWs) and the OPL processor [14]. © Reprinted by permission of IEEE.

gate switch to drive based upon information from the OCL and sends the control signal to the switches. For example, when the auto-correlation output appears from Port 6 of the decoder, the first optical gate is opened to pass the packet with OCL 6, and packets 1 and 11 are blocked by the optical gate switches. The streak camera traces of OCL 1, 6, and 11 after 50 km transmission in a dispersion-compensated single-mode fiber link show well time-resolved 16-chip OCL, Figs. 10.13(i)–(k). Their waveforms are deteriorated to some extent during the propagation, but the code detection capability remains almost unchanged. The ACP, defined in Section 7.3.1 as the peak intensity of auto-correlation/maximum intensity of cross-correlation, is theoretically infinite because of the absence of cross-correlations. It is noted that the data bit in the payload goes through the decoder and is time spread, and thus it produces a negligible level of noise.

10.2.2 Multi-granularity optical path switching

In a conventional optical path network, a wavelength path is treated as the minimum data granularity. Any sub-wavelength path having finer data granularity than a wavelength path has to be electrical, for example, TDM slots or electrical packets, and hence the data

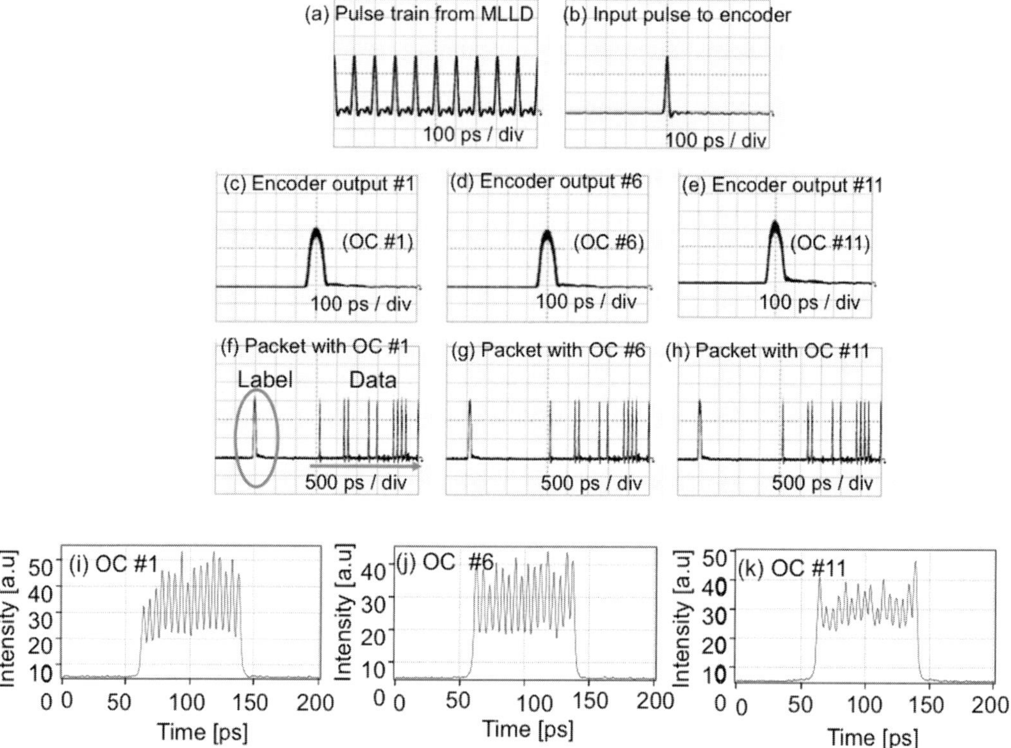

Figure 10.13 Temporal waveforms of optical code labels and the packets [1]. © Reprinted by permission of IEEE.

transparency has to be terminated each time data are transferred via a sub-wavelength path. Between a pair of source and destination nodes, therefore, no multiple transparent connection with smaller data granularity than a single wavelength can be established simultaneously using a single wavelength. However, there is a demand for transparent sub-wavelength optical paths. Optical code division multiplexing (OCDM) allows one to establish and multiplex fine granularity optical paths on a single wavelength, a so-called OC path, in which each bit is encoded by a distinct code sequence, and the OC embedded in a bit serves as an identifier of the OC path [15]. As shown in Fig. 10.15, a single 10 Gb/s wavelength path can be used for five 2 Gb/s OC paths by using five different optical codes. This technique provides sub-wavelength paths in the optical domain, and contrasts with an electrical sub-wavelength path using a time slot. As shown in Fig. 10.16, each bit is encoded by the optical encoder at the transmitter, followed by multiplexing, and at the destination node five OC paths are demultiplexed by splitting the optical power and passing through optical decoders. The auto-correlation appears at the output of the decoder [16]. The cross-correlation due to the presence of undesired OC paths has to be thresholded, and the original bit of the OC path is restored and forwarded to the switch.

The OC path routing can be performed as shown in Fig. 10.17. Here, connection A's traffic is routed from node 1 to node 5; connection B's traffic is routed from node

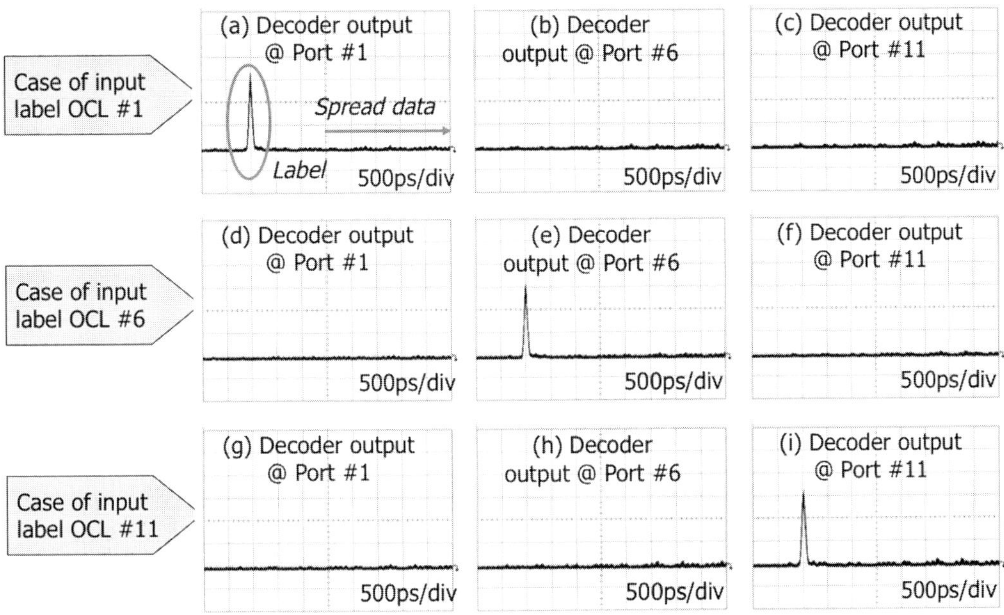

Figure 10.14 Output signals from different output ports of a multiport decoder [14]. © Reprinted by permission of IEEE.

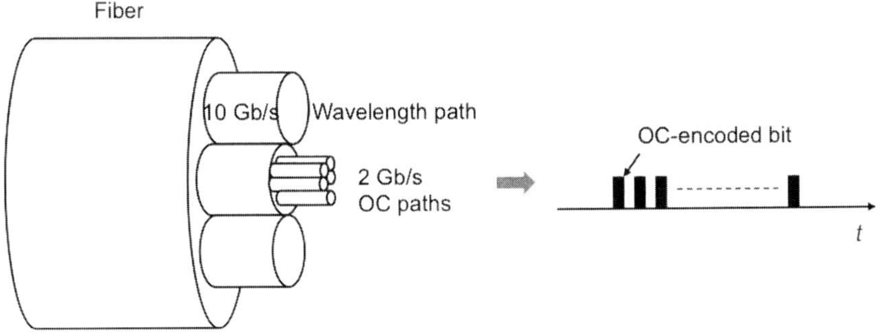

Figure 10.15 Wavelength path and a bundle of OC paths [16]. © Reprinted by permission of IEEE.

3 to node 7; and connection C's traffic is routed from node 3 to node 6. Note that two connections B and C cannot share a path on the link from node 3 to node 2 but they have to be provided with two different path identifiers. As shown in the routing table stored in node 2 in Table 10.6, the traffic of connections A and C, identified by OC 1 and OC 3, respectively, is routed through node 2 without any conversion, while the connection B, which is denoted OC 1, has to be converted to OC 2 at node 2 to avoid contention with the connection A. The OC conversion is described in Fig. 10.19 below.

The optical cross-connect (OXC) switch at a node routes the incoming path to the desired output port. In Fig. 10.18, the architecture of a 2×2 multi-granularity OXC

Table 10.6 OC-based routing table [16], © Reprinted by permission of the Optical Society of America

	Incoming		Outgoing	
Connection	Port number	OC	Port number	OC
A (1,5)	1	1	1	1
B (3,7)	2	1	1	2
C (3,6)	2	3	2	3

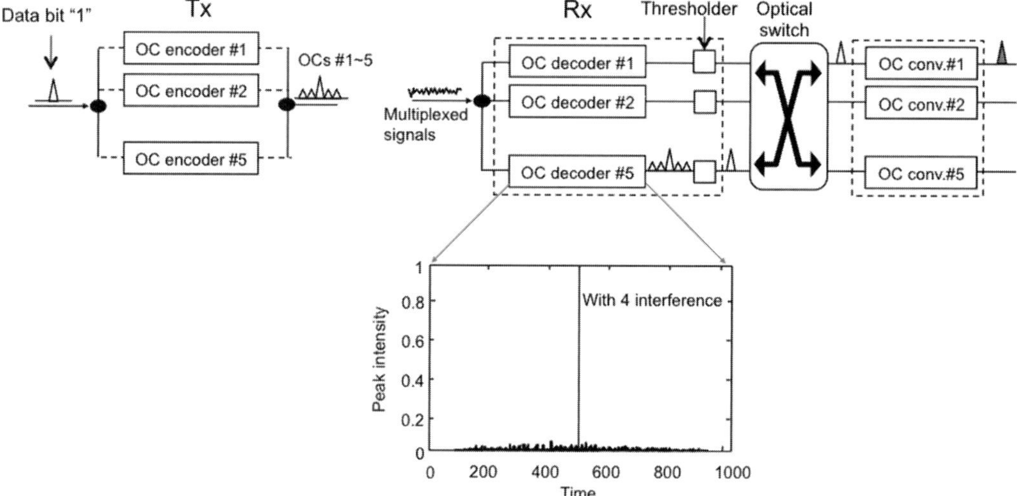

Figure 10.16 Generation and reception of an OC path [17]. © Reprinted by permission of the Optical Society of America.

switch is illustrated [17]. It consists of a wavelength demultiplexer/multiplexer, K OC path switch planes, and a controller, where K is the number of wavelengths per input port. The controller only takes responsibility for the setup of the optical switch. The cross-connect switching of the OC path involves four steps.

Step 1: Multiple paths are first wavelength demultiplexed before entering the OC path switching plane.

Step 2: OC path discrimination is performed in the OC decoder, which consists of N decoders. Here, N is the number of OC paths per wavelength.

Step 3: The controller sets up the switching state of the optical switch according to the processed optical code sequence of the incoming path. If there is a contention at the output port, the OC converter resolves this by converting the incoming optical code to another code.

Step 4: Signals passing through OC converters are multiplexed by the OC multiplexer and are guided to the wavelength-multiplexer at the target output port.

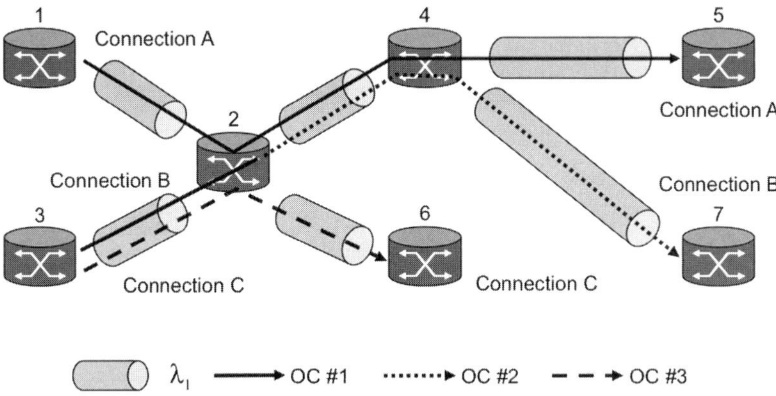

Figure 10.17 OC path routing [16]. © Reprinted by permission of IEEE.

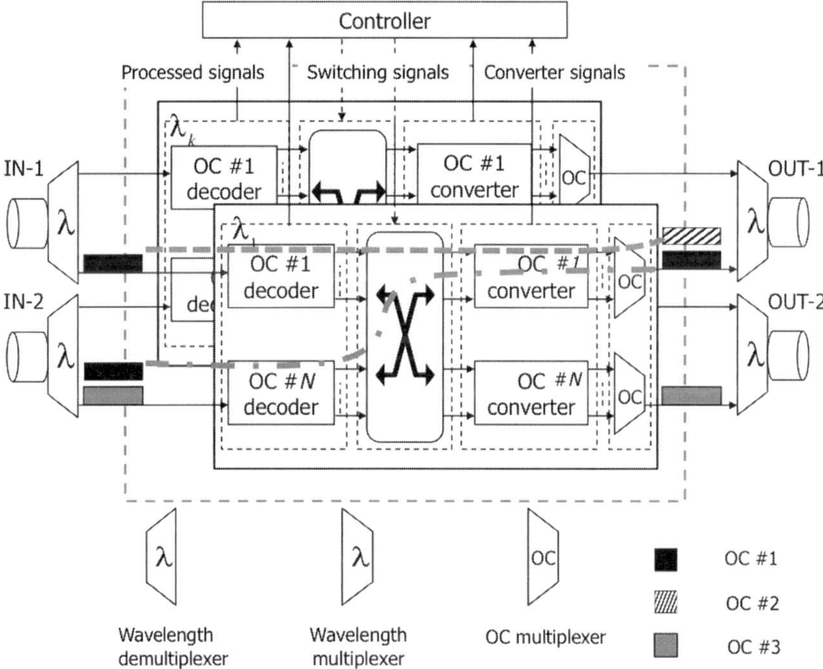

Figure 10.18 Multi-granularity optical cross-connect (OXC) switch [17]. © Reprinted by permission of the Optical Society of America.

The OC conversion can resolve the contention when the same two OC paths on the same wavelength are directed to the same output port. Suppose that both optical path $(\lambda_1, \text{OC } 1)$ from IN-1 and optical path $(\lambda_1, \text{OC } 1)$ from IN-2 are going to be switched to OUT-1 at the same time [15]. Then, optical path $(\lambda_1, \text{OC } 1)$ from IN-1 experiences OC conversion to, for example, $(\lambda_1, \text{OC } 2)$ as indicated by the dotted lines. The OC convertor, consisting of a code selector and generator, is shown in Fig. 10.19. To perform conversion from code OC i to OC j, the multiport decoder output of OC i is detected by

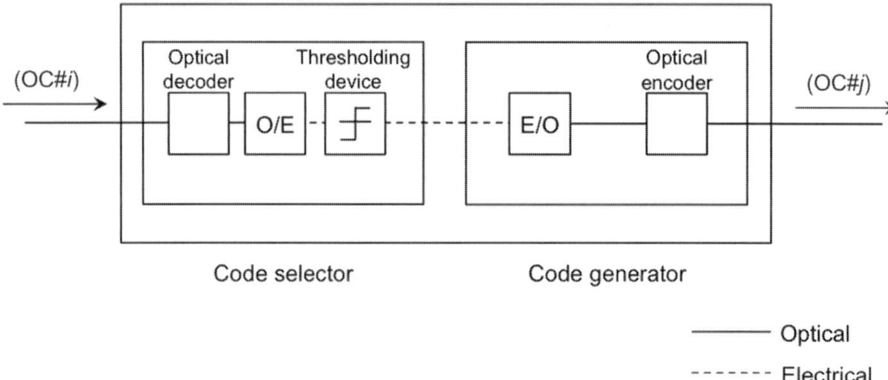

Figure 10.19 Optical code (OC) convertor [15]. © Reprinted by permission of IEEE.

the photodetector, and the electrical signal after thresholding drives the code generator. In the code generator the optical pulse is launched in the desired input port of the multiport encoder. Note that all-optical code conversion can be performed in such a way that an optical thresholding device such as an NOLM described in Section 5.4 regenerates a short pulse, and the short pulse after the optical amplification is launched in the optical encoder [6].

10.3   Faults and troubleshooting

10.3.1   Possible scenario of faults in PON

The operating company of a PON system checks the facility information and alarm conditions of the operations system continuously. When it discovers faults such as uplink-error-rate deterioration and service immobilization, it instigates troubleshooting procedures. For example, if a fault occurs at an ONU, the ONU will be disconnected from the system by turning off its power or unplugging its optical cord and restored. Two possible faults in PONs will be considered [18]. The first case is that only one subscriber cannot receive the service. Three potential faults are probable, as shown in Fig. 10.20:

(1) fault in the drop cable between the subscriber and the closest splitter,
(2) fault in the ONU equipment,
(3) fault in the subscriber's home wiring.

The second case is that all subscribers dependent on the same OLT are affected. Three potential faults are probable, as shown in Fig. 10.21:

(1) fault in the splitter closest to the OLT,
(2) fault either in the feeder cable or in the distribution cable of the ODN,
(3) fault in the OLT equipment.

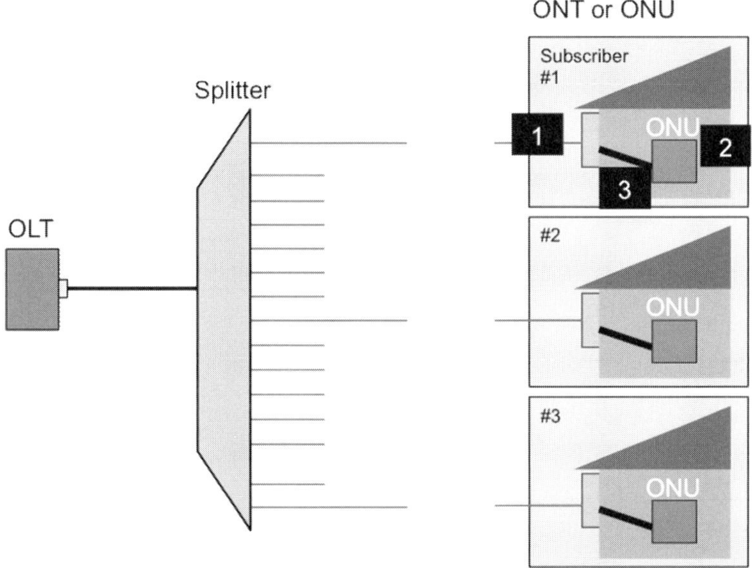

Figure 10.20　Case where only one subscriber is affected [18]. © Reprinted by courtesy of JSDU.

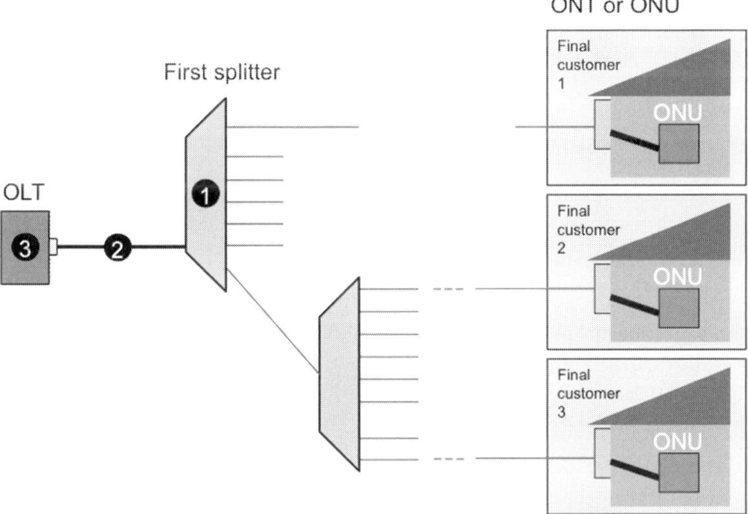

Figure 10.21　Case where all the subscribers are affected [18]. © Reprinted by courtesy of JSDU.

10.3.2　Testing and testing equipment

One of the most important factors in ensuring proper transmission is controlling the power losses in the network against the link-power budget specifications from the ITU-T Recommendation and IEEE standard, which is done by establishing a total end-to-end power budget with sufficient margin. Optical loss is defined as the difference in power level between the transmitting source and the receiving power meter. The total optical

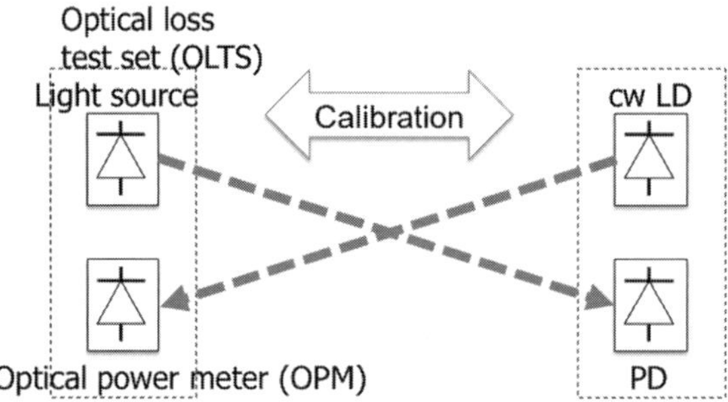

Figure 10.22 Optical loss test set (OLTS).

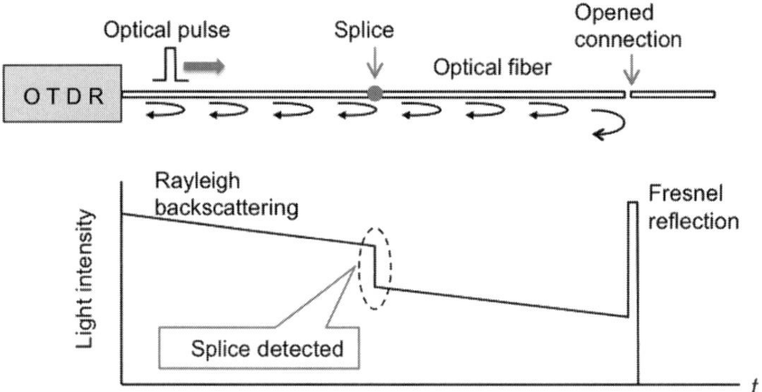

Figure 10.23 Back-reflected light optical time-domain reflectometer (OTDR).

system/link loss is the sum of the insertion loss of the connectors at the OLT and ONU, WDM coupler, splices, fiber attenuation, and splitter. The insertion loss is the loss of optical energy resulting from the insertion of a component or device in an optical path.

The optical loss test requires a sequence of two measurements. The optical loss test set (OLTS) shown in Fig. 10.22 consists of a light source and an optical power meter (OPM) combined in one unit [19]. To begin with the loss measurement, two OLTS units, one at the CO and the other at a remote site, are first referenced together using their individual light sources. Then, each OLTS sends a calibrated power value from its light source over the section under test to the other OLTS, which measures the received power and calculates the loss.

During PON installation, it is important to ensure that each cable section meets or exceeds the cable specifications. This can best be accomplished by using an optical time-domain reflectometer (OTDR). The OTDR provides a detailed map of all of the section losses, allowing users to locate and characterize every individual element in the link, including connectors, splices, splitters, couplers and faults along with the local attenuation of the fiber. The OTDR operates by launching a high-power light pulse into the fiber and measuring the back-reflected light as shown in Fig. 10.23. The

Table 10.7 Requirements for an OTDR

Case	Test location	Direction	Fault location	Wavelength
One subscriber	ONU	Uplink	ONU ~ the closest splitter	Out-of-band
All subscribers	OLT	Downlink	OLT ~ feeder cable	In-band

back-reflection is induced by Rayleigh scattering, inherent to the amorphous nature of a fiber having particles of size much smaller than the wavelength of the light. Fresnel reflection also causes back-reflection at the refractive index discontinuity. The shorter the pulse becomes, the less energy the pulse carries. As a consequence, resolution of the position improves but the pulse can only travel for a short distance due to loss caused by attenuation, splices, etc. along the fiber. Therefore, the pulse width has to be carefully determined so that the back-scattered light from the target point returns to the input end with an intensity above the noise floor.

An issue to be recalled is the wavelength used for the OTDR measurement. The wavelengths that carry the traffic in PONs are 1310 nm, 1490 nm, and 1550 nm, from Fig. 2.4. The use of light at these wavelengths in the OTDR would cause interference with the data signals if the OTDR is used in-service. As a consequence, the data signal and the measurement precision of the OTDR deteriorate at the same time. To avoid these mutual disturbances, an in-service OTDR uses an out-of-band wavelength of 1625–1650 nm, far away from the wavelengths bearing data signals. Let us consider the two cases where only one subscriber is affected, as in Fig. 10.20, and where all the subscribers are affected, as in Fig. 10.21. The former requires use of in-service OTDR for the continuity of services, which uses out-of-band wavelength. The latter can use a conventional OTDR using in-band wavelength. The requirements are summarized in Table 10.7.

10.3.3 Case studies of faults [20]

Case I: Fault at an ONU

An example of a problem with an ONU is that the transmitter continues to emit continuous wave (cw) light on the uplink at 1310 nm. This cw signal at a power level of +2 dB m affects the other ONUs administered by the same PON OLT interface package. Another similar fault is abnormal optical digital signal emission on the uplink from an uncontrolled ONU, which outputs a 1310 nm optical digital signal continuously. Examples of the temporal waveforms of an abnormal cw signal and abnormal digital signal of a faulty ONU are shown in Fig. 10.24. This abnormal uplink digital signal is overlaid on the normal uplink signals from other ONUs as shown in Fig. 10.25. This will prevent the OLT from identifying all ONUs and will result in BER deterioration and service immobilization.

The countermeasure in this case is simply to locate the faulty ONU and replace the equipment at the fault. Troubleshooting of a GE-PON accommodating 32 ONUs is conducted at the OLT in the following manner. There are four paths from the four-branch

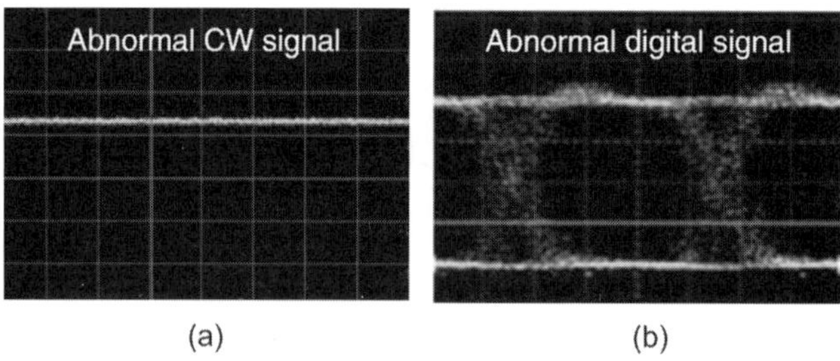

Figure 10.24 Temporal waveforms of (a) abnormal cw light and (b) abnormal digital signal from a faulty ONU [20]. © Reprinted by permission of NTT.

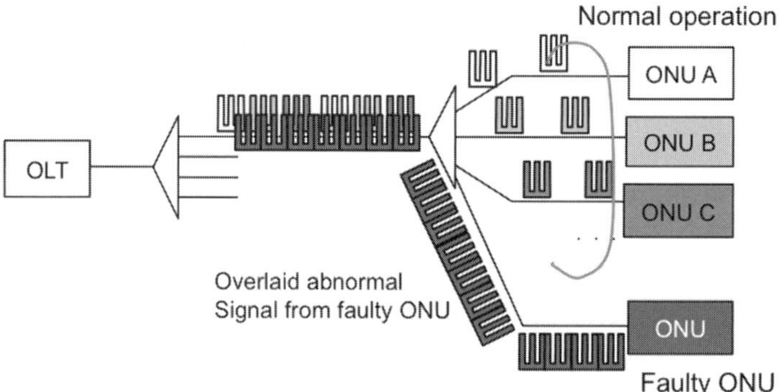

Figure 10.25 Abnormal uplink signal from a faulty ONU overlaid onto other normal signals [20]. © Reprinted by permission of NTT.

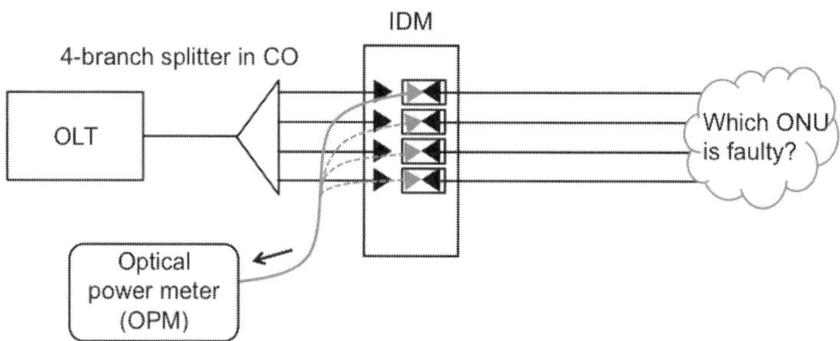

Figure 10.26 Troubleshooting procedure in the central office: optical fiber connectors for the four paths at the integrated distribution module (IDM) are disconnected and the power is monitored [20]. © Reprinted by permission of NTT.

splitter at the OLT in the CO to the integrated distribution module (IDM) as shown in Fig. 10.26. The optical fiber connectors at the IDM are disconnected one by one, and the power is monitored by the OPM. The connector where the faulty ONU is connected must be identified. Under these circumstances, normally operating ONUs are disconnected from the OLT and cannot launch signal light. The absence of received optical power, therefore, would indicate normal ONU operation. By contrast, the presence of received

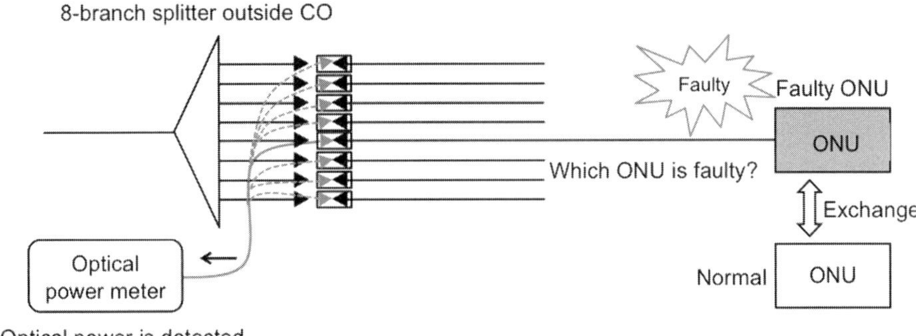

Figure 10.27 Troubleshooting procedure at an outside plant at an eight-branch splitter [20]. © Reprinted by permission of NTT.

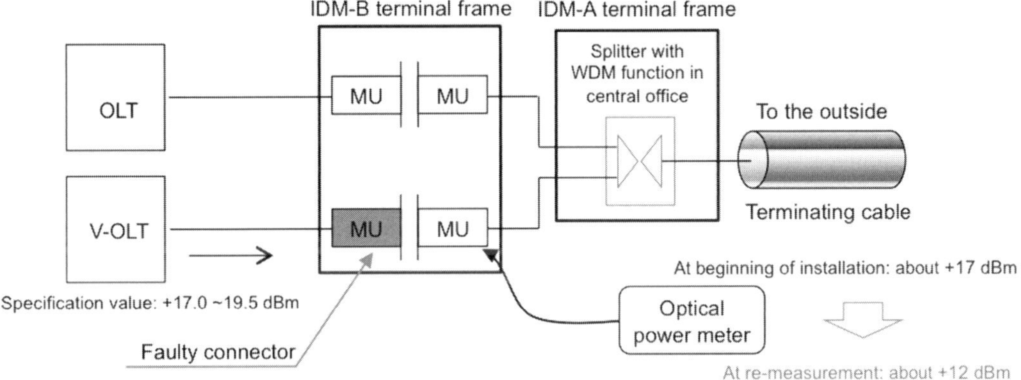

Figure 10.28 Miniature unit (MU) connector on an integration distribution module (IDM) [20]. © Reprinted by permission of NTT.

optical power would indicate that the faulty ONU with abnormal signal light is involved. Next, one must disconnect one by one the eight optical fiber connectors on the outside eight-branch splitter and check whether optical power is received from that ONU. Finally, once the faulty ONU is located, it can be exchanged with a normal one to complete the repair procedure shown in Fig. 10.27. If the optical splitter is cascaded, as in the case shown in Fig. 10.21, the next step is to go to the outside optical splitter, and one must disconnect the optical fiber connectors in turn and identify and check each connector for received optical power from that ONU using an optical power meter, as described above. Finally, once the faulty ONU is located, the equipment can be replaced or repaired.

Case II: Fault at the OLT

Another case study involves a miniature unit (MU) connector on an IDM used for video delivery services. At the beginning of installation of intra-office equipment at a video OLT (V-OLT), shown in Fig. 10.28, the received power level at the MU connector is as high as 17 dBm. When it is measured again on connection, the received power level

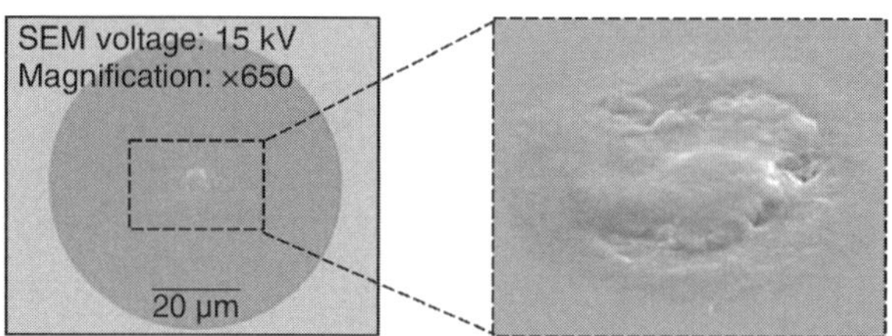

Figure 10.29 Fatal damage caused by silica melting in the core section [20]. © Reprinted by permission of NTT.

goes down by a few dB due to the increase in insertion loss at the MU connector. The insertion loss of faulty MU does not improve even after the optical fiber end surface of the MU connector is carefully cleaned.

After the connector is dismantled, the connector ferrule end surface is inspected using a scanning electron microscope (SEM). There is fatal damage caused by silica melting in the core section as shown in Fig. 10.29. An experiment is conducted to reproduce the fault: light from the V-OLT is injected into the connector and the tip of the MU connector is exposed to various types of contamination from oily hands, clothes, dust, etc. It is found that fatal damage caused by silica melting occurs in the core section and that the loss increases in the case of all these types of contamination. It is therefore deduced that the high optical power of the V-OLT equipment is converted into heat by contamination on the optical fiber end surface of the connector and that this leads to silica melting in the core section.

10.4   Safety to the human body and systems

10.4.1   Safety to the human body

Safety in the installation of optical fibers in optical communication systems includes avoiding exposure to invisible light radiation carried in the fiber, proper disposal of fiber scraps produced in cable handling and termination, and safe handling of hazardous chemicals used in termination, splicing or cleaning. Damage such as to the retina of the eye and the ingestion of glass fragments will be focused on. To avoid damage to the retina of the eye, the input power to an optical fiber has to be limited. It is fortunate that FTTH systems use the wavelength regions of 1300 nm and 1550 nm, where the energy of a photon is smaller than that of photons in the ultraviolet and visible light ranges. However, it is still better to inspect carefully whether the PON systems are safe with respect to these potential threats to the human body and the fibers.

Optical power from lasers is classified by the International Electrotechnical Commission (IEC). Depending on the potential danger, IEC 825 requires that all laser equipment be classified into one of the following classes: 1, 2, 3a, 3b, or 4. Because the minimum power limits for class 4 lasers are not used in telecommunications, they are excluded. For the other classes of lasers, the power limitations and the accompanying safety

Table 10.8 IEC 825-1 and 825-2 classes of lasers, power limits and safety requirements

| Laser class | Maximum power level | | Safety requirements |
	1310 nm	1550 nm	
Class 1	9.4 dBm	10.0 dBm	Inherently safe • Protective housing to prevent higher than classified emission • Safety interlock in the housing to prevent access to non-classified emission levels • Classification labels on the product and in the promotion literature • Caution labels on service panels, interlocked or not • User safety information in operator and service manuals
Class 3a	13.8 dBm	17.0 dBm	Safe unless viewing aids are used Additional requirements to all of the above • Key control • Beam stop to automatically disable the laser if access is required • Audible or visible "laser-on" warning
Class 3b	27.0 dBm	27.0 dBm	Additional requirements to all of the above • Remote control switch to allow disabling of the laser by a door circuit • Aperture label to indicate the location of radiation output

requirements are summarized in Table 10.8 [21]. Class 2 is used for visible laser products emitting wavelengths from 400 to 700 nm; these requirements are not considered relevant to PONs. Most emitters fall within class 1 where the only safety precautions necessary are safety labels. Only in the RF video transmission system is the signal power level significantly greater than the SBS limit described in Section 4.8, and the SBS limit restricts the maximum power to about 8 dBm to maintain the CNR around 50 dB. The SBS suppression techniques, including phase modulation, increase the launch power up to 20 dBm and 16 dBm for transmission distances of 20 km and 50~60 km, respectively, in an EPON system of 32-split ratio. Output power from EDFAs exceeding 10 mW will require class 3a or 3b special treatment, such as automatic shutoff in the case of fiber breakage, visible laser-on warnings and physical access controlled behind lock and key.

As for the problem of fiber fragments, they are essentially glass needles, which can be quite painful if stuck in one's skin and even life-threatening if ingested. The only formal standard [NECA/FOA-301] that addresses this problem simply sets several common sense rules as follows [22].

(a) Small scraps of bare fiber produced as part of the termination and splicing process must be disposed of properly in a safe container and marked according to local regulations, as it may be considered hazardous waste.

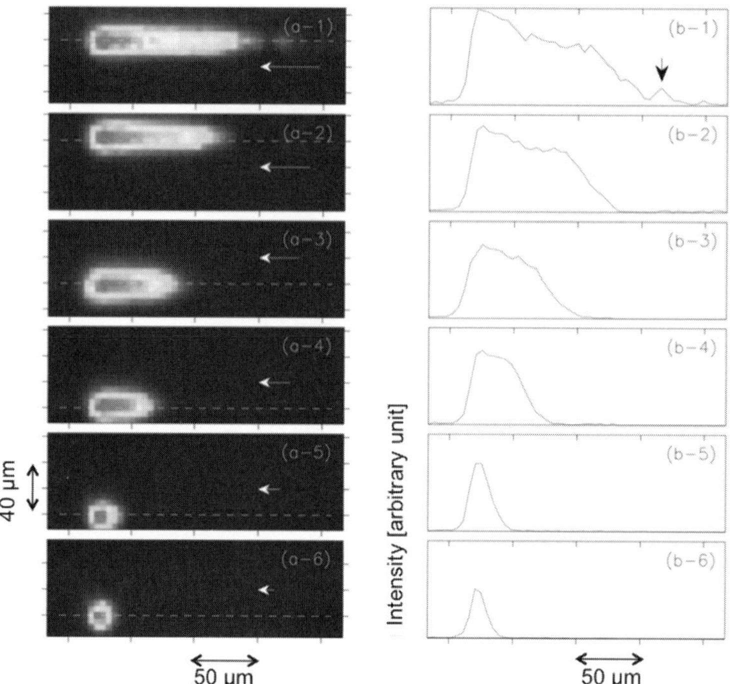

Figure 10.30 Images captured from the ultrahigh-speed video and intensity profiles along the dashed line in each image. The pump laser powers are (1) 9.0 W, (2) 7.0 W, (3) 5.0 W, (4) 3.5 W, (5) 2.0 W, and (6) 1.5 W. Each horizontal arrow indicates the distance that the optical discharge moves in 40 μs (10 frames) [24]. © Reprinted by permission of the Optical Society of America.

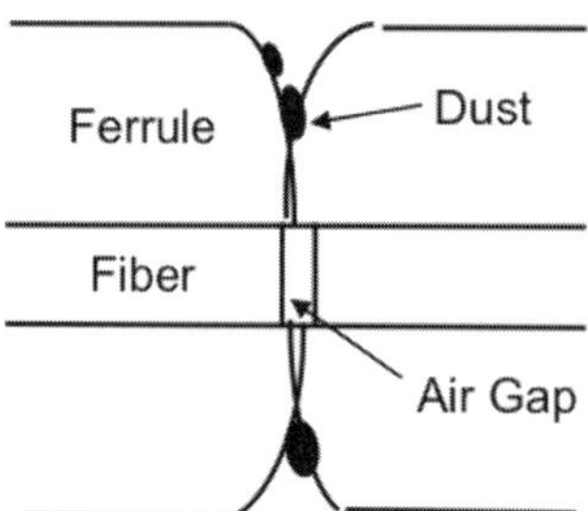

Figure 10.31 Local heating in the air gap [25]. © Reprinted by permission of IEEE.

(b) Do not drop fiber scraps on the floor where they will stick in carpets or shoes and be carried elsewhere. Place them in a marked container or stick them to double-sided adhesive tape on the work surface.

(c) Thoroughly clean the work area when finished. Do not use compressed air to clean off the work area. Sweep all scraps into a disposal container.

(d) Do not eat, drink or smoke near the working area. Fiber particles can be harmful if ingested.

(e) Wash hands well after working with fibers.

(f) Carefully inspect clothing for fiber scraps after working with fibers.

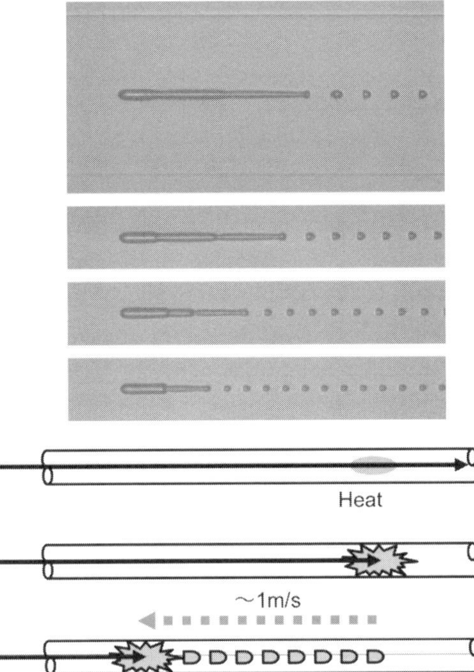

Figure 10.32 Optical micrographs showing the front part of fiber fuse damage generated in single-mode silica glass fibers. The pump laser powers are (a) 9.0 W, (b) 7.0 W, (c) 5.0 W, (d) 3.5 W, (e) 2.0 W, (f) 1.5 W, (g) ~1.3 W, and (h) ~1.2 W. The two thin lines at the top and bottom of (a) and (e) are the edges of the fiber, whose diameter is 125 μm. The height of the figures, except for (a) and (e), corresponds to 50 μm [24]. © Reprinted by permission of the Optical Society of America.

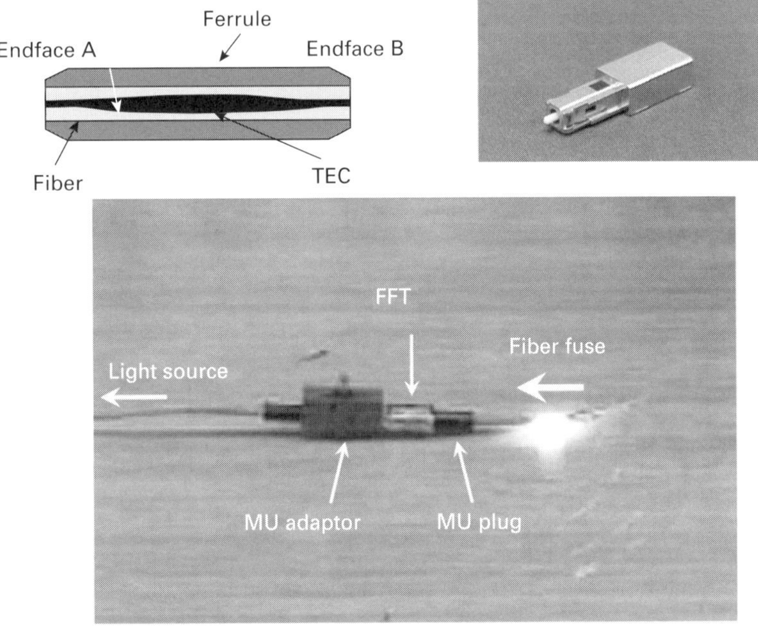

Figure 10.33 Fiber fuse terminator and the termination of fiber fuse [26]. © Reprinted by permission of IEICE.

10.4.2   Fiber fuse

Other damage is to optical fibers. The fiber fuse effect was discovered in the late 1980s [23]. The effect of fiber fuse can be disastrous in a PON where kilometers of fiber may be destroyed as a result of a failure at the input end. Fiber fuse is a photo-induced phenomenon in an optical fiber and occurs when the cw optical input power exceeds about 1 W. It is unlikely that a PON system will experience fiber fuse since the highest power level in a PON system is that for the RF video transmission system (see Section 10.3.3, Case II), and the power level of the video signal is around $10\sim20$ dB m at most. Nevertheless, it is worth reviewing fiber fuse, a phenomenon unique to optical fibers. It is recalled that the SBS will onset with a lower threshold power than fiber fuse, as described in Section 4.8. It is also likely that melting of the fiber endface will occur, as described in Section 10.3.3.

Fiber fuse is a phenomenon which occurs when the laser light propagating in the fiber is strongly absorbed by the heated part of the fiber when the silica glass fiber is heated locally to a temperature of about 1000 °C, and this heating induces the absorption of light, increasing its temperature to approximately 10^4 K. As a result, plasma is generated in the high-temperature region, shown as a bright white spot in Fig. 10.30, and it moves at the velocity of about 1 m/s along the fiber toward a source of laser radiation, causing damage to the fiber core [24]. The plasma exhausts the energy supplied by the launched light during the propagation. The reason why the fiber is not cut at 10^4 K is that the plasma moves at a faster speed than the heat which transmits from the core toward the cladding. The heating can be induced in an air gap between the fiber connector in the presence of dust, as shown in Fig. 10.31 [25]. In some cases this results in the formation of periodic and bullet-shaped voids, as shown in Fig. 10.32. However, the void formation mechanism is not fully understood. A fiber fuse terminator which is capable of terminating the propagation of fiber fuse has been developed [26]. It uses a thermally diffused expanded core (TEC) fiber whose mode field diameter is enlarged to 20 μm, double that of the standard single-mode fiber. It is encapsulated in a ferrule, and the ferrule is built in a miniature unit (MU) plug with the MU adaptor shown in Fig. 10.33. The bottom photograph demonstrates a fiber fuse from east to west about to terminate by the terminator. It has been confirmed that a fiber fuse can be terminated with an onset power up to 2 W.

References

Chapter 1

1. Wellbrock G. and Xia R. J. (2010). The road to 100G deployment, *IEEE Commun. Mag.*, March, S14–S18.
2. Cisco White Paper, June 2011 (www.cisco.com/en/US/solutions/collateral/ns341/ns525/ns537/ns705/ns827/white_paper_c11-481360.pdf).
3. Disclosure of Quarterly Data concerning Competition Review in the Telecommunications Business Field: 2288; First quarter of FY 2011, Ministry of Internal Affairs and Communications (www.soumu.go.jp/menu_news/s-news/01kiban04_02000033.html in Japanese).
4. Sakano T. (2009). Intelligent optical networking to achieve customer satisfaction – Challenges and direction toward the future, *OFC2009*, NWB (Los Angeles, CA).
5. Vanier F. (2011). World Broadband Statics: Q1 2011, POINT topic.
6. Nagel S. R. (1987). Optical fiber – the expanding medium, *IEEE Commun. Mag.*, 25; 4, 33–43.
7. Kao K. C. and Hockham G. A. (1966). Dielectric-fibre surface waveguides for optical frequencies, *Proc. IEE*, 113; 7, 1151–1158.
8. Suematsu Y. and Iga K. (2006). *Introduction to Optical Fiber Communications* (in Japanese), Tokyo: Ohmsha.
9. Kapron F. P., Keck D. B. and Maurer R. D. (1970). Radiation losses in glass optical waveguides, *Appl. Phys. Lett.*, 17; 10, 423–425.
10. Itaya Y. (2008). Photonic Challenges Toward Future Broadband Society, *ACP 2008*, Plenary talk (Shanghai, China).
11. Morioka T., Jinno M., Takara H. and Kubota H. (2011). Revolutional optical transport technology in the future (in Japanese), *NTT Technol. J.*, 23; 3, 32–36.
12. Morioka T., Awaji Y., Ryf R., Winzer P. and Richardson D. (2012). Enhancing optical communications with brand new fibers, *IEEE Commun. Mag.*, February, S31–S42.
13. Takara H., Sano A., Kobayashi T., Kubota H., Kawakami H., Matsuura A., Miyamoto Y., Abe Y., Ono H., Shikama K., Goto Y., Tsujikawa K., Sasaki Y., Ishida I., Takenaga K., Matsuo S., Saitoh K., Koshiba M. and Morioka T. (2012). 1.01-Pb/s (12 SDM/222 WDM/456 Gb/s) Crosstalk-managed Transmission with 91.4-b/s/Hz Aggregate Spectral Efficiency, *ECOC2012*, Th.3.c.1 (Amsterdam).
14. Gregory G. Raleigh and Cioffi J. M. (1998). Spatio-temporal coding for wireless communication, *IEEE Trans. Commun.*, 46; 3, 357–366.
15. Randel S., Sierra A., Mumtaz S., Tulino A., Ryf R., Winzer P. J., Schmidt C. and Essiambre R. J. (2012). Adaptive MIMO Signal Processing for Mode-division Multiplexing, *OFC2012*, OW3D.5 (Los Angeles, CA).

16. Koester C. J. and Snitzer E. (1964). Amplification in a fiber laser, *Appl. Opt.*, 3, 1182–1186.

17. Stone J. and Burrus C. A. (1973). Neodymium-doped silica lasers in end-pumped fiber geometry, *Appl. Phys. Lett.*, 23, 388–389.

18. Mears R. J., Reekie L., Poole S. B. and Payne D. N. (1985). Neodymium-doped silica single-mode fibre laser, *Electron. Lett.*, 21, 738–740.

19. Roberts K., Beckett D., Boertjes D., Berthold J. and Laperle C. (2010). 100G and beyond with digital coherent signal processing, *IEEE Commun. Mag.*, July, 62–69.

20. Okoshi T. and Kikuchi K. (1980). Frequency stabilization of semiconductor lasers for heterodyne-type optical communication schemes, *Electron. Lett.*, 16; 5, 179–181.

21. Favre F. and LeGuen D. (1980). High frequency stability of laser diode for heterodyne communication systems, *Electron. Lett.*, 16; 18, 709–710.

22. Yamazaki E., Yamanaka S., Kisaka Y., Nakagawa T., Murata K., Yoshida E., Sakano T., Tomizawa M., Miyamoto Y., Matsuoka S., Matsui J., Shibayama A., Abe J., Nakamura Y., Noguchi H., Fukuchi K., Onaka H., Fukumitsu K., Komaki K., Takeuchi O., Sakamoto Y., Nakashima H., Mizuochi T., Kubo K., Miyata Y., Nishimoto H., Hirano S. and Onohara K. (2011). Fast optical channel recovery in field demonstration of 100-Gbit/s Ethernet over OTN using real-time DSP, *Opt. Express*, 19; 14, 13179–13184.

23. Shannon C. E. (1949). Communication in the presence of noise, *Proc. IRE*, 37; 2, 10–21.

24. Duand L. B. and Lowery A. J. (2010). Improved single channel backpropagration for inta-channel fiber nonlinearity compensation in long-haul optical communication systems, *Opt. Express*, 18:2, 17075–17088.

25. Li L., Tao Z., Liu L., Yan W., Oda S., Hoshida T. and Rasmussen J. C. (2010). Nonlinear polarization crosstalk canceller for dual-polarization digital coherent receivers, OFC 2010, OWE3 (San Diego, CA, March 2010).

26. Mitra P. P. and Stark J. B. (2001). Nonlinear limits to the information capacity of optical fibre communications, *Nature*, 411, 1027–1030.

27. Essiambre R.-J. and Tkach R. W. (2012). Capacity trends and limits of optical communication networks, *Proc. IEEE*, 100; 5, 1035–1055.

28. Roberts P. J., Couny F., Sabert H., Mangan B. J., Williams D. P., Farr L., Mason M. W., Tomlinson A., Birks T. A., Knight J. C. and Russell P. S. J. (2005). Ultimate low loss of hollow-core photonic crystal fibres, *Opt. Express*, 13; 1, 236–244.

29. Collins B. C. (200). Wavelength Selectable Switches and Future Photonic Network Applications, *PS 2009*, FrII2–4 (Pisa, Italy).

30. Baran P. (2002). The beginning of packet switching: some underlying concepts, *IEEE Commun. Mag.*, July, 42–48.

31. Miyahara H. (2001). *Osaka University-NTT Meeting*.

32. Recommendation ITU-T G.694.1 (02/2012). Figure I.1.

33. Reduction of CO_2 by the ICT in 2020. Working group for environmental problem, Ministry of Internal Affairs and Communications in Japan, May 2010 (www.soumu.go.jp/main_content/000065258.pdf).

34. Hogari K., Yamada Y. and Toge K. (2010). Design and performance of ultra-high-density optical fiber cable with rollable optical fiber ribbons, *Opt. Fiber Technol.*, 16, 257–263.

35. Muller S., Bechtolsheim A. and Hendel A. (2007). HSSG Speeds and Feeds Reality Check, *IEEE 802.3 Higher Speed Study Group Meeting*, January.

Chapter 2

1. Kani J., Bourgart F., Cui A., Rafel A., Campbell M., Davey R. and Rodrigues S. (2009). Next-generation PON – Part I: technology roadmap and general requirements, *IEEE Commun. Mag.*, November, 43–49.
2. Chanclou P., Cui A., Geihardt, Nakamura H. and Nesset D. (2012). Network operator requirements for the next generation of optical access networks, *IEEE Networks*, March/April, 8–14.
3. ITU-T G.983.1 (2001). Broadband Passive Optical Networks (B-PON).
4. ITU-T G.984.1 (2008). Gigabit-Capable Passive Optical Networks (G-PON): General Characteristics.
5. IEEE 802.3ah (2004). Ethernet for the First Mile.
6. IEEE 802.3av (2009) 10Gb/s Ethernet Passive Optical Network.
7. ITU-T G987, 10 Gb/s-Capable Passive Optical Network (XG-PON) Systems.
8. ITU-T G.984.3 (2008). Gigabit-Capable Passive Optical Networks (G-PON): Transmission Convergence Layer Specification.
9. ITU-T G.984.2 (2003). Gigabit-Capable Passive Optical Networks (GPON): Physical Media Dependent (PMD) Layer Specification.
10. IEEE Standard 802.3, 2005 Edition, Carrier Sense Multiple Access with Collision Detection (CSMA/CD) access method and physical layer specifications.
11. Effenberger F., Cleary D., Haran O., Kramer G., Ding R., Oron M. and Pfeiffer T. (2007). An introduction to PON technologies, *IEEE Commun. Mag.*, March, S17–S25.
12. Kani J. and Suzuki K. (2009). Standardization of next-generation 10G-PON systems, *NTT Technol. J.*, 21; 9, 90–93 (in Japanese).
13. Tanaka K., Agata A. and Horiuchi Y. (2010). IEEE 802.3av 10G-EPON Standardization and Its Research and Development Status, *J. Lightwave Technol.*, 28; 4, 651–661.
14. Miki N. and Kumozaki K. (2007). Chapter 5, in Lam C. ed., *Passive Optical Networks: Principles and Practice*, Academic Press.
15. Basic Lecture Series (2005). GE-PON Technology, *NTT Technol. J.*, 17; 10 (in Japanese).
16. Bosco G., Carena A., Curri V., Poggiolini P. and Forghieri F. (2010). Performance limits of Nyquist-WDM and CO-OFDM in high-speed PM-QPSK, *IEEE Photonic Technol. Lett.*, 22; 15, 1129–1131.
17. Chandrasekhar S. and Liu X. (2012). OFDM based supperchannel transmission technology, *J. Lightwave Technol.*, 30; 24, 3816–3823.
18. Yan M., Tao Z., Li L., Hoshida T. and Rasmussen J. C. (2012). Experimental Comparison of No-Guard-Interval-OFDM and Nyquist-WDM Superchannels, OFC/NFOEC 2012, OTh1B.2 (Los Angeles, CA, March 2012).
19. Iwatsuki K. and Kani J. (2009). Applications and technical issues of wavelength-division multiplexing passive optical networks with colorless optical network units, *J. Opt. Commun. Netw.*, 1; 4, C17–24.
20. Choi K-. M., Baik J-. S. and Lee C.-H. (2005). Broad-band light source using mutually injected Fabry–Perot laser diodes for WDM-PON, *IEEE Photonic Technol Lett.*, 17; 12, 2529–2531.
21. Ji H.-C., Yamashita I. and Kitayama K. (2008). Bidirectional transmission of downstream broadcast and upstream baseband signals over a single wavelength in WDM-PON using mutually injected FPLDs and RSOA, *IEEE Photonic Technol. Lett.*, 20; 20, 1709–1711.

22. ITU-T G.987 (2013). 40-Gigabit-capable passive optical networks (NG-PON2): General requirements.

23. Luo Y., Zhou X., Yan X., Peng G., Qian Y. and Ma Y. (2013). Time- and wavelength-division multiplexed passive optical network (TWDM-PON) for next-generation PON stage 2 (NG-PON2), *IEEE Lightwave Technol.*, 31: 4, 587–593.

24. Nakamura H., Tamaki S., Hara K., Kimura S. and Hadama H. (2011). 40Gbit/s λ-tunable stacked-WDM/TDM-PON using dynamic wavelength and bandwidth allocation, OFC2011, OThT4 (Los Angeles, CA).

25. Chang R. W. (1966). Synthesis of band-limited orthogonal signals for multi-channel data transmission, *Bell Syst. Tech. J.*, 45, 1775–1796.

26. Nakajima S., Arita T. and Higuchi K. (2012). *How Mobile Phones Work* (in Japanese), Fig. 6.3, Tokyo: Nikkei BP.

27. Shieh W. and Djordjevic I. (2010). *OFDM for Optical Communications*, Academic Press.

28. Yoshida Y., Ishii K., Maruta A., Akiyama Y., Yoshida T., Suzuki N., Koguchi K., Nakagawa J., Mizuochi T. and Kitayama K. (2012). Experimental Demonstration of 2xONU 30Gbps Digitally-Supported-Coherent IFDMA-PON Uplink, OFC2012, OW3B.5

29. Ministry of Internal Affairs and Communications, Japan, Taskforce report (in Japanese) (www.soumu.go.jp/main_content/000065258.pdf).

30. Baliga J., Ayre R., Hinton K. and Tucker R. S. (2011). Energy consumption in wired and wireless access networks, *IEEE Commun. Mag.*, June, 70–77.

31. Ishii K., Akiyama Y., Suzuki K and Nakagawa J. (2011). Low-power consumption optical receiver technology toward next-generation PON (in Japanese), *IEICE Society Conference*, Symposium BI-5, entitled Future Perspectives of Photonic Switching Node (Sapporo, Japan, September 2011).

32. ITU-T Series G Supplement 45, GPON Power Conservation, 05/2009.

33. Kubo R., Kani J., Fujimoto Y., Yoshimoto N. and Kumozaki K. (2010). Adaptive power saving mechanism for 10 gigabit class PON systems, *IEICE Trans. Commun.*, 2; E93.B, 280–288.

Chapter 3

1. Kao K. C. and Hockham G. A. (1966). Dielectric-fibre surface waveguides for optical frequencies, *Proc. IEE*, 113; 7, 1151–1158.

2. Nagayama K., Kakui M., Matsui M., Saitoh I. and Chigusa Y. (2002). Ultra-low-loss (0.1484 dB/km) pure silica core fibre and extension of transmission distance, *Electron. Lett.*, 38; 20, 1168–1169.

3. Roberts P. J., Couny F., Sabert H., Mangan B. J., Williams D. P., Farr L., Mason M. W., Tomlinson A., Birks T. A., Knight J. C. and Russell P. S. J. (2005). Ultimate low loss of hollow-core photonic crystal fibres. *Opt. Express*, 13; 1, 236–244.

4. Hecht J. (2002). *Understanding Fiber Optics*, fourth edn., Colombus, OH: Prentice Hall.

5. Kikuchi K. (1997). *Fundamentals of Optical Fiber Communications* (in Japanese), Tokyo: Shokodo.

6. Gloge D. (1971). Weakly guiding fibers, *Appl. Opt.*, 10;10, 2252–2258.

7. ITU-T Recommendation G.652 (06/2005).

8. ITU-T Recommendation G.655 (03/2003) and (11/2009).

9. ITU-T Recommendation G.657 (12/2006).

10. Marcuse D. (1977). Loss analysis of single-mode fiber splices, *Bell Syst. Tech. J.*, 56; 5, 703–718.

11. Snyder A. W. and Love J. (1983). *Optical Waveguide Theory*, Chapter 36, Springer.

12. Matsuo S., Ikeda M., Kuwaki H. and Himeno K. (2005). Low-bending-loss and low-splice-loss single-mode fibers employing a trench index profile, *IEICE Trans.*, E88; 5, 889–895.

13. Sakai J., K. Kitayama K., Ikeda M., Kato Y. and Kimura T. (1978). Design considerations of broadband dual mode optical fibers, *IEEE Trans. Microwave Theory Technol.*, 26, 658–665.

14. Kitayama K., Kato Y., Seikai S. and Uchida N. (1981). Structural optimization for two-mode fiber: theory and experiment, *IEEE J. Quantum Electron.*, 17; 6, 1057–1063.

15. Kitayama K., Kato Y., Seikai S., Uchida N. and Akiyama M. (1982). Transmission bandwidth of the two-mode fiber link, *IEEE J. Quantum Electron.*, 18; 11, 1871–1876.

16. Qian D., Huang M.-F., Ip E., Huang Y.-K., Shao Y., Hu J. and Wang T. (2011). 101.7-Tb/s ($370 \times$ 294-Gb/s) PDM-128QAM-OFDM Transmission over $3 \times$ 55-km SSMF Using Pilot-Based Phase Noise Mitigation, *OFC2011*, PDPB5, 2011 (Los Angeles, CA).

17. Nakazawa M., Yoshida Y., Maruta A. and Kitayama K. (2012). On the Computational Complexity of MIMO Processing in Mode Division Multiplexing Transmission over 2-mode Fiber, *OECC2012* (Busan, Korea, July 2012).

18. Kato Y., Kitayama K., Seikai S. and Uchida N. (1982). Modal equalization for two-mode fibre link using a step-index fibre, *Electron. Lett.*, 18: 9, 356–358.

19. Maruyama R., Shoji T., Kuwaki N., Matsuo S., Sato K. and Ohashi M. (2013). Design and fabrication of long DMD maximally flattened two-mode optical fibres suitable for MIMO processing, *ECOC* 2013, Mo.4.A.3, (London, September 2013).

20. Okamoto K. and Okoshi T. (1976). Analysis of wave propagation in optical fibers having core with a-power refractive-index distribution and uniform cladding, *IEEE Trans. Microwave Theory Technol.*, 24; 7, 416–421.

21. The Japan Society of Applied Physics, 1987. *Fundamentals of Laser Diodes* (in Japanese), Tokyo: Ohmsha.

22. Malitson I. H. (1965). Interspecimen comparison of the refractive index of fused silica, *J. Opt. Soc. Am.*, 55, 1205–1209.

23. ITU-T Recommendation G.656 (03/2003) and (07/2010).

24. Agrawal G. P. (2001). *Nonlinear Fiber Optics*, third edn., Chapter 3, San Diego, CA: Academic Press.

25. Grüner-Nielsen L., Wandel M., Kristensen P., Jørgensen C., Jørgensen L. V., Edvold B., Pálsdóttir B. and Jakobsen D. (2005). Dispersion-compensating fibers, *J. Lightwave Technol.*, 23; 11, 3566–3579.

26. Yamazaki E., Yamanaka S., Kisaka Y., Nakagawa T., Murata K., Yoshida E., Sakano T., Tomizawa M., Miyamoto Y., Matsuoka S., Matsui J., Shibayama A., Abe J., Nakamura Y., Noguchi H., Fukuchi K., Onaka H., Fukumitsu K., Komaki K., Takeuchi O., Sakamoto Y., Nakashima H., Mizuochi T., Kubo K., Miyata Y., Nishimoto H., Hirano S. and Onohara K. (2011). Fast optical channel recovery in field demonstration of 100-Gbit/s Ethernet over OTN using real-time DSP, *Opt. Express*, 19; 14, 13179–13184.

Chapter 4

1. Kikuchi K. (1997). *Fundamentals of Optical Fiber Communications* (in Japanese), Chapter 5, Tokyo: Shokodo.

2. Green Jr. P. E. (1993). *Fiber Optical Networks*, Chapter 8, Prentice Hall.

3. Okoshi T. and Kikuchi K. (1988). *Coherent Optical Fiber Communications*, Boston, MA: Kluwer Academic.

4. Gnauck A. H. and Winzer P. J. (2005). Optical phase-shift-keyed transmission, *J. Lightwave Technol.*, 23; 1, 115–130.

5. Winzer P. J. and Essiambre R.-J. (2006). Advanced modulation formats, *Proc. IEEE*, 94; 5, 952–985.

6. Essiambre R.-J., Cramer G., Winzer P. J., Foschini G. J. and Goebel B. (2010). Capacity limits of optical fiber networks, *J. Lightwave Technol.*, 28; 4, 662–701.

7. Kametani S., Sugihara T. and Mizuochi T. (2009). 6-QAM modulation by polar coordinate transformation with a single dual drive Mach-Zehnder Modulator, *OFC2009*, OWG6 (San Diego, CA).

8. Agrawal G. P. (2001). *Nonlinear Fiber Optics*, third edn., Chapter 9, San Diego, CA: Academic Press.

9. Cotter D. (1983). Stimulated Brillouin scattering in monomode optical fiber, *J. Opt. Commun.*, 4; 1, 10–19.

Chapter 5

1. Green Jr. P. E. (1993). *Fiber Optical Networks*, Chapter 3, Prentice Hall.

2. Rawson E. G. (1978). Star couplers using fused biconically tapered multimode fibres, *Electron. Lett.*, 14; 9, 274–275.

3. Hida Y., Inoue Y., Hanawa F., Fukumitsu T., Enomoto Y. and Takato N. (1999). *IEEE Photonics Technol. Lett.*, 11; 1, 96–98.

4. Yariv A. (1975). *Quantum Electronics*, second edn., Chapter 11, New York: John Wiley & Sons.

5. Morioka T., Kawanishi S., Mori K. and Saruwatari M. (1994). Nearly penalty-free, 4 ps supercontinuum Gbit/s pulse generation over 1535–1560 nm, *Electron Lett.*, 30; 10, 790–791.

6. Okuno T., Onishi M. and Nishimura M. (1998). Generation of ultra-broad-band supercontinuum by dispersion-flattened and decreasing fiber, *IEEE Photonic Technol. Lett.*, 10; 1, 72–74.

7. Mori K., Takara H. and Kawanishi S. (2001). Analysis and design of supercontinuum pulse generation in a single-mode optical fiber, *J. Opt. Soc. Am. B*, 18; 12, 1780–1791.

8. Sotobayashi H., Chujo W. and Kitayama K. (2004). Highly spectral-efficient optical code-division multiplexing transmission system, *IEEE J. Select. Topics Quantum Electron.*, 10; 2, 250–258.

9. Miyoshi Y., Namiki S. and Kitayama k. (2112). Performance evaluation of resolution-enhanced ADC using optical multiperiod transfer functions of NOLMs, *IEEE J. Select. Topics Quantum Electron.*, 18; 2, 779–784.

10. Jiang Z., Seo D. S., Yang S. D., Leaird D. E., Roussev R. V., Langrock C., Fejer M. M. and Weiner A. M. (2005). Four-user 10-Gb/s spectrally phase-coded O-CDMA system operating at 30 fJ/bit, *IEEE Photonics Technol. Lett.*, 17, 705–707.

11. Wang X., Hamanaka T., Wada N. and Kitayama K. (2005). Dispersion-flattened- fiber based optical thresholder for multiple-access-interference suppression in OCDMA system, *Opt. Express*, 13; 14, 5499–5505.

12. Miniscalco W. J. (1991). Erbium-doped glasses for fiber amplifiers at 1500 nm, *J. Lightwave Technol.*, 24; 25, 888–890.

13. Kikuchi K. (1997). *Fundamentals of Optical Fiber Communications* (in Japanese), Chapter 5, Tokyo: Shokodo.

14. Ono H., Yamada M. and Ohishi Y. (1997). Gain-flattened Er3+-doped fiber amplifier for a WDM signal in the 1.57–1.60-mm wavelength region, *IEEE Photonic Technol. Lett.*, 9; 5, 596–598.

15. Nogawa M., Katsurai H., Nakamura M., Kamitsuna H. and Ohtomo Y. (2011). IC Technology for 10 Gb/s Burst-mode receiver, *NTT Tech. J.*, 1, 31–35 (in Japanese).

16. Yoshima S., Tanaka Y., Kataoka N., Wada N., Nakagawa J. and Kitayama K., Full-duplex, extended-reach 10G-TDM-OCDM-PON system without en/decoder at ONU, *J. Lightwave Technol.*, in press.

17. Nakagawa J., Nogami M., Suzuki N., Noda M., Yoshima S. and Tagami H. (2010). 10.3-Gb/s burst-mode 3R receiver incorporating full AGC optical receiver and 82.5 GS/s over-sampling CDR for 10 G-EPON systems, *IEEE Photon. Technol. Lett.*, 22; 7, 471–473.

18. Takesue H. and Sugie T. (2008). Wavelength channel data rewrite using saturated SOA modulator for WDM networks with centralized light sources, *J. Lightwave Technol.*, 26; 1, 99–107.

19. Kim S. Y., Jun S. B., Takushima Y., Son E. S. and Chung Y. C. (2007). Enhanced performance of RSOA-based WDM PON by using Manchester coding, *J. Opt. Networking*, 6; 6, 624–630.

20. Ji H.-C., Yamashita I. and Kitayama K. (2008). Bidirectional transmission of downstream broadcast and upstream baseband signals over a single wavelength in WDM-PON using mutually injected FPLDs and RSOA, *IEEE Photonic Technol. Lett.*, 20; 20, 1709–1711.

Chapter 6

1. Dixon R. C. (1994). *Spread Spectrum Systems with Commercial Applications Spread Spectrum*, Chapter 2, New York: John Wiley & Sons.

2. Shannon, C. E. (1949). Communication in the presence of noise, *Proc. IRE*, 37; 2, 10–21.

3. Kitayama K., Sotobayashi H. and Wada N. (1999). Optical code division multiplexing (OCDM) and its applications to photonic networks (Invited), *IEICE A.*, E82-A, 2616–2626.

4. Tur M. (1997). Private communications, Tel-Aviv University.

5. Sampson D. D., Pendock G. J. and Griffin R. A. (1997). Photonic code-division multiple-access communications, *Fiber Integrated Optics*, 16; 2, 129–157.

6. Delisle C. and Cielo P. (1975). Application de la modulation spectrale la transmission de l'information. *Can. J. Phys.*, 53, 1047–1053.

7. Prucnal P. R., Santoro M. A. and Fan T. R. (1986). Spread spectrum fiber-optic local area network using optical processing, *J. Lightwave Technol.*, 4; 5, 547–554.

8. Scott R. P., Cong W., Hernandez V. J., Li K., Kolner B. H., Heritage J. P. and Yoo S. J. B. (2005). An eight-user time-slotted SPECTS O-CDMA testbed: demonstration and simulations, *J. Lightwave Technol.*, 23; 10, 3232–3240.

9. Wang X. and Kitayama K. (2004). Analysis of beat noise in coherent and incoherent time-spreading OCDMA, *J. Lightwave Technol.*, 22; 10, 2226–2235.

Chapter 7

1. Davies P. A. and Shaar A. A. (1983). Asynchronous multiplexing for an optical fiber local area network, *Electronic Lett.*, 19, 390–392.

2. Dixon R. C. (1994). *Spread Spectrum Systems with Commercial Applications Spread Spectrum*, Chapter 3, New York: John Wiley & Sons.

3. Prucnal P. R., Santoro M. A. and Fan T. R. (1986). Spread spectrum fiber-optic local area network using optical processing, *J. Lightwave Technol.*, 4; 5, 547–554.

4. Goh T., Yasu M., Hattori K., Himeno A., Okuno M. and Ohmori Y. (1998). Low-loss and high-extinction-ratio silica-based strictly nonblocking 16x16 thermooptic matrix switch, *IEEE Photon. Technol. Lett.*, 10; 6, 810–812.

5. Wada N. and Kitayama K. (1999). A 10 Gb/s optical code division multiplexing using 8-chip optical bipolar code and coherent detection, *J. Lightwave Technol.*, 17; 10, 1758–1765.

6. Petropoulos P., Ibsen M., Ellis A. D. and Richardson D. J. (2001). Rectangular pulse generation based on pulse reshaping using a superstructured fiber Bragg grating, *J. Lightwave Technol.*, 19; 5, 746–752.

7. Wang X., Matsushima K., Nishiki A., Wada N. and Kitayama K. (2004). High reflectivity superstructured FBG for coherent optical code generation and recognition, *Opt. Express*, 12; 22, 5457–5468.

8. Wang X., Matsushima K., Kitayama K., Nishiki A., Wada N. and Kubota F. (2005). High-performance optical code generation and recognition by use of a 511-chip, 640 Gchip/s phase-shifted superstructured fiber Bragg grating, *Opt. Lett.*, 30; 4, 355–357.

9. Hill K. O. and Meltz G. (1997). Fiber Bragg grating technology fundamentals and overview, *J. Lightwave Technol.*, 15; 8, 1263–1276.

10. Cincotti G., Wada N., Kataoka N. and Kitayama K. (2006). Characterization of a full encoder/decoder in the AWG configuration for code-based photonic routers. Part I: modeling and design, *J. Lightwave Technol.*, 24; 1, 103–112.

11. Wada N., Cincotti G., Yoshima S., Kataoka N. and Kitayama K. (2006). Characterization of a full encoder/decoder in the AWG configuration for code-based photonic routers. Part II: experiments and applications, *J. Lightwave Technol.*, 24; 1, 113–121.

12. Wang X., Wada N., Cincotti G., Miyazaki T. and Kitayama K. (2006). Demonstration of over 128-Gb/s-capacity (12-User X 10.71-Gb/s/User) asynchronous OCDMA using FEC and AWG-based multiport optical encoder/decoders, *IEEE Photonics Technol. Lett.*, 18; 15, 1603–1605.

13. Yoshima S., Nakagawa N., Kataoka N., Suzuki N., Noda M., Nogami M, Nakagawa J. and Kitayama K. (2010). 10 Gb/s-based PON over OCDMA uplink burst transmission using SSFBG encoder/multi-port decoder and burst-mode receiver, *J. Lightwave Technol.*, 28; 4, 365–371.

14. Omichi K., Nomura R., Matsumoto R., Shimizu S., Terada Y., Sakamoto A., Yamauchi R., Wada N. and Kitayama K. (2012). Superstructured FBG based optical encoder/decoder for highly-confidential 40 Gbps telecommunication network, OFS 2012, *SPIE*, 8421, 1–3 (Beijing, China).

15. Kataoka N., Wang X., Wada N., Cincotti G., Terada Y. and Kitayama K. (2009). 8 X 8 full-duplex demonstration of asynchronous, 10Gbps, DPSK-OCDMA system using apodized SSFBG and multi-port en/decoder, *ECOC 2009*, 6.5.5 (Vienna, Austria, September 2009).

16. Kataoka N., Wada N., Wang X., Cincotti G., Sakamoto A., Terada Y., Miyazaki T. and Kitayama K. (2009). Field trial of duplex, 10 Gbps X 8-user DPSK-OCDMA system using a single 16 X 16 multi-port encoder/decoder and 16-level phase-shifted SSFBG encoder/decoders, *J. Lightwave Technol.*, 27; 3, 299–305.

Chapter 8

1. Kitayama K., Sasaki M., Araki S., Tsubokawa M., Tomita A., Inoue K., Harasawa K., Nagasako Y. and Takada A. (2011). Security in photonic networks: threats and security enhancement, *J. Lightwave Technol.*, 29; 21, 3210–3222.
2. Zeltsan Z. (2005). ITU-T recommendation X.805 and its application to NGN, *ITU/IETF Workshop on NGN*.
3. Kodama T. (2011). Studies on Secure M-ary Optical Code Division Multiplexing Using a Single Multi-port Encoder/Decoder, Doctoral dissertation.
4. Wang X., Wada N., Miyazaki T., Cincotti G. and Kitayama K. (2007). Asynchronous multiuser coherent OCDMA system with code-shift-keying and balanced detection, *IEEE Select. Topics Quantum Electron.*, 13; 5, 1463–1470.
5. Wang X., Wada N., Miyazaki T. and Kitayama K. (2006). Coherent OCDMA system using DPSK data format with balanced detection, *IEEE Photonics Technol. Lett.*, 18; 7, 826–828.
6. Kataoka N., Wada N., Cincotti G., Kitayama K. and Miyazaki T. (2007). A novel multiplexed optical code label processing with huge number of address entry for scalable optical packet switched network, *ECOC2007*, Tu3.2.6 (Berlin).
7. Cincotti G., Sacchieri V., Manzacca G., Kataoka N., Wada N. and Kitayama K. (2008). Physical layer security: all-optical cryptography in access networks, *ICTON2008* (Athens, Greece).
8. Cincotti G., Manzacca G., Sacchieri V., Wang X., Wada N. and Kitayama K. (2008). Secure OCDM transmission using a planar multiport encoder/decoder, *J. Lightwave Technol.*, 26; 13, 1798–1806.
9. Shake T. (2005). Security performance of optical CDMA against eavesdropping confidentiality performance of spectral-phase-encoded optical CDMA, *J. Lightwave Technol.*, 23; 2, 655–670.
10. Wu B. B., Prucnal P. R. and Narimanov E. E. (2006). Secure transmission over an existing public WDM lightwave network, *IEEE Photonic Technol. Lett.*, 18; 17, 1870–1872.
11. Menendez R., Agarwal A., Toliver P., Jackel J. and Etemad S. (2007). Direct optical processing of M-ary code-shift keyed spectral phase encoded OCDMA, *J. Opt. Networks*, 6; 5, 442–450.
12. Kodama T., Nakagawa N., Kitayama K., Kataoka N., Wada N., Cincotti G., Wang X. and Miyamazaki T. (2010). Secure 2.5Gbit/s, 16-ary OCDM block-ciphering with XOR using a single multi-port en/decoder, *J. Lightwave Technol.*, 28; 1, 181–187.
13. Kodama T., Kataoka N., Wada N., Cincotti G., Wang X., Miyazaki T. and Kitayama K. (2010). High-security 2.5 Gbps, polarization multiplexed 256-ary OCDM using a single multi-port encoder/decoder, *Opt. Express*, 18; 20, 21376–21385.
14. Kodama T., Kataoka N., Wada N., Cincotti G., Wang X. and Kitayama K. (2011). 4096-Ary OCDM/OCDMA system using multidimensional PSK codes generated by a single multiport en/decoder, *J. Lightwave Technol.*, 29; 22, 3372–3380.

Chapter 9

1. Wang X., Wada N., Cincotti G., Miyazaki T. and Kitayama K. (2006). Demonstration of over 128-Gb/s-capacity (12-User X 10.71-Gb/s/User) asynchronous OCDMA using FEC and AWG-based multiport optical encoder/decoders, *IEEE Photonics Technol. Lett.*, 18; 15, 1603–1605.
2. Shieh W. and Djordjevic I. (2010). *OFDM for Optical Communications*, Academic Press.
3. Yoshima S., Nakagawa N., Kataoka N., Suzuki N., Noda M., Nogami M., Nakagawa J. and Kitayama K. (2010). 10 Gb/s-based PON over OCDMA uplink burst transmission using SSFBG encoder/multi-port decoder and burst-mode receiver, *J. Lightwave Technol.*, 28; 4, 365–371.
4. Cincotti G., Kataoka N., Wada N., Wang X., Miyazaki T. and Kitayama K. (2009). Demonstration of asynchronous, 10Gbps OCDMA PON system with colorless and sourceless ONUs, *ECOC 2009*, 6.5.7 (Vienna).
5. Yoshima S., Tanaka Y., Kataoka N., Wada N., Nakagawa J. and Kitayama K., Full-duplex, extended-reach 10G-TDM-OCDM-PON system without en/decoder at ONU, *J. Lightwave Technol.*, in press.
6. Kodama T., Tanaka Y., Yoshima S., Kataoka N., Nakagawa J., Shimizu S., Wada N. and Kitayama K. (2013). Scaling the system capacity and reach of 10G-TDM- OCDM-PON system without en/decoder at ONU, *J. Opt. Commun. Networks*, 5; 2, 134–143.
7. Kitayama K., Wang X. and Wada N. (2006). OCDMA over WDM PON: A solution path to gigabit-symmetric FTTH, *J. Lightwave Technol.*, 24; 4, 1654–1662.
8. Wang X., Wada N., Cincotti G., Miyazaki T. and Kitayama K. (2007). Field trial of 3-WDM X 10-OCDMA X 10.71 Gbps, asynchronous, WDM/DPSK-OCDMA using hybrid E/D without FEC and optical thresholding, *J. Lightwave Technol.*, 25; 1, 207–215.
9. Kataoka N., Cincotti G., Wada N. and Kitayama K. (2011). Demonstration of asynchronous, 40Gbps X 4-user DPSK-OCDMA transmission using a multi-port encoder/decoder, *Opt. Express*, 19; 26, B965–970.
10. Kataoka N., Cincotti G., Wada N. and Kitayama K. (2011). 2.56 Tbps (40-Gbps X 8-wavelength X 4-OC X 2-POL) asynchronous WDM-OCDMA-PON using a multi-port encoder/decoder, *ECOC 2011*, Th.13.B.6 (Geneva, September 2011).
11. Omichi K., Nomura R., Matsumoto R., Shimizu S., Terada Y., Sakamoto A., Yamauchi R., Wada N. and Kitayama K. (2012). Superstructured FBG based optical encoder/decoder for highly- confidential 40 Gbps telecommunication network, OFS 2012, *SPIE*, 8421, 1–3 (Beijing, China). Matsumoto R., Kodama T., Shimizu S., Nomura R., Omichi K., Wada N. and Kitayama K. (2013). Apodized SSFBG en/decoder for 40G-OCDM-PON system, OECC/PS 2013, MP1-6 (Kyoto, Japan).
12. Matsumoto R., Kodama T., Shimizu S., Nomura R., Omichi K., Wada N. and Kitayama K. (2013). Cost-effective, asynchronous 4 X 40Gbps full-duplex OCDMA demonstrator using apodized SSFBGs and a multi-port encoder/decoder, OFC2013, OW4D.7 (Anaheim, CA).
13. Kitayama K. (1994). Novel spatial spread spectrum based fiber optic CDMA networks for image transmission, *IEEE J. Select. Areas Commun.*, 12; 4, 762–772.
14. Kitayama K., Nakamura M., Igasaki Y. and Kaneda K. (1997). Image fiber-optic two-dimensional parallel links based upon optical space-CDMA: experiment, *J. Lightwave Technol.*, 15; 1, 202–212.

15. Nakamura M. and Kitayama K. (1998). System performances of optical space code-division multiple-access-based fiber-optic two-dimensional parallel data link, *Appl. Opt.*, 37; 14, 2915–2924.

16. Nakamura M., Kitayama K., Igasaki Y., Shamoto N. and Kaneda K. (2002). Image fiber optic space-CDMA parallel transmission experiment using 8 X 8 VCSEL/PD arrays, *Appl. Opt.*, 41; 32, 6901–6906.

17. Nakamura M. and Kitayama K. (2001). Two-dimensional erbium-doped image fiber amplifier (EDIFA), *J. Select. Topics Quantum Electron.*, 7; 3, 434–438.

Chapter 10

1. Kaneko S., Kataoka N., Miki N., Kimura H., Wada N. and Kitayama K. (2011). Optical code-division-multiple access: thorough comparison with TDM- and DWDM-PONs for future PON systems toward 100Gbit/s/ONU, unpublished work.

2. Kataoka N., Cincotti G., Wada N. and Kitayama K. (2011). Demonstration of asynchronous, 40Gbps X 4-user DPSK-OCDMA transmission using a multi-port encoder/decoder, *ECOC2011*, Tu.5.C.4 (Geneva).

3. Wang X., Wada N., Kataoka N., Miyazaki T., Cincotti G. and Kitayama K. (2007). 100 km field trial of 1.24 Tbit/s, spectral efficient asynchronous 5 WDM × 25 DPSK-OCDMA using one set of 50 × 50 ports large scale en/decoder, *OFC2007*, PDP14 (Anaheim, CA.).

4. Tucker R. S. (2011). Green optical communications – part 1: Energy limitations in transport, *IEEE J. Select. Topics Quantum Electron.*, 17; 2, 245–260.

5. Cisco, CRS-3 data sheet (2010). (www.cisco.com/en/US/prod/collateral/routers/ps5763/data_sheet_c78-408226.html).

6. Kitayama K., Wada N. and Sotobayashi H. (2000). Architectural considerations for photonic IP router based upon optical code correlation (Invited), *IEEE J. Lightwave Technol.*, 18; 12, 1834–1844.

7. Nozaki K., Shinya A., Matsuo S., Segawa T., Sato T., Kawaguchi Y., Takahashi R. and Notomi M. (2012). Ultralow-power all-optical RAM based on nanocavities, *Nature Photonics*, 26; February, 1–5.

8. Nakahara T., Suzaki Y., Urata R., Segawa T., Ishikawa H. and Takahashi R. (2011). Enhanced multi-hop operation using hybrid optoelectronic router with time-to-live-based selective forward error correction, *Opt. Express*, 12; 19, B301–307.

9. Takushima Y. and Kikuchi K. (1994). Photonic switching using spread spectrum technique, *Electron. Lett.*, 30, 436–438.

10. Vaughn M. D. and Blumenthal D. J. (1997). All-optical updating of subcarrier encoded packet headers with simultaneous wavelength conversion of baseband payload in semiconductor amplifiers, *IEEE Photon. Technol. Lett.*, 9, 827–829.

11. Way W. I., Lin Y.-M. and Chang G.-K. (2000). A novel optical label swapping technique using erasable optical single-sideband subcarrier label, *OFC1999*, WD6 (Baltimore, MD).

12. Cardakli M. C., Gurkan D., Havstad S. A. and Willner A. E. (2000). Variable-bit-rate header recognition for reconfigurable networks using tunable fiber-Bragg-gratings as optical correlators, *OFC2000*, TuN2 (Baltimore, MD).

13. Kitayama K. and Wada N. (1999). Photonic IP routing, *IEEE Photonic Technol. Lett.*, 11, 1689–1691.

14. Wada N., Cincotti G., Yoshima S., Kataoka N. and Kitayama K. (2006). Characterization of a full encoder/decoder in the AWG configuration for code-based photonic routers. Part II: experiments and applications, *IEEE/OSA J. Lightwave Technol.*, 24; 1, 113–121.

15. Kitayama K. (1998). Code division multiplexing lightwave networks based upon optical code conversion, *IEEE Select. Areas Commun.*, 16, 1309–1319.

16. Huang S., Baba K., Murata M. and Kitayama K. (2006). Variable-bandwidth optical paths: Comparison between optical code-labeled path and OCDM path, *IEEE/OSA J. Lightwave Technol.*, 24; 10, 3563–3573.

17. Huang S., Baba K., Murata M. and Kitayama K. (2006). Architecture design and performance evaluation of multigranularity optical networks based on optical code division multiplexing, *J. Opt. Networking*, 5; 12, 1028–1042.

18. Maintenance & Troubleshooting of a PON Network with an OTDR, JDSU.

19. FTTx PON Guide: Testing Passive Optical Networks, third edition, EXFO (www. 3-edge.de/export/sites/3EDGE/de/main/solutions/FTTx-PON-Networks/Content-Misc/ FTTx-PON-Reference-Guide.pdf).

20. NTT East Japan (2012). Case studies of faults and countermeasures in a passive optical network system, *NTT Technical Review*, 10; 7, 31–35.

21. Technical reference 73603 (1999). Unbundled dark fiber (UDF) Technical specifications, February (www.docstoc.com/docs/2140807/Unbundled-Dark-Fiber-(UDF)-Technical-Specifications#).

22. Green P. E. Jr. (2006). *Fiber To The Home, The New Empowerment*, John Wiley & Sons.

23. Kashyap R. and Blow K. J. (1988). Observation of catastrophic self-propelled self-focusing in optical fibres, *Electron. Lett.*, 24; 1 47–49.

24. Todoroki S. (2005). Origin of periodic void formation during fiber fuse, *Optic Express*, 13; 17, 6381–6389.

25. Shuto Y., Yanagi S., Asakawa S., Kobayashi M. and Nagase R. (2004). Fiber fuse generation in single-mode fiber-optic connectors, *IEEE Photonics Technol. Lett.*, 16; 1, 171–176.

26. Yanagi S., Asakawa S., Kobayashi M., Shuto Y. and Naruse R. (2004). Fiber fuse terminator (in Japanese), Technical Report of IEICE, OPE2004–178.

Index